THE COMMON MIND

An Essay on Psychology,
Society, and Politics

Philip Pettit

New York   Oxford
OXFORD UNIVERSITY PRESS
1993

Oxford University Press
Oxford New York Toronto
Delhi Bombay Calcutta Madras Karachi
Kuala Lumpur Singapore Hong Kong Tokyo
Nairobi Dar es Salaam Cape Town
Melbourne Auckland Madrid

and associated companies in
Berlin Ibadan

Published by Oxford University Press, Inc.,
200 Madison Avenue, New York, New York 10016

Oxford is a registered trademark of Oxford University Press

Library of Congress Cataloging-in-Publication Data
Pettit, Philip, 1945-
The common mind : an essay on psychology, society, and politics /
Philip Pettit.
p. cm. Includes bibliographical references and index.
ISBN 0-19-507818-7
1. Intentionality (Philosophy) 2. Intentionality (Philosophy)—
Social aspects. 3. Social psychology—Philosophy. 4. Political
science—Philosophy. I. Title.
B105.I56P48 1993 128—dc20 92-16184

9 8 7 6 5 4 3 2 1

Printed in the United States of America
on acid-free paper

For Eileen

Preface

This book has been years in the making. But that's not a moan, for the making of it has been great fun. It has pushed me into thinking and writing about a variety of topics that I would otherwise have left untouched. It has drawn me into a range of challenging dialogues and exchanges. And it has involved me in a number of very enjoyable and profitable collaborations on publications related to the theme of the book. Without those collaborations this book might have been completed earlier but it certainly would not have been completed in the present form.

I should begin with an acknowledgment of my indebtedness to relevant collaborators: to Frank Jackson, with whom I share the theory of intentionality and causal relevance presented here; to Michael Smith, who has an equal stake in the depiction of human subjects as deliberative agents; to Geoffrey Brennan, with whom I have made connected investigations into the nuances of rationality in choice and of feasibility in institutions; and to John Braithwaite, with whom I have tried to elaborate the sort of republican theory of politics, in particular of criminal justice, that is supported here. I should also mention my indebtedness to Peter Menzies, with whom I am currently working on themes connected with holism and anthropocentricism. I have drawn on other collaborative work to a lesser extent but I should not fail to record at least my debts to those with whom my joint work is cited in the text: Alan Hamlin, Chandran Kukathas, Huw Price, and Bob Sugden.

These collaborators come from economics, politics, and sociology, as well as from philosophy. With their help I have absorbed interdisciplinary lessons and insights that would not otherwise have come my way. With the encouragement of their collaboration I have tried to write a book that will engage with debates that occur within and across a variety of disciplines.

The nature of the book is easily explained. It begins with questions of psychology: questions to do with what it means to be an intentional agent and, in particular, what it means to be an agent with the capacity for thought. Having sketched an overall view of the intentional, thinking agent, it then goes on to explore the difference that social life makes to the mentality of such agents.

And, having developed a picture of mind in society—of the common or social mind—it turns, finally, to the lessons of this picture for the pursuit of social and political theory: for the explanation of what happens on the social scene and for the evaluation of the different ways in which that scene may be structured.

The book is in three parts, and each part corresponds to a stage in the enterprise just described. The first part deals with topics in what is variously known as the philosophy of mind, philosophical psychology, or the philosophy of psychology; the second deals with questions in social ontology or metaphysics, questions in the theory of how psychological subjects relate socially to one another and to the social entities they constitute; and the third looks at issues of social explanation and political evaluation, issues of social and political theory. The first part of the book gives a picture of mind; the second, of mind and society; and the third, of mind, society, and this associated sort of theory.

The architecture of the book is dictated by a pair of dependence claims. The first is that we cannot hope to chart the ways of social explanation or political evaluation unless we have an overall view of people and society: unless we have a guiding picture of the nature of individuals and of the difference that social life makes to individuals. The second claim is that equally we cannot hope to develop such an overall view of people and society unless we have a picture of their psychological make-up, a picture of what it is to be psychologically equipped in the manner of human beings. Seven years ago I committed myself to writing a book on social and political theory. The first dependence proposition meant that the book would also have to deal with social ontology, with the theory of mind and society. The second dependence proposition ensured that it would also have to cover philosophical psychology, the theory of mind as such.

The main propositions defended in the book are summarised in the previews offered at the beginning of each of the three main parts. These summaries are best avoided in the course of reading the full text. I hope that they will give some sense of the direction of the argument for someone who skims and that they will help as a record of the stages of its development for someone who does not. The summaries may also make it possible for a reader to dip into the book at different points, with at least an overall picture of where the part that interests her fits with other pieces. It would be nice to think that the book is a seamless web, a fabric in which there are no joints, and that it has to be clasped as a whole if it is to have the desired effect on a reader. But I am no master of that sort of art or craft. The book is a simple piece of architecture and I am sure that there will be no serious loss, at least no serious aesthetic loss, in surveying it with a selective eye.

Perhaps it will help the reader further if I draw attention to the main points of novelty in the book. The principal novelty in the first part is the distinction that I draw between intentional systems that can merely believe and desire and intentional systems that can also think: intentional systems, as I put it, that can act intentionally with a view to having beliefs and desires that satisfy certain rationality constraints. The treatment of merely intentional systems in the first chapter follows fairly familiar lines; the new departures are that it offers a distinctive account of desire and deliberation and introduces the program model of causal relevance to account for the impact of intentional states. The treatment of thinking systems on the other hand, and more generally the discussion of rule-following subjects, does not fit any standard mould; indeed it provides the basis for much that is new in later parts of the book.

The main novelty in the second part is the distinction between the question that divides individualists and collectivists, as I describe them, and the question that separates atomists and holists. The first issue has to do with whether our ordinary intentional psychology is undermined by collective agencies: aggregate regularities and forces. The second has to do with whether we are individually dependent on our relations with one another for the possession of essential human capacities, in particular the capacity to think. In arguing for individualism, the chapter on the first question covers fairly well explored territory; if there is novelty here, it is in the organisation of the material and in the setting out of the different issues involved. The chapter on the second question breaks new ground, however, in defending and developing a holistic perspective. The defence is novel in building on the account of thinking systems presented in the first part of the book, arguing that under certain plausible conditions the capacity for thought requires community with others. The development is novel in showing that the holism involved does not require a retreat to any sort of relativism: that while it certainly entails an anthropocentrism, this doctrine is consistent with a realist vision of inquiry.

Finally, the main novelty in the third part of the book probably consists in the connections drawn between the positions argued there and the holistic individualism emerging from earlier discussions. On the explanation side—the side of social theory—I show that the approach developed earlier generates an interesting view of interpretation but then I argue that the fact that human beings are interpretable in this manner is consistent with the pursuit of other strategies of social explanation also; I provide room in this way both for rational choice explanation and for explanation of a more structural or historicist stamp. On the evaluation side I show that the holistic individualism impacts in various ways on political theorising. The doctrine is methodologically significant, impacting negatively on fashionable, contractarian approaches to the discipline. It is significant in the construal that it motivates for various values, underpinning a republican as distinct from a liberal conception of negative liberty. And it is significant in pointing us towards a particular strategy of feasibility analysis: a particular strategy for assessing the institutional feasibility of different values and policies.

So much for the scope and structure of the book. Perhaps I should also comment on its style. Although the book aspires to find an interdisciplinary readership, I do not pretend to have found or forged a new interdisciplinary style. I am myself a philosopher and I bear the marks of my profession. I love the accumulation of distinctions, the development of cut-and-dried taxonomies, the pursuit of common-or-garden argument. I have a taste for straightforward theses rather than formal theorems or elusive motifs. I get excited by substantial problems, and not by technical puzzles or tantalizing enigmas. I go for deflationary prose in preference to diagrams, matrices, and tables or, for that matter, to high-flying rhetoric or cascading metaphor. In any case I am no good in the ways of formalism and figure, however helpful I may find the writing of those who embrace such modes. So, for better or worse, the book has the aspect of a work of philosophy, in particular a work of analytical philosophy.

But philosophy can be great fun and I hope that non-philosophical readers will not be put off. Philosophy is irreverent and adventurous and it offers a fine prospect of real, intellectual excitement about familiar, human things: things like the nature of mind, the character of social relations, the possibilities of politics; things like the topics covered in this book. Intellectual excitement is real to the extent that it can survive the project of increasing intelligibility. And few disciplines are committed in the manner of analytical philosophy to looking at the potentially exciting aspects of human and social life, without any compromise of intelligibility. Of the other disciplines that focus on similar concerns, too many only look to technically tractable matters. And of those that target exciting concerns, too many seem content just to savour and celebrate the themes in question; it is as if the only point were to roll them round the tongue, like old brandy.

Perhaps these remarks should be used, not to align myself with analytical philosophy, but rather with analytical theory. For analytical philosophy often falls short of the ideals I have just charted, sticking too narrowly to topics that belong in the charmed circle of the traditional curriculum, and using only well-tested patterns of argument in the service of their illumination. And, on the other side, the disciplines and traditions that compete with analytical philosophy are not by any means dichotomised, in the manner my remarks may have suggested, between the excessively technical and the overly rhetorical. The style of research and writing with which I identify—the style of analytical theory—may belong equally to practitioners of a variety of disciplines and traditions, even if I associate it mainly with analytical philosophy. It is a no-nonsense, no-inhibition style of careful analysis and theory-building and, while this work will undoubtedly fall short of the demands of that style, my aim is to do the best I can with it. Do not look in this book for the compelling mathematical proofs that will leave you no room for dissent. Do not look for the passages of overwhelming literary power that will bring you to your knees in commitment. The enterprise is pursued in a different, more pedestrian key.

Postmodernists dominate many of the humanities and social sciences and they will be surprised at anyone's publishing an analytical treatment of the topics covered in this book (Rose 1991). Enthusiasts of the postmodernist movement will find the self-confidence with which I approach my topics amusing and anachronistic. They will be shocked at my failure to keep a distance from the discourse I pursue: at my failure to hold the presuppositions that distinguish the discourse in focus, and to keep a continual eye on the rival modes of articulation with which it may be confronted. They will see me as a sort of fossil from the eighteenth-century Enlightenment, for it is often said in their circles that the distinguishing mark of the Enlightenment project was an insensitivity to the parochial character of its own modes of thought and expression: an assumption that whereas unenlightened thought seeks reality in the refractive languages of different traditions, enlightened thought can gain an undistorted view of what there is.

The postmodernist challenge is a difficult one to confront. Perhaps I should begin with acknowledging the points on which I agree with postmodernists and then explain why I do not go along with their models of intellectual discussion.

Postmodernists will generally disavow doctrine. But this is disingenuous, for there are a number of propositions that distinguish their commitment. They deny that the human subject is the self-transparent, self-directive entity imagined in the Cartesian tradition, and resurrected in recent existentialist thought. They hold that the subject is essentially dependent on social relations and cultural resources, in particular the resources of language and tradition, for the realisation of its distinctive capacities. They hold, in particular, that reason is not a universal given: that what is reasonable, what is theoretically or practically compelling, is a function of the local culture or cultures within which people operate. This leads them to think, on the other side, that reality is not a universal given either. They deny that theoretical regularities and practical norms are a promised land to which different traditions may lead from different directions; they portray traditions as each having their own reality in view, with the real corresponding to what is found compelling within the local habits of theoretical and practical thought.

This doctrinal sketch is rough and ready, and allows of immense variations of emphasis and application. But the striking thing is that the propositions involved can all be interpreted in such a way that the pages that follow constitute a defence. I argue in this book that human beings are characterised by the ability to think, that thinking requires social relations and cultural resources, and that this means that naive realism is put in question. True, I hold that the compromise to realism involved in admitting the social character of thought need not be of relativistic proportions. I hold that the lesson is, not that each tradition of discourse constructs its own reality, but that whether a discourse directs us to reality is a question that is always *sub judice*: in particular, *sub judice conversationis*, under the judgment of intellectual exchange. The answer to the question is contingent on whether the claims that distinguish the discourse prove their worth in intellectual negotiation between different individuals and cultures. But the rejection of relativism does not cast me beyond the postmodernist pale, for relativism is also rejected by many heroes of the movement (MacIntyre 1988, chap. 18; Rorty 1980).

So much for points of agreement. The points of disagreement are equally striking. Postmodernist writing is characterised, not by the presentation of any sort of theory, but by the pursuit of a kind of therapy. Given their theory—a theory that often remains unvoiced—postmodernists seem to assume that any unselfconscious involvement in a discourse, for example any unselfconscious pursuit of historical or psychological or philosophical inquiry, is problematic. They take unselfconsciousness to betray an ignorance of the fact that every discourse is distinguished by particular assumptions and that those assumptions can always be challenged in rival modes of thought. Thus they abstain from such unselfconscious discoursing themselves and they devote their energies to the so-called deconstruction of that type of discoursing on the part of others. The deconstruction may take many forms but the general aim is to reveal to the practitioners of the discourse that they are not driven by the uniquely compelling call of the real in cleaving to the paths they pursue: that there are choices implicit in many of the judgments they make, choices that become salient in the light of alternative discursive models. The approach might take as its motto that delightful quip: anything you can do, I can do meta. Postmodernists refuse to get involved, refuse to take any ground-level discourse seriously. They remain sceptical and detached, even supercilious, about the efforts of those who would commit themselves wholeheartedly to such a discourse. They devote themselves to higher-level commentary on those efforts, and they devote themselves in particular to demonstrating the discursive naivete of those who would let such efforts absorb their souls. Their speciality is the arched brow and the quizzical eye.

Although I am sympathetic to the distinctive postmodernist doctrines, at least under certain versions, I think that the deconstructive stance is deeply misconceived. It is a mistake to think, as postmodernists appear to think, that any wholehearted discursive commitment amounts to an announcement that the discourse in question is unproblematic, that it is free of controversial presuppositions. There is a difference between announcing that a discourse is not problematic and not announcing that it is, as there is a difference between telling someone she is not a friend and not telling her that she is. Discursive involvement or commitment does not necessarily reflect a naivete about the matters postmodernists belabour; it does not necessarily manifest an ignorance about the cultural roots and references on which postmodernists like to dwell. All that it may reflect is a desire to get on and do something other than rehearse the postmodernist verities: a desire to break the spiral of meta-commentary and try to make a contribution in the terms—albeit, the limited and situated terms—of some ground-level debate.

This desire should not be derided by the postmodernist. The eye that tries to keep itself in the picture cannot see anything else; the condition under which vision operates is that the eye looks beyond itself. A similar condition applies in the case of intellectual work. It is impossible to participate in a discourse, working at the resolution of problems generated by its assumptions, if one is intent on remaining aloof and self-aware. It is impossible to advance a discourse, exploring its logic and its limits, if one refuses to sink oneself in it. As the enjoyment of a play requires the suspension of disbelief, so discursive participation requires the suspension of scepticism. Indeed the point is made, and perhaps even exaggerated, by another hero of the postmodernists: the historian of science, Thomas Kuhn (1970). He argues that one of the distinctive features of the successful scientific advance is the dogmatism with which practitioners attach themselves to the new paradigm of research, a dogmatism that often flouts all the familiar criteria of reasonable assent. If his observations are granted, then the detachment postmodernists affect must be seen as an inhibitor of serious intellectual inquiry.

I have mentioned two characteristics of postmodernism: its distinctive doctrines and its deconstructionist methodology. This book is far removed from the postmodernist models of discussion, not because of a downright rejection of the distinctive doctrines, but because of a refusal to follow deconstructionist, and ultimately self-defeating, patterns of debate. But it is also distanced from postmodernism in regard to a third characteristic of the movement and perhaps I should mention this too. That characteristic is a preference for a rhetoric in which the subjects are abstractions like discourses, practices, formations, genres, languages, traditions, even knowledges. These subjects are described as interacting with one another in moments of conjunction, interpolation, contestation, and the like, and they are represented as producing, transforming, locating, shaping, and constructing human beings and human actions. The metaphor at work has a glacial character. The discourses present themselves like ice sheets that move across and against one another and that carry us along, now on this trajectory, now on that.

As I reject deconstructionist methodology, so I renounce this inflationary rhetoric. At times the rhetoric is harmless enough. It may make a useful point to say that a certain language constructs someone's speech rather than saying that she speaks that language. But at other times the rhetoric raises serious worries. It projects a story under which one of the main questions debated in this book is begged. This is the question between individualists and collectivists, as to whether our ordinary psychological sense of ourselves can survive the recognition of social regularities and forces. Despite arguing that the ability to think requires social relations and resources, despite being in this respect a holist rather than an atomist, I maintain that all the evidence is against collectivism. I hold that socially resourced though we are, our ordinary psychological self-image is fundamentally sound: the agency recognised in that image is not compromised by the existence of social regularities. It would not serve my purposes in staking out this position—this holistic individualism, as I describe it—to introduce the rhetoric of postmodernism, for that rhetoric makes for a constant temptation to think in collectivist terms. Put the cart before the horse in your habits of speech and your habits of thought are all too likely to follow.

It remains for me only to express appropriate words of thanks. I could not have written the book in a more fitting interdisciplinary setting than the Research School of Social Sciences at the Australian National University. Most of the co-authors mentioned above are currently colleagues here; the remainder have been visiting fellows at the school or have been colleagues elsewhere in the university. I am grateful to the keepers of the school for their work in sustaining it: to my colleagues in general, who maintain the spirit and standing of the place; to Paul Bourke, who has been a fine director and defender during his five years of office; and to the various reviewers who have found such gratifying things to say about us in a period of unrelenting scrutiny. I began actively thinking about the book as an Overseas Fellow at Churchill College, Cambridge, in 1986 and I also worked on it during a period in Oxford in 1989, when I was a Visiting Fellow of Corpus Christi College and a Visitor at Nuffield College. I am grateful to those three Colleges for their support.

There are many individual colleagues who have given of their time in discussing some of the typescript, or some closely related papers, and I am delighted to thank them, with the usual qualification that they should not be incriminated by association with any of my errors. John Bishop, Simon Blackburn, John Braithwaite, Geoff Brennan, John Campbell, Greg Currie, Martin Davies, Bob Goodin, Stein Helgeby, Barry Hindess, Frank Jackson, Peter Menzies, Graham Oddie, Christopher Peacocke, Huw Price, Ian Ravenscroft, Geoff Sayre McCord, Jack Smart, Michael Smith, Dan Sperber, Kim Sterelny, Bob Sugden, Janna Thompson, Tim Williamson, Andrew Woodfield, and Crispin Wright have all been a great help. I am very happy, finally, to be able to thank my wife, Eileen McNally, who continues with such grace to tolerate my obsessions and even to find them amusing. I dedicate the book to her. She will be even happier than me to see the thing in print.

Canberra   P. P.
February 1992

Contents

III   MIND, SOCIETY, AND THEORY

I
MIND

Preview

Intentional states are those mental states which, like perception and belief and desire, are about things: they are about the things, as we say, that figure in the contents of the states. Such states serve to connect an agent with its world, letting it have a representation of what there is and of what opportunities there are for action. They give it the bearings that it needs to intervene successfully in the external order. The two chapters in this first part of the book are both concerned, in different ways, with intentional states: with the intentional resources of human subjects.

The first chapter deals with basic intentionality, with intentionality in the sense in which it need not involve thought, whereas the second considers what is required for the appearance of intentional subjects that are also able to think. One of the main departures from orthodoxy in this section of the book is the introduction, at least in the discussion of intentionality, of the distinction between non-thinking and thinking subjects. The distinction is usually ignored in treatments of intentionality and when it is introduced, it is not drawn in the terms that I use here.

The treatment of basic intentionality in the first chapter follows a fairly well-known track. The questions addressed are these. What makes a system into an intentional agent? What sorts of entities are intentional states? And how does the picture presented here fit with the pictures developed in the literature?

On the question of what makes a system an intentional agent, I take an easy-going line. To be an intentional system, to be a system that has beliefs and desires and the like, is to be exposed perceptually to a certain sort of environment and to interact with that environment in a way that makes sense, and makes sense non-accidentally, in belief-desire terms. We assume that there are regularities characteristic of beliefs and desires, regularities that dictate both the effect of certain sorts of evidence on what beliefs and desires are maintained, and the effect of certain sorts of belief-desire profiles on what responses are evinced. Those regularities will typically identify what it is to be evidentially rational in the attitudes one holds and responsively rational in the responses one makes, though they may also point us towards obstacles that get in the way of rational performance. A system will count as an intentional agent to the extent that its interactions with its environment, or at least some of its interactions, are governed by such regularities.

For all we have said, the intentional system may be a device rigged up with the environment, whose only responses are the formation of beliefs: say, beliefs represented in displays on a screen. But most familiar systems are active and indeed deliberative agents: systems that intervene in the environment and that do so, roughly, because of the desirable properties they register in the chosen option, because of the properties they register in the option, which engage with their desires. There is a system, say a robot, that is poking about in the grass. If it is an intentional system, then we will be able to see its intervention as being governed by a regularity relating certain beliefs and desires to the sort of behaviour displayed, where it is independently plausible—that is, plausible in the light of other regularities—to ascribe those beliefs and desires to the system. We will be able to see the system as displaying intentionally intelligible behaviour in the light of intentionally intelligible beliefs and desires. The robot is a gardener, we may find, and it is trying to aerate the lawn.

So much for the crucial question as to what makes a system an intentional agent. The next question discussed is what sorts of entities intentional states are. Unsurprisingly, I reject the traditional, Cartesian view that they are entities composed of a special non-naturalistic stuff. I argue that beliefs and desires and the like are higher-order naturalistic states. Consider the elasticity of an eraser: the state associated with the eraser's capacity to bend under pressure and regain its shape afterwards. This is not an ordinary naturalistic state like the molecular make-up of the eraser but a higher-order state which can be identified in this way: the eraser is so configured naturalistically—it has such a molecular structure—that it bends under suitable pressure. I say in parallel that to be in the state of believing that p—to be, like the robot, in the state of believing that raking aerates the grass—is to be in a state identified as follows: the system is so configured naturalistically— the robot has such an electronic profile—that it behaves in the belief-that-p way; the robot behaves in a way that more or less compels us to ascribe to it the belief that raking aerates grass.

This line of argument raises a well-known difficulty. We think of beliefs and desires as causally relevant to what an agent does and it is not clear how higher-order naturalistic states can enjoy causal relevance. When the eraser bends, it seems that all the required causal work is done by the state that underlies the elasticity—by the molecular structure of the eraser—and that the elasticity itself is inert. And by parallel it may appear that on our picture, beliefs and desires are causally inert as well: they are not causally relevant to actions, since any actions will be entirely accounted for in terms of more basic, electronic or neural states.

In response to this problem, I introduce the program model of causal relevance, which I have developed in joint work with Frank Jackson. The model is important, not just in the first chapter, but also in later discussions of social causality and explanation. It is true that the elasticity of the eraser does not produce the bending in the same way that the molecular structure does; it does not collaborate on all fours with the molecular structure, either as an earlier input or as a collateral, simultaneous element. Yet the elasticity is causally relevant, by our ordinary intuitions in these matters. So how does it attain such relevance? The answer suggested by the program model is that it is causally relevant so far as its realisation ensures that there will be some molecular structure in place—maybe this, maybe that—which is suited to produce the bending effect. The elasticity does not produce the bending in the same way as the molecular structure. It arranges non-causally for the presence of such a productive state; it ensures or otherwise raises the probability that there is such a state in place. And it is in that sense that it too is causally relevant to the bending.

The elasticity example points us towards other cases where programming obtains, and obtains in a more interesting fashion. The fact that the water in a closed flask is boiling programs for the cracking of the flask, though the cracking is actually produced by this or that vibrating molecule breaking a bond in the surface. The fact that the peg is square, the hole round, programs for the obstruction of the peg, though the obstruction is actually produced by this or that overlapping part. I argue that many states are causally relevant through arranging in this way for a factor that produces the effect in a more basic sense; indeed I later generalise the model, in Chapter 3, so that most states are seen to have causal relevance in this fashion.

The program model enables us to understand the causal relevance of intentional states like beliefs and desires. Suppose that my desire that p and my belief that I can make it the case that p by X-ing are causally relevant to my X-ing. The action will be produced in a more basic sense, at least on the non-Cartesian picture of intentional states, by my neural profile at the time. But still we will be able to think of the belief-desire complex as programming for the action. The presence of that complex more or less ensures, given that I am a properly constituted human subject, that there will be some neural profile realised within me—maybe this, maybe that—which will produce a suitable action.

Having discoursed on intentional agents and intentional states, the final question that I address in the first chapter is how the picture presented here fits within the constellation of pictures on offer in the philosophy of mind. I need not summarise that discussion here, since it will only be of interest to initiates. For the record it may be worth mentioning that my position amounts to a form of functionalism about intentional states, in particular a broad or externalist form of functionalism that allows contextual factors to figure in the constitution of such states. There are problems raised in the literature for such a position, usually problems related to an alleged indeterminacy of content: related to the claim that on the functionalist account it often remains indeterminate as to what exactly are the contents of an agent's intentional states, what exactly are the propositions believed and the states of affairs desired. I take a relaxed view of this family of problems, on the grounds that they disappear in the perspective of the second chapter; they are resoluble in the case of thinking subjects.

The second chapter begins with the distinction between non-thinking and thinking intentional subjects. Every intentional subject will instantiate intentional states and every intentional subject will do things on the basis of the states it realises: it will draw inferences, for example, and perform actions. The actions it performs will be intentional in a distinct sense from that in which the states are intentional; they will be performed, roughly, under the control of the agent's desires. Now the distinction between the thinking and non-thinking subject, as I understand it, is determined by the range of things that the subject is capable of doing intentionally. The subject will be a thinking subject if and only if, among the things that it can do intentionally, it can do things that are designed to promote the prospect of its meeting various constraints of rationality: to promote the prospect of its having beliefs that are indeed true, for example, or the prospect of its performing actions that are indeed desirable. The thinking subject can cogitate and ratiocinate, deliberating in the light of this or that consideration as to what it is right to believe or best to do. The non-thinking subject cannot. It may display a coherent pattern of beliefs and desires, it may generally respond to things in a rational way, but it will not do any thinking; it will be a relative automaton.

Perhaps the best way of capturing this contrast is to see that while the thinking subject must have a desire to be rational—for example, a desire to have beliefs that are more likely to be true than false—the non-thinking subject will have no such desire in its make-up. It may be designed generally to adjust, say in the light of new evidence, so as to have true beliefs; it may in that sense be a truth-seeking system. But it will not have a desire to have true beliefs that will lead to action on the basis of this or that channeling belief: say, the belief that by taking time, by investigating further, or by listening to others, it can increase the expectation of having true beliefs. It will not have a desire for truth, or more generally a desire for rationality, in the common-or-garden sense of desire. It counts as a believer but, by contrast with the thinking subject, it is a blind believer, not a conscious one.

This contrast is salient to anyone who begins to think about the intentional character of human beings, since it is obvious that even if robots and non-human animals are intentional subjects, they contrast with human beings in lacking the capacity for thought. The contrast has slipped from view in recent philosophy of mind, because of the focus on topics generated by artificial intelligence and cognitive science. These disciplines are not particularly concerned with human beings, only more generally with representational or intentional systems, and within the perspective that they nurture there is little interest in the things that distinguish human beings among intentional systems.

The second chapter is devoted to building up a picture of what it is like for an intentional system to be not just intentional, but capable of thought: to be a subject like one of us. I identify two crucial requirements that must be fulfilled by any thinking system and most of the chapter is given to how these are met in the human case. I describe the first requirement as that of intentional ascent, the second as that of rule-following.

The requirement of intentional ascent, briefly, is the requirement that the subject be able to have not just beliefs with certain contents, but beliefs about the contents, actual and potential, of other beliefs. The subject must be able to have beliefs about the proposition that p, as well as being able to have the belief that p: for example, it must be able to have beliefs to the effect that it is likely that p, given that q; that it is impossible that p and that q; and the like. Unless it is able to have such beliefs, then it will be unable to orientate itself in the project of promoting the prospect that its beliefs are true. How do human beings come to have such beliefs about contents or propositions? It turns out that the problem is quite a tricky one, as the more obvious models do not work: for example, propositions cannot be exemplified by truth-makers, in the way in which properties may be exemplified by instances, for thinkers must have access to false propositions as well as true ones. I argue that the best way to make naturalistic sense of the ability is by recourse to a model under which propositions are presented by proxies or signs as potential objects of belief. The use of language would make appropriate signs available but I emphasise that the signs whereby propositions are presented need not be intended communicatively and need not even be produced by the subjects themselves.

The other crucial requirement for thought is that the person be able not only to have beliefs about propositions, but to treat propositions and the elements out of which they are constructed as rules of thought. The idea is that the subject must be able to recognise that a proposition dictates for each of an indefinitely large number of possible ways the world may be, that it should be believed or not believed under that eventuality; and that, recognising this, the subject must be able to make efforts to see that its own belief-forming responses honour the dictate of the proposition. In short, the subject must be able to treat the proposition as a rule, identifying a constraint that it represents for belief-forming practice and setting itself intentionally to form beliefs in accordance with the constraint. If the proposition is 'It is raining on that hill', then the subject must be able to grasp the condition that makes it appropriate at the relevant time and place to hold a belief with that content and must be able, furthermore, to try to abide by that condition.

This problem of rule-following is at the centre of a contemporary philosophical debate, though it is not always related to the possibility of thought, and I defend here a line of response that I have already presented elsewhere (Pettit 1990a, 1990b). We follow the rule associated with a proposition through following more basic rules associated with elements of propositions. Thus I follow the rule for belief dictated by the proposition 'That is a box' by following the rule associated with the property of being a box; this more basic rule makes it appropriate to group some things together as boxes, other things not. In order to present my line on rule-following, let us begin then by considering how the rule associated with the concept of a box might be presented to me, as something that dictates responses in an indefinite range of cases and as something that I can identify and try to respect in those cases.

In order to dictate responses in an indefinite range of cases, the rule must be unlike anything finite and concrete, such as a paradigm or image or formula, that can be presented to our finite minds. It cannot reduce to a set of paradigm boxes, apparently, since it will always be possible to treat these as instances of a wider or wayward class, such as boxes-or-bicycles, boxes up to a certain time, bicycles after, and so on: there will be no necessity linking the exemplars to any particular class. But on the other side the rule must be of a kind with such more or less tangible things, such concrete exemplars, if it is to be identifiable and followable by a finite mind like mine. The problem is how to reconcile these conflicting demands: how to give an account of rules and rule-following that makes rules look objectively satisfactory, being normative over an indefinite range of cases, while being subjectively accessible, being the sorts of things that anyone of us can intentionally try to track.

The essence of my solution is this. Certain paradigm boxes can exemplify the box-rule to me, so far as I am naturally inclined to go on from those exemplars in a certain way, associating other boxes, albeit boxes of different shapes, sizes, and colours, with the exemplars, and definitely not associating bicycles with them. I use the exemplars to fix the rule in question: pointing at them, as it were, I say to myself that it is that kind of thing that I mean by 'box' and that it is of instances of that kind, and only of such instances, that I intend to hold box-identifying beliefs. But the kind is fixed by my extrapolative inclination, my inclination to go on from the original paradigms. And doesn't that mean that I cannot be wrong in any future belief that something is a box? Doesn't it mean that anything I am inclined to call a box is a box? No, not necessarily.

I use the exemplars to fix the relevant rule and kind, with the extrapolative inclination making that project possible. But it may be that I follow this practice, only subject to the satisfaction of certain conditions. Suppose that I assume, perhaps as a matter of my very nature, perhaps as part of a standard convention, that if my inclination leads me in what by my lights are different directions at different times, or that if my inclination leads me in what by my lights is a different direction from that followed by another, then there is something amiss. Suppose I am disposed in such cases to reserve commitment until a perturbing condition has been identified, a condition that means that the inclination that it affects should be discounted. In that case, and it surely has a certain plausibility, the rule and kind that I should be said to identify on the basis of the initial exemplars is not whatever kind is fixed by my actual extrapolative inclination, but whatever kind is fixed by that inclination as it operates under circumstances that do not occasion interpersonal or intertemporal inconstancy: by that inclination, as it operates under those favourable conditions that survive negotiation with myself across time or with other individuals.

I describe this sort of story about rule-following as 'ethocentric'; the materials it deploys are habits of response—extrapolative inclinations—and practices of negotiation and the word 'ethos' can refer to either element, habit, or practice. A rule like the box-rule is presented to me in the concrete exemplars that I initially confront. The rule extends to yield judgments on new cases so far as it finds a voice in my extrapolative inclination: boxes are things of that kind, as I say in pointing at the exemplars, where the kind are fixed via the inclination. But still, the rule is not something over which I am an authority, something in the application of which I cannot go wrong. For the rule and kind that I identify on the basis of the exemplars are fixed via the inclination, only as the inclination operates under those conditions, identified in negotiation, where convergence across times and individuals proves to be attainable. On any occasion I must see myself as less than an authority on the rule, as a fallible employer of the rule, for the rule to which I address my efforts of construal is authoritatively presented only in a voice that is beyond my control: the concerted voice of intertemporally or interpersonally consistent responses.

There are many questions to be raised about this ethocentric story and I try to consider the most telling ones in my discussion. I argue that the ethocentric materials—habits of response and practices of negotiation—are reasonable posits for the story to invoke; that the work to which the story puts those materials, as it suggests that they make rule-following possible, is of a sort to which they are fitted; and that the consequences of the ethocentric story are not as counterintuitive as they may at first seem. I believe that the ethocentric account is the only way of making naturalistic sense of rule-following and, in particular, of thought.

The upshot of this discussion is that we human beings can succeed in being higher-grade intentional systems, can succeed in being thinkers, only so far as we are equipped with certain extrapolative inclinations and are able to defer to the operation of those inclinations in others, or in ourselves at other times, in identifying the rules to which we try to remain faithful when we form our beliefs. The self isolated in time and society, the solitary, momentary self, cannot follow a rule, cannot try to be faithful to any constraint, because it has nothing to play off against. We human beings can follow rules, in particular follow the rules of thought, so far as we interact with ourselves across time or, more plausibly, interact with one another. Such interpersonal, or at least intertemporal, community is essential for the project of thought. This result connects with themes that we shall explore further in the second part of the book.

1

Intentionality

Intentional subjects are a species of agent. Roughly, they are agents that engage with their environment in such a way that we ascribe beliefs and desires to them; by contrast with stimulus-response automatons, they act on the basis of how they construe their situation and how they feel about it. Thinking subjects are in turn a species of intentional agent. They not only act on the basis of beliefs and desires, they act with a view to having beliefs and desires that meet certain desiderata: they act, say, with a view to having beliefs that answer to the facts and desires that answer to whatever values they countenance.

Human beings are intentional and thinking subjects and the first of the three parts of this book will be given to examining what those features involve. In this chapter we look at the requirements of intentionality, in the next at the requirements of thought. This chapter in turn breaks down into a number of sections. First, we investigate how certain agents get to count as intentional. And then we examine the place of intentional states—beliefs and desires—in the physical world and the role they play in the production of action. Finally, we situate the view of intentionality emerging from these discussions in the context of some rival contemporary theories.

One of the main themes of the book is that thought makes a great difference to intentionality; the difference made should begin to emerge in the next chapter. In this chapter we will be looking at intentionality without any assumption that it is accompanied by thought or of course by the linguistic expression of thought. We will be looking at intentional agents and intentional states in the sense in which they may be exemplified well beyond the human sphere: in higher mammals, for example, and perhaps even in robotic intelligences. This focus means that we postpone certain problems that arise distinctively with human beings until the next chapter (see Garfield 1988, Rudder Baker 1987).

Intentional Agents

Definition

There are three conditions that it seems reasonable to impose on any system that is to count as an agent, whether the agent be intentional or not. First, the system must be embedded in an environment: a distinct, surrounding context. Second, it must be subject to perceptual inputs from that environment: inputs that are transmitted via recognised sensory or sensory-like organs. Third, the system must respond to those inputs under the control of certain regularities; its responses must be reliably driven by the regularities. When I speak of regularities, I mean lawlike as distinct from accidental regularities. I prefer to speak of regularities rather than laws, because the focus is on presumptive patterns in nature, not on the formulae in which such patterns are articulated by theories, and the word 'regularity' is the more suitable for this purpose; the word 'law' is used both of natural patterns and of theoretical formulae.

The third condition rules out a number of scenarios. It will not do if the relation between inputs and outputs is utterly random. It will not do if the system's responses are a regular function of those inputs together with certain randomly varying independent factors. And it will not do, finally, if the responses conform accidentally to the regularities: if they conform, because things have been rigged so that in all actual circumstances—but not in all the intuitively relevant circumstances that might have come about—the subject behaves as it would do if the regularities were in control.

No agency without interaction, then: specifically, no agency without regular, law-driven interaction with a perceptual environment. With this account of the requirements of agency in hand, we can distinguish between types of agent on the basis of the sorts of regularities that govern their interaction with the environment. In particular, we can demarcate intentional agents from non-intentional ones. Intentional agents interact with their environment under the control of intentional regularities. Non-intentional agents do not.

Intentional regularities are regularities that involve intentional properties, as chemical regularities are regularities that involve chemical properties. But what are intentional properties? The Latin word *intentio* means a directing towards a target and, as the etymology suggests, intentional properties are target-directed properties. It is an intentional property of a perceptual input that it is a perception that p; of a behavioural output that it is an attempt to bring about q; of a state of an agent that it is a belief that r or that it is desirable that r. The input, output, and state are each identified by reference to a proposition that serves as its target or, in a much-used phrase, as its content: as what it bears upon. The notion of a content, the notion of a proposition, and indeed also the notion of a property, will receive further commentary in the next chapter; for the moment we may take them as given.

The lesson then is that intentional agents are driven by regularities involving intentional properties, as they interact with their environment. The intentional regularities that are most commonly identified are rational regularities: the sorts of regularities exhibited by an agent that is minimally rational in theoretical and practical matters (Cherniak 1986). In normal circumstances, and absent other relevant considerations, the fact that it appears to be the case that p induces in the agent the belief that p (inductive rationality); the belief that p conjoins with the belief that if p, then q to produce the belief that q (deductive rationality); the belief that q interacts with the belief that it is undesirable that q to produce an attempt to make it the case that not q (practical rationality); and so on.

In discussing intentional regularities I concentrate on regularities that involve contents that can be articulated in ordinary language: contents that bear on familiar objects like trees and birds and familiar properties like being a eucalypt or a parrot. It is worth mentioning, in passing, that any intentional system that is governed by such regularities is likely to be simultaneously governed by intentional regularities involving contents that can only be articulated in the language of scientific psychology: contents that bear, for example, on subliminal perceptual cues rather than on familiar objects. No system is likely to satisfy a regularity involving the belief that that tree is a eucalypt without simultaneously satisfying a regularity involving a corresponding cue-centred belief; presumably an intentional system will be sensitive to the object only through being sensitive, however subliminally, to cueing features (see Jackson and Pettit 1992a). I propose, for the moment, to ignore unfamiliar beliefs and unfamiliar regularities and to focus on the common-or-garden kinds: the kinds directed to contents that can be readily articulated in ordinary language. I return to the topic again at the end of the next section.

I hope that this is enough to give an initial idea of what it is for a system to be an intentional agent. The intentional agent is a system that is designed in such a way, whether by nature, nurture, or engineering, that it embodies certain intentional regularities in its environmental interaction: say, that it embodies the regularities characteristic of minimal rationality, at least in circumstances that we can count as normal. That a system is intentional in this sense allows it also to conform to regularities of other kinds in its interaction. If it is an engineered system—say, a robot—then it will conform to the regularities of the relevant engineering theory. If it is a biological system like one of us, then it will conform to the regularities of biology. And in either case it will conform to the regularities of micro-physics. The important point is that apart from conforming to such regularities, it is so constructed as also to conform to regularities that are intentional.

How broad is the category of intentional agents? It may appear to extend deep into the animal and indeed mechanical worlds, as we can describe even the behaviour of termites and central heating systems in intentional terms: we can think of different categories of termites as pursuing different social tasks, we can think of the central heating system as looking after the business of keeping the house above a certain temperature. But this appearance is misleading. I said that if a system is to be governed by certain regularities, then it is not enough for things to be rigged so that in all actual circumstances its responses fall under them. It must also conform in all intuitively relevant circumstances. But, however we demarcate intuitively relevant circumstances, it is clear that they go well beyond the conditions under which termites or central heating systems comply with intentional regularities. We know, or so I assume, that termites only comply provided various chemical conditions are just right: that things go quite awry, for example, if we interfere with the chemical secretions given off by the queen. And we know, of course, that the central heating system ceases to do its job if the thermostat is manipulated in any of a variety of ways. Thus we can see termites and heating systems only as mechanical simulacra of truly intentional subjects. The intentional patterns we seem to discern in their responses are not resilient enough over a range of plausible disturbances to constitute governing laws.

But if the category of intentional agents does not extend deep into the animal and mechanical worlds, if it does not extend as far as termites and heating systems, I do think that it extends in some measure. In particular, I think that it is hard to avoid taking more or less adaptable animals and artifices as intentional subjects. Think of the cat in pursuit of its prey or the chess-playing robot in pursuit of check-mate. The intentional regularity manifested in this sort of response, and the other intentional regularities it presupposes, are borne out more or less robustly in the behaviour of the systems in question. Move the cat's quarry and the cat will change its tactics, alter your play against the robot and the robot will adjust its moves. In such cases we naturally think of the subjects as being driven by intentional regularities, not just as accidentally conforming to them (Dennett 1987, chap. 7). We tend to take the behavioural evidence of adaptability in the pursuit of ends as evidence that here we have intentional systems, here we have systems that act on the basis of intentional states. I think that this tendency is well placed.

There has been one very influential tradition in psychology, methodological behaviourism, which would deny that we should be impressed in this way by evidence of adaptability. Behaviouristic psychology tries to show that the interactions of various agents with their environment, even the interactions of adaptable agents, are not evidence of intentionality. It argues that they are reliable only under such non-intentional inputs, and only for such non-intentionally specifiable responses, as it provides a heuristic for recognising: that they are reliable only so far as they are conditioned responses of a non-intentionally identifiable type to non-intentionally describable stimuli (Sober 1985). Were a behaviouristic story available for any system, then it might be said that that system's behaviour conforms to intentional regularities, not in all intuitively relevant conditions, but only in conditions where the conditioning applies. Thus it might be said that the cat isn't reliably pursuing its prey out of a desire to catch it, because it only displays the prey-pursuing pattern under circumstances that connect with its conditioning. But we ought not to be disturbed by the behaviouristic challenge, for the behaviouristic program has conspicuously failed to fulfill its promise. It is still reasonable practice to argue from the adaptability of behaviour to the presence of intentionality.

The upshot of this discussion is a generous sense of which agents are intentional, which non-intentional. An agent is intentional just in case it interacts with its perceptual environment under the control of intentional regularities and the presumption is that wherever there is evidence of behavioural adaptability, there is intentionality at work. This claim will be resisted by at least two groups. Many will say that intentionality also requires a certain sort of internal make-up in the agent: say, a certain neural or electronic organisation. This position is unattractive, for we attribute intentionality in our everyday dealings without a knowledge of internal make-up; and moreover, we do so with an authority which it is hard to see how to challenge (Jackson and Pettit 1990a). I discuss the position a little further in the last section of this chapter.

But while I resist the first group of likely opponents of my generous theory of intentional agents, I have more sympathy with a second. These are the philosophers who will say that any system capable of having beliefs and being intentional must also itself have the concept of belief; thus it must be vastly more complex than cats or robotic chess-players appear to be (Davidson 1984, essay 11, and 1985). I hope that those who take this view will bear with me. There are two different ways of being an intentional system, the thinking and the non-thinking way. In the next chapter I will introduce that distinction and try to show what intentionality with thought involves. I believe that the divide these philosophers want to mark is best cast, not as the distinction between intentional and non-intentional agents, but rather as the distinction between thinking and non-thinking intentional subjects. Thus I go along with the spirit, if not the letter, of their proposal.

Examples

But it is one thing to introduce a definition of what it is to be an intentional agent. It is another to give that definition life, illustrating how it applies in simple cases and then in cases that are progressively more complex. I shall now attempt to give the definition life in this way, generating a sense of how we come to think of certain systems as intentional agents. I begin with a simple artifice—an imagined inferring machine that is not much more complex than the familiar adding machine—and show first how that machine, and then how more complex variations on it, can count as intentional agents; the more complex variations get to be closer and closer to intentional subjects like you and me.

Imagine an inferring machine, the analogue of a simple adding device, which has a pad as input mechanism—a perceptual organ, so to speak—and a display screen as the medium of its behavioural output. The device accepts as input any of a number of simple declarative sentences—let us represent them schematically as 'p', 'q', 'r', and so on—and any sentence constructed from them with the help of logical operators like 'and', 'or', and 'not'. Depending on the entry made on the input pad, the output screen displays different responses. The output screen will display 'Yes' if the sentence is consistent in propositional logic with sentences already given a 'Yes' response, 'No' if it is not. Is the device imagined an intentional agent? Does it interact with the environment mediated by the pad under the control of intentional regularities? Perhaps not, since it reliably connects the intentionally characterised inputs with the intentionally characterised outputs only so far as they assume specifiable inscriptional—and therefore non-intentional—forms. But let us ignore that feature for the moment: let us imagine that inputs, and indeed outputs, may vary indefinitely in inscriptional form. If we take this line, then it is hard to see how we can deny that the inferring machine is a rudimentary, intentional agent.

If we treat the inferring machine as an intentional system, then we will naturally ascribe beliefs and other intentional states on the following pattern.

1. The inferrer is presented with evidence that p—it appears to it that p—for any sentence 'p' that is entered on the pad.
2. The inferrer believes any proposition to which it gives a 'Yes' response and we may assume that it disbelieves the negation of the proposition.
3. The inferrer disbelieves any proposition to which it gives a 'No' response and we may assume that it believes the negation.
4. The inferrer may believe propositions that express the propositional rules of inference it follows in awarding its verdicts; alternatively, it may just follow those rules as a matter of habit, there being nothing involved that earns the title of belief. The inferrer embodies propositional logic but it is not clear whether embodying a logic means having certain beliefs.
5. The inferrer does not believe or disbelieve any propositions with which it has not been presented.

In ascribing beliefs after this pattern, we follow a three-stage procedure. First, we identify the inputs, outputs, and existing states of the system that can be non-arbitrarily tagged with propositions, non-arbitrarily assigned intentional properties; the task is trivial in this case, since we take our cue from the sentences mentioned in the original specification of the device. Next, we identify the system's states of belief and disbelief and the habits of inference it follows; such states affect the system's behaviour, helping to determine which propositions it will accept in the future. Third, we distinguish among beliefs between those whose contents are explicitly represented on the pad and those structural beliefs, if any are admitted—see stage 4—whose contents are reconstructed from the machine's inferential behaviour. But never mind about the procedure followed, you may say. The main question is why we ascribe such beliefs at all. Why, as I have alleged, would we find the pattern of belief-ascription so natural?

Consider a parallel question. Why do we find it natural to characterise a certain sort of machine as an adding machine? The most plausible answer is that the machine simulates the adding function (Cummins 1989). The adding function takes two or more numbers as its inputs or arguments and yields their sum as ouput or value. The machine simulates that function in the sense that we can non-arbitrarily tag its inputs and outputs with numbers and under normal functioning the number that tags an output always turns out to be the sum of the numbers tagging the inputs. The adding machine is an analogue of our inferring machine. As the adding machine simulates the adding function, so we can say that the inferring machine simulates the belief or acceptance function, at least relative to propositional logic. The acceptance function takes an input proposition—a proposition for which there is some evidence—plus the propositions it already believes as its arguments and it goes to acceptance or rejection of the original proposition as value: or, if you prefer, it goes to that proposition or the negation of that proposition as value. The inferring machine simulates this function in the obvious way. We can nonarbitarily tag the inputs, the outputs, and certain existing states of the system with propositions, and we find that the output proposition in every case, or at least in every case where the machine functions properly, corresponds with the input proposition if and only if the proposition is consistent in propositional logic with the propositions already believed; it corresponds with the input proposition in just those cases where that proposition is believable under propositional logic.

It should be clear how easily we might come to think of the simple inferrer as an intentional system, specifically a system with the particular beliefs and other intentional states mentioned. It remains now to specify various ways in which the machine might be more complex—and might approximate agents like you and me—without our tendency to treat it as an intentional system being undermined. If we can specify such extra complexities, in particular complexities that do not subvert the system's status as an intentional system, then we shall have shown how easy and natural it is to think of complex machines and organisms as intentional agents.

1. The simple inferrer is equipped with a pad on which sentences that express the input propositions are written. A first development in complexity would keep that pad or at least an analogue of the pad; an analogue might be the receipt of information from the testimony of others. This development would supplement that indirect information-gathering device with a sensor for directly recording what we can continue to regard as input propositions. With this development the system would certainly cease to be constrained by the inscriptional regularities mentioned earlier and would have a firmer claim on the status of an intentional subject. Of course it will not always be clear what a sensor is recording. Moreover, any sensor is likely to fire occasionally at the wrong time and not to fire always at the right; similar malfunctioning possibilities were present with the simple inferrer but were not so salient. So with such a system we as interpreters would need to develop a view as to what it is sensitive to, when it is likely to be reliable, and what it is likely to be recording or missing when it is unreliable. We would need a sense of the circumstances that are normal for the system, the circumstances under which it works properly. Otherwise we would be plagued with problems of interpretation.

2. The inferrer as characterised is the executor of propositional logic, the logic of operators like 'and', 'or', and 'not'. Obviously it might execute a much richer logic than this in determining what is consistent with what. It might even be sensitive to informal and perhaps defeasible relations, respecting the sort of implication that binds 'bachelor' to 'unmarried', 'generous' to 'desirable', and 'gold' to 'glitters'. This extra complexity in the logic of the system, like the extra complexity in the inputting device, would not put the intentional status of the system in question in any way. But like the other complexity, it would introduce more problems of interpretation. We would have to struggle to identify precisely the relations for which the system embodies corresponding habits of inference.

3a. Moving from inputting device and logic to the system's outputting device, we can see a whole range of possible complexities that might be introduced here. Suppose that the inferrer finds an input proposition consistent with what it already holds. Instead of directly endorsing that proposition by way of output, it might endorse a proposition that follows from what it already holds together with the input proposition. For example, if it is faced with deciding whether p or not p, and it follows that p from what it already holds together with an input proposition 'q', then it might endorse 'p' by way of output rather than directly endorsing 'q'. The sentence 'p' might appear in the output display.

3b. Again, the inferrer only endorses a proposition by displaying 'Yes', by displaying 'No' to its negation or, as in this last case, by some close variations. A further step in complexity at the output end, complementing the extra complexity at the input, would have the system endorse propositions, not just by means of such articulation, but also by means of inarticulate action. The system might be equipped robotically to intervene in its perceptual environment and a certain intervention—say, G-ing—might count as the endorsement of a proposition: say, directly, the proposition that G-ing is desirable; or, indirectly, the proposition that p, where 'p' is an input proposition and combines with what is already believed to imply that G-ing is desirable. This development would mean that the inferrer intervenes in the environment with which the input pad is assumed to connect it. The inferrer becomes a truly active agent for the first time. More on this when we discuss desire.

3c. Yet another development in complexity at the output end would loosen up the requirements for endorsing a proposition even further. It would mean that immediate articulation or action was not the only means of endorsement. The development would allow the system to endorse a proposition merely by becoming disposed to evince appropriate articulation or action in certain circumstances. The relevant change in the system would be hidden at the time of input; it would appear only later, in the circumstances mentioned.

3d. All of these extra complexities at the output end are consistent, like the earlier complexities, with the system retaining its status as an intentional agent. But all of them, again like the earlier complexities, make the intentional interpretation of the system, the identification of its intentional states, more and more difficult. The point hardly needs to be laboured. A final complexity at the input end would also sustain the system's intentional status, while exacerbating this problem of interpretation. Take the case where the simple inferrer rejects a proposition that is inconsistent with what it already believes. Under the final modification, the inconsistency of the input proposition might lead the system, not necessarily to reject that proposition, but instead to reject one of the propositions already believed, a proposition whose rejection achieves the desired consistency. It might reject this proposition by any of a variety of types of articulation or action; the possibilities will ramify as with modes of endorsing a proposition.

As we pursue these possible developments of our simple inferrer, it should be clear that we are approaching familiar terrain. The fully developed inferrer displays much of the complexity found in higher mammals, ourselves included. Creatures like us are exposed to input propositions via testimony and via sensors—our perceptual organs—and we endorse propositions by way of output in articulation or action. So at least it is possible to interpret our interaction with the environment in which we are embedded. Furthermore, there generally seems to be a logic at work in the pattern displayed by the novel inputs and outputs and the propositions that we can reasonably be assumed to believe already. There are circumstances—abnormal circumstances, we may hope—in which we go awry, as through carelessness and compulsion, hypnosis and breakdown. But generally we are interpretable as developed versions of the simple inferrer. If it is to count as an intentional agent, then so surely should creatures like us.

Desires

There is one crucial task that needs to be discharged, before I can claim to have defined and motivated a suitable concept of intentional agency. I need to show where desires are supposed to figure in my picture of the intentional agent. In most philosophical discussions, and perhaps even in our common sense reflections, the idea of the intentional subject is equated with the subject whose actions issue in a suitable way from its beliefs and desires. But I have said nothing about desires in my account of what it is to be an intentional agent, or in the examples of intentional agency that I have developed. So where does desire fit in?

On the picture presented, there are two sorts of central intentional states that help to determine action: one is the explicit belief, the belief in a proposition that has been suitably represented at input or output end; the other is the habit of inference, a habit that may or may not be associated—we left this open— with a structural belief in the corresponding logical principle. Habits of inference come in a variety of kinds, given that the system may be faithful to quite a rich formal logic and even to informal and defeasible relations of implication. At one end of the spectrum we have the *modus ponens* habit of inferring that q from 'p' and 'If p, then q'. At the other end, we have the habits of inferring 'unmarried' from 'bachelor', 'desirable' from 'generous', and 'glitters' from 'gold'. In each case the habit causally links certain input beliefs to an output belief; the input propositions appear to support the output proposition; and, to gesture at the need to rule out flukes, it is because of this appearance that the causal linkage holds (Cummins 1983, chap. 3).

If we are to see where desires should fit into this picture of beliefs and habits of inference, we must get a clearer idea on what exactly desires are. I take a view of desires under which they are marked by two features: they are goal-seeking and belief-channelled states. That they are goal-seeking marks them off from beliefs, and that they are belief-channelled marks them off from habits of inference.

Beliefs are required, roughly, to fit the world; it is inappropriate, under familiar intentional regularities, to stick by beliefs that inaccurately reflect how things are. Desires contrast with beliefs to the extent that the world is required to fit them, rather than the other way around. If they are to be satisfied, if they are to have their way with the agent, then the world must be brought into line with desires, not desires with the world (Anscombe 1957, Searle 1983, M. Smith 1987). In this sense desires are goal-seeking. They dispose the agent to try to bring about a certain state of the world—the goal at which they point—in demanding to be satisfied. The contrast between beliefs and desires can be brought out by the observation that whereas evidence that not p will tend to disturb the belief that p, it will have no such effect on the desire that p (M. Smith 1987). The belief that p is required to fit the world, so that such evidence will conflict with the belief. But the desire that p requires the world to fit it, so that the evidence may serve only to sustain and sharpen the desire.

The second feature of desires is that not only are they goal-seeking, they are also belief-chanelled states. If a desire leads to a certain type of response—say, my moving my hand forward towards this cup—that is because the agent believes that that sort of response is designed to make the world fit the desire: it realises my goal of having a sip of coffee. The belief about the suitability of the response for realising the goal channels the impact of the desire. In this respect desires contrast with habits of inference rather than with beliefs: with habits such as that which takes a subject from the beliefs that if p, q and that p to the belief that q, or from the belief that someone is a bachelor to the belief that he is unmarried. Such habits require the world to fit them—they require the agent to have appropriate beliefs—but they can issue in a suitable response without the agent having any belief about a connection between that response and the goal in question. My habit of inferring 'unmarried' from 'bachelor' can issue in the belief that someone is unmarried without my registering that by forming that belief I satisfy the inferential goal: I believe something entailed by my belief that the person is a bachelor. But my desire for an option cannot lead me to perform it without my believing that by acting in such and such a way—a way for which I may not have any independent description or desire—I do in all likelihood realise the option.

It should be no surprise, though the point is usually neglected, that desires contrast with habits of inference in the manner characterised. With the choices that desires serve to generate there are different responses that an agent can make by way of realising the goal in question. If my goal is to drink that coffee, I may realise it by reaching out with my left hand or my right hand, in this manner or in that. If my goal is something non-behavioural—say, to recall a poem or to pray for someone's welfare—then I may realise it by summoning up this or that image, or rehearsing this or that formula. The desire produces a response in each case, so far as the agent has a belief that connects the type of response produced with the goal that it is after. With the responses that habits of inference serve to produce, on the other hand, the agent is not capable of realising the goal in question—the forming of a suitable belief—in different ways. Thus there is no need for the response to be channelled by a belief that that type of response is one that is likely to realise the goal.

So much for the sorts of states that desires are. The question before us is whether we can find room for them within the picture that we have painted so far. I say that we can find room for desires, indeed that we must find room for desires, at two different places in the picture. We must posit desires for the options agents choose and we must posit desires for the properties of options, or of the outcomes of options, that guide agents in their choices. We must ascribe both option-desires and property-desires.

The picture we have described makes room for systems that act on an independent environment, believing that this or that option is desirable: desirable *simpliciter*, desirable in the sense of being the thing to do. Some intentional systems may not act in this way: the simple inferrer and many of its developments will provide examples, for making responses on a display screen is not acting on the independent environment mediated by the input pad. But familiar systems like you and me certainly do act on an independent environment and we may call these active, as distinct from passive, intentional systems. Now it turns out that we must ascribe option-desires to any active, intentional system and that we must ascribe property-desires to any active system that is, as I shall put it, deliberative.

It makes sense to ascribe beliefs that certain options are desirable to an active intentional system only if in general the system is disposed to do what it believes desirable. However the notion of desirability is to be analysed, the belief that an option is desirable must make it appropriate for a subject to perform it. This condition would be fulfilled, for example, if something counts as desirable in a given context just in case the agent would desire it for that context, were the agent to be operating according to some independently suitable standard: were the agent to be rational or whatever (Smith 1992). But if the agent is disposed to perform an option that it believes desirable, as it will be when it responds appropriately to the desirability-belief, then by our account it has a desire for that option. It is in a goal-seeking state, which is satisfied only if the option is performed; and it is in a state that can lead it towards that goal, so we may assume, only via a belief that the response it disposes the agent to produce is a way of performing the option. This is desire in the sense in which to desire an option is to be fixed on it or to be set on enacting it: if you like, to have the intention of performing it. I shall call it option-desire. An agent desires an option in this sense if and only if it is disposed to choose it from among the options that it recognises as feasible.

Options are best understood as the basic items of agent control: as possibilities such that when an agent acts on a desire for an option, it does so without antecedently forming a desire for some means of realising it. Of course when an agent acts on an option-desire it will produce a response that it believes will realise the option; the desire will be belief-channelled, as we put it. But the agent will not have formed an antecedent desire for that response; antecedently, it will have conceived of the response only as its implementing the option: as reaching for the coffee, as trying to recall the poem, or whatever. It is crucial that an active intentional system come to form desires vis-à-vis options, since otherwise it would never act on beliefs as to what option is desirable; its beliefs and habits of inference would leave it suspended in a world of inactive contemplation and autonomic response.

If option-desires have to be recognised in any agent that acts on the belief that an option is desirable, there are other sorts of desires that have to be recognised in any agent that forms its beliefs about what option is desirable on the basis of registering the desirable properties of that option and its possible outcomes, together with the probability of those outcomes. We have pictured the intentional subject as operating in this fashion: as it forms beliefs on the basis of registered evidence, so it forms decisions on the basis of the valued properties, which it registers in the different options and their different possible outcomes. Such an agent will naturally be represented as deliberative: when it behaves up to par, then it comes to the belief that an option is desirable by inference, however implicit and defeasible, from the belief that the option and its likely outcomes have such and such desirable properties; recognition of the presence of those properties causally engenders the belief and it does so because their presence defeasibly supports the belief.

If we suppose that an agent is deliberative, then we must take it to be generally disposed, for any property that it believes desirable, to choose a bearer of the property. The disposition will not be sure-fire, of course, since other options may have different desirable properties. It will be sure-fire, at best, when the agent chooses from among options that leave it indifferent in other regards. If the belief that a property was desirable did not make it appropriate to have a disposition of this kind, then it would make no sense to ascribe such beliefs to an agent. But if the agent is disposed to choose the bearer of any property that it believes desirable, then it is in a goal-seeking, belief-channelled state relative to that property. It is in a state that is satisfied, for appropriate scenarios, only if the bearer of the property is chosen; and it is in a state that presumably leads the agent towards that choice, towards that goal, only via an appropriate, channelling belief. I describe the sort of state involved as a property-desire.

An option-desire is a disposition to choose one particular option from among the various alternatives that are recognised as feasible in any situation. It is a state of being bent on a particular option, as we put it earlier. A property-desire is a disposition to choose a bearer of the property, at least from among any set of alternatives that otherwise leave the agent indifferent. It is, in more ordinary parlance, a state of cherishing or prizing the instantiation of the feature in question. What distinguishes these desires, under our picture, is their relative status and their distinctive objects. Property-desires are more basic in status than option-desires, since the role they play requires them to determine which option-desires an agent will form. This difference in status corresponds to a difference in the objects of the two sorts of desire. The option-desire focuses on a sort of way the world may be made to be: namely, the sort of way it will be in the event of a certain choice. The property-desire focuses on a feature that is perceptually or otherwise salient to the agent in making such a choice: a feature that plays an important role in how the agent sees or thinks about the options it faces.[1]

Have we made room for all the desires that there are? The fact that both property- and option-desires are dispositionally characterised may raise doubts. Some desires in this dispositional sense, in particular some property-desires, often assume a richer profile. As a subject is moved by them, as they become activated in its psychology, the desires have a phenomenal presence to consciousness; like hunger or lust or a hankering for a smoke they become felt tensions that the subject finds itself anxious to relieve. I shall not discuss such desires explicitly but our schema would appear to allow them a place. To mention just one possibility, we might identify phenomenal desires as dispositional desires that are conditionally tied to the presence of an associated sensation. Thus we might identify hunger as a dispositional desire for food that is conditionally tied to the presence of hunger-pangs. Hunger on this rendering would be the desire for food-to-relieve-hunger-pangs rather than the desire for food as such. It would be the sort of desire that is not satisfied, at least not in the fullest possible way, if food comes along only after the hunger-pangs have disappeared (Pettit and Smith 1990).

I began this section with a worry about the view of intentionality developed earlier: the worry that the view does not give explicit recognition to desires, only to beliefs and habits of inference. I hope that my remarks about option-desires and property-desires will have dispelled that worry. Although the picture does not explicitly include desires, still it leaves room for them to be recognised as elements that are essential if the picture is to represent active and deliberative intentional systems.

We have stressed all along that there are variations in how any intentional agent can be interpreted, variations in the states that may be ascribed by way of making sense of the agent's performance. In introducing desire into the image of the intentional agent, I should emphasise that this openness of interpretation applies here also. Do we really say that intentional agents both believe that options and outcomes have certain desirable properties and entertain corresponding property-desires? Do we say that they both hold that options are desirable and form corresponding option-desires? Or do we look for a greater economy, perhaps reducing some of these elements further? This sort of question I would prefer to leave open, as I leave open the question of whether we should associate certain habits of inference with structural beliefs. Not every feature of the picture developed here, in particular not every feature associated with the role of desire, is to be regarded as mandatory.

That said however, I would like to add that if intentional agents in general are to be represented as operating on lines similar to those on which we human beings work, then there is good reason to find a place for both the sorts of desires I have distinguished and both of the corresponding desirability-beliefs. It appears that as we find our way to action, not only do we form intentions to do the things we do, we form such intentions on the deliberative basis of recognising properties that we cherish and usually expect others to cherish (Pettit 1991a, Pettit and Smith, 1990, 1992; also Morton 1990, chap. 3, and Schick 1991). Those properties, however implicitly recognised, constitute the sorts of reasons for our choices that we might quote in self-justification— though perhaps only very weak self-justification—to others. They are grounds such that their general appeal may make our choices reasonable or at least understandable—there may be other considerations that we are culpable for neglecting—to those other people. Furthermore, they are properties such that their hold on us explains the regret we often feel for having missed out on a particular option—the bearer of cherished and now unrealised properties—even when we think that we chose the most desirable option overall (Hurley 1989, Jackson 1984).

There is one further word I want to add to this discussion of the place of desire in our representation of intentional agents. One of the most lively debates in contemporary moral philosophy is over the question of whether desires are cognitive states or not: roughly, over whether there are facts—say, facts of objective desirability or rightness—such that whether desires are suitable or not depends on whether they correspond to those facts (see Collins 1988, Lewis 1988, Price 1988, Smith 1987 and 1988).[2] Desire will be cognitive in this sense if it is on a par with what is admitted to be factual belief—the paradigm of the cognitive state—or if its presence is necessitated by the presence of such a cognitive state. Thus if the presence of certain property-desires is necessitated by an understanding of the properties involved, or by a belief that they are desirable, then if we assume that those states of understanding and belief are cognitive, the property-desires will be cognitive too. And equally if the presence of an option-desire is necessitated by the belief that the option is desirable, and that belief is a cognitive state, then it will also count as cognitive (Pettit 1987a, Pettit and Price 1989).

The picture developed here suggests that desires become appropriate—if you like, rational—in the light of corresponding beliefs about desirability. Each sort of desire is governed by a desirability-belief and, if things are working properly, then a subject should have a property-desire or an option-desire if and only if it sees the property or option as desirable. But what I want to stress, in conclusion, is that the picture does not commit us to the claim that desires are cognitive states. There are at least two different ways in which we might break the link with cognitivism.

First, we need not think that the property-desires and option-desires mentioned spring by necessitation from the states that make them appropriate. We can represent the association with the other states as something weaker than necessitation; we can think of the desires as appearing more or less contingently—by a contingency of agent, culture, or species—in the wake of those states. Believing that a property or option is desirable may incline one, other things being equal, to desire it but there may be nothing strictly incoherent in the notion that one has that belief without the corresponding desire (see Smith 1987). Some will argue that even the possibility of practical unreason, in which desires fail to match beliefs about what is desirable, is sufficient to establish this result. (On practical unreason, see Pettit and Smith forthcoming.)

The other way in which we might break the link between our presentation of desire and the cognitivist doctrine would allow us to think that the states that trigger desire do so by necessitation; it would allow us to think that just to understand what kindness is, or just to believe that kindness is desirable, means having a corresponding property-desire; that just to believe an option desirable is to desire it. While maintaining such things, we can break the link with cognitivism by denying that the triggering state is properly a cognitive one. We can say that though it behaves like a cognitive state so far as it appears in inference and the like, still the state implicitly involves the presence of a desiderative factor. And thus we can hold that it is no wonder if the presence of the state necessitates the presence of a corresponding desire (Blackburn 1984a, chap. 4).

Intentional States

Ontological Character

To instantiate an intentional property is to be in an intentional state or to undergo an intentional event. To instantiate a property like that of believing that p or desiring that q is to be in the corresponding belief or desire state. To instantiate a property like that of noticing that r is to undergo the corresponding perceptual event. The state is the concretisation of the one sort of property, the event of the other. Where the property is something abstract, something that may be realised here or there, by this or that subject, the state or event is a concrete thing, an entity that is, or is part of, an ostensible substance.

If we are to understand intentionality fully, then one thing that we need to be clear about is the standing that intentional states and events have within intentional subjects. I restrict myself to states in my discussion here, as the extension to events is unproblematic; in any case it is generally possible to treat events as changes in respect of states, so that the extension may not even be necessary. I also restrict myself, for the moment, to intentional states with contents involving trees and birds and such familiar objects: intentional states, more generally, with contents that can be articulated easily in ordinary language. For most of the discussion I ignore intentional states with contents that bear on items like perceptual cueing features: intentional states with contents that may only lend themselves to being articulated in the language of scientific psychology. I return to this topic at the end of the section.

Every intentional subject with which we are familiar is characterised, not just by intentional states, but also by more common, naturalistic ones. By naturalistic states I mean the sorts of states recognised in the natural sciences: physics, chemistry, biology, and so on. Some of these naturalistic states will be molar, such as the subject's weight or height or number of parts. Others will be molecular, consisting of the having of parts that are in certain states: arms that are strong, a brain that is normally functioning, neurones that are in such and such a configuration. What we need to be clear about in particular with intentional states is how they relate to naturalistic states of these kinds.

The reason why it is essential to be clear about the relation between intentional states and naturalistic states is that both sorts of state are relevant to behaviour. Consider any action that we trace to states of belief and desire in the agent. That same action will be a naturalistic event that can be traced to certain of the agent's naturalistic states, in particular to its neural or electronic configurations. Thus we see that an agent's intentional states are naturalistically demanding; they require the agent to be naturalistically such as to produce certain naturalistic responses. It is necessary to be clear about the relation between intentional and naturalistic states, because it is necessary in particular to understand how intentional states can be naturalistically demanding in this way.

Two plausible assumptions have dominated modern discussions of the relation between intentional and naturalistic states and I am prepared to go along with them. The first is a negative, non-definability assumption; the second is a positive, supervenience one.

The negative assumption is that it is not possible to define an intentional property such as the property of believing that p in terms of naturalistic properties. It is not possible to define it even in the weak sense of identifying a lawlike, biconditional correlation—one that is not logically necessary—between an intentional property and naturalistic properties. A suitable correlation would go like this: x believes that p, at least in naturalistic worlds like ours, if and only if x displays this or that or the other naturalistic configuration. The point frequently made against the possibility of such a correlation is that there seems to be no end to the different naturalistic ways in which a believer that p may be. The believer that p may vary indefinitely in height or weight, of course. But it may also vary, it would seem, in the precise configuration of its neurones and indeed, given the possibility of robotic intelligences, in the sort of physical stuff of which it is composed. Take any believer that p and specify exactly what its naturalistic states are or could be, so far as current evidence goes on such believers. There will always be some further variation imaginable that is consistent with its remaining a believer that p. Thus there seems to be no hope of defining what it is to be a believer that p in naturalistic terms.

The positive assumption that I endorse is the complement of this negative one; where the negative assumption denies a certain linkage between naturalistic and intentional properties, this assumption asserts a connection. Imagine that we are presented with a naturalistic system that has a certain intentional profile: a certain configuration of beliefs and desires. Can we identify certain naturalistic conditions satisfied by this system such that given those conditions—given the scenario they specify—the system could not have differed intentionally? The supervenience assumption is that, yes, we can.

First scenario. Suppose we keep the system itself naturalistically unchanged in its internal make-up, neural or electronic. Suppose that it remains the same system—though perhaps with parts that change in the way our cells change—and that it satisfies exactly the same general naturalistic properties. Will this scenario ensure that the system is not of a different intentional profile? No, it will not. A subject believes that a particular thing in its environment is of a certain category: say, that that tree is a eucalypt. Imagine now that there was no tree there, only an illusion, but that the system was identical in its naturalistic properties. Would the subject still have believed that that tree was a eucalypt? Clearly not (McGinn 1989, pp. 36–43; Salmon 1982). The subject might have believed that there was a tree there and that it was a eucalypt. But it would not have believed that that tree—that very tree—was a eucalypt. The difference in the external, naturalistic environment would make for a difference in the subject's intentional profile. The property of believing that that tree is a eucalypt is only displayed, so we think, by a subject that has a particular relation: it satisfies a relational property vis-à-vis this particular tree.

Second scenario. Not only do we keep the system itself naturalistically unchanged, we also keep its environment—for safety, we keep the rest of the things in the world, past and present—naturalistically unchanged. Not only do we keep the system unchanged in regard to its naturalistic properties, we also keep its naturalistic relations fixed: we fix its relational properties vis-à-vis naturalistic objects—they are a sub-class, we may take it, of its naturalistic properties—and we also fix the naturalistic objects involved in those relations. Will this provide a guarantee of intentional identity? Again, no.[3] There is a further problem, which was also relevant for the first scenario: though the system and its world remain naturalistically unchanged, the laws or regularities governing the naturalistic may differ, in which case the behaviour of the system will surely be different too. And if the behaviour of the system is different, the intentional profile cannot remain the same.

Third scenario. The system and its environment are naturalistically unchanged and obey the actual naturalistic laws. That is to say, we keep the system unchanged in regard to its naturalistic properties, its naturalistic relations, and the naturalistic regularities. Does this mean that the system will necessarily display the same intentional profile, the same set of beliefs and desires? Can we imagine a variation at the intentional order consistently with this degree of fixture at the naturalistic? The supervenience assumption is that no such variation is possible; or, strictly, that no such variation is possible provided that there is no non-actual sort of stuff added to the subject in the process: in particular, no purely mental stuff of the kind that Descartes envisaged at the source of intentionality.

The assumption is intuitively plausible. How could I come to be believing or desiring different things without any change in my naturalistic make-up, in the make-up of the naturalistic world, or in the laws that govern the naturalistic universe? It is as difficult to imagine as it is to imagine a painting remaining unchanged in its naturalistic properties while altering in regard to its aesthetic: becoming less beautiful without changing in any naturalistic manner. If we do not rule out Cartesianism as logically incoherent—as we probably should not—then we may be able to imagine me changing intentionally through the infusion of the sort of thinking substance that Descartes envisaged but it is agreed that no such intervention occurs. Thus it seems inevitable under the scenario given that I will retain the same beliefs and desires as before.

But if the assumption is so plausible, is there anyone who would deny it? There is. Suppose that not only do we think that Cartesianism is logically coherent, obtaining in some possible world; we are Cartesians and think that it is true at the actual world: we think that intentional states involve configurations in non-naturalistic stuff. If we believe this, then we will deny the assumption in question. We will believe that without any naturalistic change in properties, relations or laws, and without the addition of any non-actual stuff, an intentional system might be different in regard to its beliefs or desires: it might change in regard to the stuff that actually fixes its intentional states. There might be less of that stuff around, it might be differently organised, or it might obey different (non-naturalistic) laws.

The assumption I make about the intentional in relation to the naturalistic is a familiar sort of supervenience assumption. Suppose we take two possible worlds that both obey (all and only) the actual, naturalistic laws and in neither of which non-actual stuff figures. My assumption is that those worlds may involve intentionality and that if they are naturalistically identical, they will be intentionally identical too. Naturalistic identity ensures intentional identity so long as comparisons are not disturbed by the presence of non-actual stuff or by a divergence from the naturalistic laws that actually govern the universe. Same naturalistic properties and individuals, same intentional profile; different intentional profile, different naturalistic properties or individuals. (On the varieties of supervenience, see Horgan 1984 and Kim 1990.)

There are three features of this supervenience assumption philosophers will wish to note; others may prefer to skip. The first is that it postulates a global form of supervenience; it says, subject to the provisos, that naturalistically identical worlds are going to be intentionally identical too (Currie 1984). This formula ensures that for any system, we will count its naturalistic relations as well as its naturalistic properties in the supervenience base of its intentional states; keep the world naturalistically identical and these are bound to remain the same. Some philosophers are bolder in their supervenience claims, arguing for the local supervenience of intentional profile on the neural or electronic— in effect, the intrinsic, naturalistic—constitution of the subject in question. But such a supervenience thesis is highly dubious, for a reason already noted: with a change in the objects that populate the environment, there may well be a change in the things the subject has beliefs about and therefore, by most counts, a change in the beliefs themselves. Here the subject believes that that building is tall; there he cannot have such a belief, for there is no building to connect with. I prefer to avoid such objections and to stick to a formula that it is hard to imagine people rejecting.

The second feature of our assumption worth noting is that the supervenience postulated is a contingent form of supervenience (D. Lewis 1983). If we said that any two worlds that were naturalistically identical—and perhaps obeyed the actual naturalistic laws—were going to be intentionally identical too, then we would be defending a necessary supervenience doctrine. The thesis, if true, would be true at every possible world. But we assert the supervenience relationship only of pairs of worlds in which there is no non-actual stuff to complicate things: only of pairs of worlds that satisfy a condition independent of the supervenience itself.[4] Thus our thesis, even if it is true, is not true at every possible world, only at every world in which there is no non-actual stuff. Are there necessary supervenience truths? Plausibly, yes. We can say of any two possible worlds that if they contain matter distributed in space at exactly corresponding points then they will contain the same shapes. We shall speak of supervenience again in this book, but in general our concerns will be with contingent forms of the supervenience claim.

The third feature that I want to note about our supervenience assumption is that it is unusually demanding in the supervenience base and, consequently, that it is a relatively weak assumption. It does not require the pairs of worlds addressed in the assumption just to be naturalistically indiscernible, to be qualitatively identical in naturalistic respects. It requires them to be naturalistically identical, involving the same naturalistic individuals as well as the same naturalistic properties. One general reason for this requirement is a desire to keep the assumption as uncontroversial as possible. A more specific reason is to cope with the fact that if in this world I have a belief about X but in that world a belief about Y, it is not enough for me to have the same beliefs in both worlds that X is exactly similar to Y; X must be just the same individual as Y.[5]

Back to the relation between intentional and naturalistic states. What we have postulated so far, in common with much contemporary thought, is that while intentional profiles contingently supervene on the naturalistic—on the naturalistic properties and relations of the subject—they cannot be defined in terms of the naturalistic. The non-definability and supervenience assumptions together support a view as to the standing of intentional states in intentional systems, or at least in intentional systems of the familiar naturalistic kind. They suggest that intentional states are, in a sense to be explained, higher-order naturalistic states.

Take the elasticity of an eraser. If an eraser is elastic then it is in a naturalistic state—it is of a molecular structure—which ensures that it displays elasticity behaviour: it bends when put under such and such pressure and regains its shape afterwards. We might think of the elasticity of this eraser as being the underlying naturalistic state: that is, as being the particular molecular structure that realises or fulfills the elasticity condition. Or we might think of it as the higher-order state of being in such a lower-order, realiser state—maybe this, maybe that—as fulfills the condition: call this higher-order state the role state as distinct from the realiser state (Jackson and Pettit 1988). I shall assume that the elasticity is the higher-order, role state, but nothing much will be affected by that choice. Taken in either way elasticity counts as a second-grade sort of naturalistic state and all that my assumption means is that second-grade gets interpreted as higher-order.

The elasticity of the eraser offers a distinctive model for how we should think of intentional states. The elasticity of the eraser is not definable in terms of naturalistic properties, because there may be any number of molecular or other ways in which an object may realise the elasticity condition. Again, the elasticity of the eraser is supervenient on the naturalistic properties of the eraser, for if it has the properties sufficient to fulfill the elasticity condition, then it has to be elastic. Given that intentional states satisfy corresponding non-definability and supervenience conditions, a plausible explanation is that they are essentially states of the same, naturalistically higher-order kind as the elasticity (McGinn 1982, chap. 2).

This will be a plausible explanation, at least, if we can identify a condition in the intentional case that corresponds to the elasticity condition in the other. And it turns out that we can, under the account of intentionality in the preceding section. To be elastic is to be such as to display elasticity behaviour: to bend under appropriate pressure and regain shape afterwards. And to be in a certain intentional state, according to the account in the preceding section, is to be such as to interact with a perceptual environment in accordance with appropriate regularities: regularities that strictly engage the relevant intentional property. Thus to believe that p is to display a pattern of interaction with a perceptual environment that invites, among other intentional states, the ascription of the belief that p. It is to be governed, when things go well, by certain intentional regularities that involve the property of believing that p. This will include the behaviour of forming the belief that q on learning that if p, then q; the behaviour of trying to make it the case that not p on coming to believe that it is desirable that not p; and so on.

Given this parallel, the twin assumptions of non-definability and supervenience suggest that the state of believing that p is the higher-order, role state of being naturalistically such—such in regard to ground-order naturalistic properties and relations—that the believer displays suitable belief-that-p behaviour. On such a higher-order picture, there may be indefinitely many ways of being a believer that p—as there may be such ways of being elastic—and so we can see why the property of believing that p should not be definable in basic, naturalistic terms. Again, on this picture, the fact that a system is in one of the naturalistic ways that supports belief-that-p behaviour ensures that it believes that p, as the fact that an eraser is of such a molecular structure that it bends appropriately ensures that it is elastic; and so we can also see why the property of believing that p should be supervenient on the naturalistic properties and relations of the believer. The higher-order picture is a hypothesis that neatly explains non-definability and supervenience. Moreover, it explains those features better than any rival hypothesis I can envisage. And so I propose to endorse the picture.

The analogy with the elastic eraser serves to demonstrate how naturalistically unexceptional intentional states are on the higher-order picture. But there is a great difference between a simple disposition like elasticity and an intentional state like believing that p and this is also important to note. The source of the difference is that whereas the behaviour involved in being elastic—that of bending under appropriate pressure—is relatively simple, behaviour such as belief-that-p behaviour is extremely complex. It involves a whole range of different responses, as we have seen, not just a single unvarying kind of display. Moreover, it involves responses that place requirements on the other intentional states of the agent, apart from its belief that p, and that place requirements in some cases on the external context that the agent inhabits. It introduces what we may describe as a collateral and a contextual form of complexity.

The collateral complexity amounts to a familiar sort of intrasubjective holism. On our account in the last section, a subject will display belief-that-p behaviour only so far as it simultaneously displays belief-that-q behaviour, belief-that-r behaviour, and so on. An agent cannot interact with its environment so as to display the first sort of behaviour without at the same time displaying the other sorts of behaviour too. Thus if the agent interacts in such a way that it counts as believing that p, it will interact in such a way that it counts equally as believing that q, believing that r, and so on. The belief-that-p behaviour, unlike the behaviour characteristic of elasticity, is not an atomistic sort of response and it does not serve to define an atomistic state like the elasticity disposition. There is no such thing as the solitary intentional response and no such thing therefore as the solitary intentional state.

So much for the collateral complexity of belief-that-p behaviour. Belief-that-p behaviour, unlike the elasticity response, may also involve contextual complexity. We find such complexity, for example, if 'p' is a proposition that refers directly to something in the environment: a proposition like 'that tree is a eucalypt'. When we speak of the eraser being naturalistically such as to bend under appropriate pressure, we mean being such in regard to internally realised naturalistic properties: such in regard to molecular structure. When we speak of an intentional system being such as to display belief-that-p behaviour, and 'p' is a proposition of that kind, we mean such in regard to both naturalistic properties and naturalistic relations. For a system to be such as to display belief-that-p behaviour it must not only be internally organised in a certain way, it must also have a certain relationship with that particular tree: it must inhabit a world where that tree exists and it must have made contact with that tree. To be such as to display belief-that-p behaviour is to enjoy a suitable contextual embedding as well as a suitable intrinsic constitution.

This point can do with emphasis. It is natural on the higher-order picture, particularly in view of the parallel with elasticity, to think of an intentional state like believing that p as a sophisticated kind of disposition. But a disposition may be context-bound or context-free and the point here is that where 'p' refers directly to something in the environment, being naturalistically such as to display belief-that-p behaviour—that is, believing that p—will be a context-bound disposition. Consider two familiar dispositions like being brave and being envious. A person may be brave without ever acting bravely: the occasion may not arise in the actual context. Thus someone's being brave does not require that they inhabit or have intervened in any particular context or kind of context. A person may not be envious on the other hand without having been exposed to the object of the envy and without having exhibited at least the internal response—say, a feeling—characteristic of envy. Thus someone's being envious does require that she inhabit a particular context and does require that she have interacted with that context in a particular way. If being naturalistically such as to display belief-that-p behaviour—or, more simply, believing that p—is to be compared with such dispositions, then it should be grouped with envy rather than bravery. It is a disposition that presupposes an actual relationship with context; it is not merely the disposition to form such a relationship, if an appropriate context presents itself. In a phrase, it is a context-bound, not a context-free, state. The point comes up again in the next section.

All of this is simply to repeat the point that while an intentional state like believing that p is naturalistically supervenient, the minimal supervenience base may include the naturalistic relations of the system as well as the naturalistic properties that it realises in itself: its intrinsic naturalistic states, neural or electronic. The naturalistic base for being envious is not confined within the skin of the envious person; it extends to include the historically grounded relations of that person to the object of envy. Similarly the naturalistic base for believing that p, where 'p' refers directly to something in the environment, extends to include the historically grounded relations of the believer with that object of reference as well as the neural or electronic states that play a suitable role within the subject.

But though the supervenience base for most intentional states involves relations as well as intrinsic states, we should recognise that there still is a sense in which the intrinsic base is pivotal. Suppose that a context-bound naturalistic state, a state involving a naturalistic relation of some kind to context, ensures that a subject displays belief-that-p behaviour. In any such case, there must be a context-free naturalistic state of the subject that ensures that the agent displays that behaviour; after all, the behaviour is driven, as it were, from within. Again suppose that a context-bound naturalistic state ensures that a person is envious of someone else. In such a case too, there must be a context-free state within the person that ensures that she displays envy; here also, the behaviour is produced by factors within the subject. An analogy may help to make the point more telling. Consider the context-bound naturalistic state of an object x, which consists in its being heavier than y. That state ensures that x outbalances y. But if it does this, it does it by virtue of the presence of a context-free naturalistic state in x— its mass—which ensures that x outbalances y.

But if there is a context-free naturalistic state involved in this way in believing that p, then we can express the view we have taken in a schema that privileges that state. A person is envious of another, we can say, if and only if there is a context-free naturalistic state in that person that ensures that the person displays envy and which is embedded in the actual world, specifically in a history of interaction with the object of the envy. A subject believes that p, we can say in parallel, if and only if there is a context-free naturalistic state in that subject that ensures that the subject displays belief-that-p behaviour and which is embedded in the actual world, specifically in a history of suitable commerce with any items to which 'p' directly refers. This context-free state is the neural or electronic realiser of the belief-state within the subject. We shall refer to it henceforth simply as the realiser of the belief.

We have been stressing the relative complexity of belief-that-p behaviour: first, its collateral complexity and, just now, its contextual complexity. This complexity means that even if we take intentional states as higher-order, naturalistic states like elasticity, we can leave lots of contrasts between the two sorts of states in place. There is one further respect in which belief-that-p behaviour is more complex than elasticity behaviour that I would like to mention too. This involves epistemological rather than ontological complexity and it connects with issues that will concern us later in the book.

Belief-that-p behaviour is ontologically complex, because it involves a range of collaterally and perhaps contextually complex responses. It is epistemologically complex, because it is no easy matter to see what constitutes such behaviour: to see exactly what range of responses is required. There is no plausible, reductive account—no account in non-intentional terms—of the range of responses to be expected of any subject that believes that p. As believers ourselves, we may be quite sure that this and that subject, subjects that differ in other ways among themselves, both display belief-that-p behaviour and both believe that p. But the fact that we cannot give a reductive account of belief-that-p behaviour means that we will be hard pressed to say why we attribute that belief in the two cases. I shall assume for now that this problem need not concern us, as I shall be dealing with it later in the book: most explicitly in Chapter 5, when we discuss intentional interpretation.

This completes our discussion of the ontological standing of intentional states. An intentional state like the belief that p is a higher-order, naturalistic state or, if you prefer, a naturalistic role state. Although it is significantly more complex in various ways, it relates to ordinary, naturalistic states in the same fashion as a disposition like elasticity. It is, in a word, a common-or-garden sort of thing.

Causal Relevance

Having settled on the higher-order view of intentional states, we straightaway face a problem. With an intentional system, we invoke its intentional states to explain any actions it performs and indeed to explain various other adjustments it makes—say, the adjustment of coming to believe that p under a certain perceptual input. Explaining actions and other adjustments in this way, we show that we think the explanatory intentional states are relevant to the occurrence of those changes. More particularly, we show that we think the states are causally relevant to their occurrence, for what can relevance mean in this context, if not causal relevance (Davidson 1980, essay 1; D. Lewis 1986, essay 22)? But now a problem confronts us. How can higher-order, role states be causally relevant? How can they be significant factors in ensuring that an agent acts and adjusts after a certain pattern? (For some treatments of the problem see Blackburn 1991, Block 1990a, Dretske 1988. The treatment given here follows Jackson and Pettit 1988, 1990 a, b, c, and 1992c. See also Pettit 1992.)

In asking this question, and indeed in future discussions, I do not presuppose any particular account of what constitutes causal relevance in a property or in the realiser of a property: say, in a state. Thus I leave open issues like the following: whether there are objective causal powers—say, mass or charge or spin—that are relevant in the sense of being primitively efficacious in producing things; whether it is sufficient for the causal relevance of a property that it figures in the antecedent of a regularity that we regard as causal; and whether it is sufficient for its causal relevance that the property identifies a necessary, or sufficient, or perhaps just probabilifying condition for the occurrence of something. I assume that we all think of certain properties as causally relevant, and that we therefore assign causal relevance in suitable cases to their realisers, but I do not need or wish to commit myself to any particular account of causal relevance.

Although I do not presuppose any particular account of causal relevance, however, I do make a general assumption about how higher-order properties get to be causally relevant. I assume that if a higher-order property is relevant to a certain effect, this is because of how it relates to lower-order properties that are relevant to that effect. The elasticity of the eraser is causally relevant to its bending. I assume that that property is relevant because of how it relates to the lower-order properties that are relevant: because of how it relates to the molecular structure of the eraser. This assumption is surely plausible. After all, the elasticity brings about the bending only so far as the molecular structure brings it about, not in virtue of an autonomous power. Were the molecular structure irrelevant—had the bending occurred just because of the presence of some alien force—then the elasticity would have been irrelevant too.

Now to the problem. What is the relation between a higher-order and lower-order cause in virtue of which the higher-order counts as causally relevant? The two sorts of factors do not combine as necessary parts of the same causal whole; if the molecular structure has done its work, there is no need to wait on the contribution of the elasticity, or indeed vice versa. Again, the factors do not relate as temporally different stages in the same causal chain; there is no time-lag between the contribution of the one and the contribution of the other. So how do they relate to one another? In particular, how does the higher-order relate to the lower-order in such a way that both sorts of state can count as causally relevant? Unless we can find an answer to this question, we will have to regard the causal relevance of higher-order factors as mysterious or illusory. We will have to assert that, though higher-order factors are relevant, we cannot give an account of how they are relevant; or we will have to conclude that they are not relevant in any distinctive way at all: we will have to endorse a sort of epiphenomenalism about higher-order properties and states.

One response to the problem raised will be to say that it is misconceived. The most likely suggestion on these lines will be that for any apparently higher-order factor like elasticity, we can reinterpret it as a lower-order causal influence: as a lower-order influence described, as it were, in higher-order terms. We can take the elasticity of the eraser, not to be the role state of being molecularly such as to bend under appropriate pressure, but the molecular structure itself: to be the realiser state, as we called it, which instantiates the role state. And, so the response goes, this recasting makes it unproblematic how the factor in question is causally relevant. It is relevant through being identical with the lower-order factor. The elasticity is relevant to the bending because it just is, under another name, the molecular structure that causes the bending. And the belief that p is relevant to some response that it rationalises, because it is just the neural or electronic state that causes the agent to make that response; it is the realiser state that instantiates the role state of being neurally or electronically such as to display belief-that-p behaviour, not the role state itself (D. Lewis 1983b, essay 7; also Armstrong 1968, Smart 1959).

I believe that it is possible, so far as ordinary usage goes, to take someone's belief that p, or indeed to take a disposition like elasticity, to be either the higher-order, role state or the corresponding realiser state (Jackson and Pettit 1988). I have assumed that an intentional state, like a disposition, should be identified with the role state but I recognise that this assumption is not mandatory. Thus I admit that the response under consideration here is perfectly coherent. But though it is coherent, the response is to no avail: it does not solve our problem.

When we hold that an intentional state like believing that p is causally relevant to an event, say the performance of an action, we have two things in mind: both that the state in question played a certain causal part and that it did so in virtue of the property of being a belief that p. On the response under consideration, the realiser state is given a part in the genesis of the event explained by it but it plays that part in virtue of its neural or electronic character, not in virtue of constituting or helping to constitute a belief that p. The response gives a certain causal relevance to the belief that p, but not a causal relevance that the state possesses in virtue of being the belief that p: not a causal relevance that it has *qua* belief that p. Thus the response will allow our problem to arise in a new format. How can a realiser state *qua* belief that p have causal relevance over and beyond its relevance *qua* neural or electronic configuration?

The point I am making against this sort of response can be made very nicely against a parallel response in the case of simple dispositions. Take the dispositions of electrical conductivity, ductility, and opacity. The realiser state for all three dispositions may be the same in a given metal, consisting of the cloud of free electrons that permeates the metal and holds the atoms in a solid state (Menzies 1987, pp. 566–67). Now consider an effect we might invoke the conductivity to explain: say, the shock that someone receives when an electric current is passed through the metal. What makes the conductivity causally relevant? It will do no good to say that the conductivity just is the realiser state and that it is relevant, so far as that realiser state mediates the shock. This response will do no good, because so far as this consideration goes the ductility and the opacity of the metal are causally relevant to just the same extent, contrary to all our intuitions. The response preserves the causal relevance of the conductivity only in a sense of relevance shared with the ductility and opacity. It does not make the conductivity relevant *qua* conductivity.

The point can be made directly in the intentional case, if we are prepared to grant with some theorists that one and the same neural—or presumably electronic—state might be the realiser state for different beliefs: say, the belief that p and the belief that q (Jackson and Pettit 1990c; Ramsey, Stich, and Garon 1990). The identification of someone's belief that p with that realiser state will not solve the problem of what makes the belief causally relevant to effects it is invoked to explain. It will leave us with the problem of what makes the realiser state causally relevant to those effects *qua* belief that p, rather than *qua* belief that q, or indeed *qua* neural or mechanical configuration.

Back then to our problem. We shall consider the problem in the form in which it arises for role states, and we shall offer a resolution of the problem in that form, but the solution offered can easily be adapted for someone who thinks of intentional states in the realiser mould (Jackson and Pettit 1988). What we need is a story that will show how a higher-order, role state can be just as causally relevant to the occurrence of an event as its lower-order realisation. It is common in philosophy to recognise that causal factors may relate in a coordinate way as necessary parts of a single causal whole, or in a sequential way as necessary parts of a single causal chain. We know that role states cannot collaborate with realiser states in either of these ways. What we need to do then is to identify a third kind of causal collaboration, a form of collaboration suited to factors of different orders.

There are two things to note about the sort of story we are after. First, it will assume with any higher-order and lower-order factors to which it applies that the lower-order factor is unproblematically relevant in producing the given effect; the problem will be to explain how the higher-order factor is relevant too. This assumption means that we can speak of the lower-order factor as producing the effect, without further comment on what such production involves. But it does not rule out the possibility that the lower-order factor is itself higher-order in relation to some other factor, and so on down a looming hierarchy (see Pettit 1992). The issue comes up again in chapter three.

The second thing that needs to be noted about the story we are after is that it is an ontological story, not an epistemological one. It is an account of how, in the objective arrangement of things, higher-order causes relate to their lower-order counterparts. It is not an account of how we must subjectively come to know that certain higher-order causes are causally relevant. The story will ideally give us an account, for example, of the relation between the elasticity that makes an eraser bend and the molecular structure that does so, articulating a constraint on such joint relevance. But it will not suggest that in order to recognise that the elasticity is relevant to the bending, we have to know about the relationship it enjoys to the molecular structure. Knowing about that relationship may be sufficient for recognising the relevance of the elasticity; but it will not be necessary, except perhaps in some special cases (see chapter 5).

With these points clarified, we should press on in the attempt to deal with our problem. The best-known story about how higher-order causal factors collaborate with lower-order is that they do so through being supervenient on them (Kim 1984). The idea would be that for any factor at a given order that is causally relevant to the production of a certain effect, any higher-order factor will be causally relevant in virtue of being supervenient on the relevant factor. But this idea comes to nothing.

The nature of the cloud of free electrons that permeates the piece of metal in our earlier example is causally relevant to the shock transmitted, but of the distinct dispositional states that supervene on the electrons, not all are causally relevant to the shock: the electrical conductivity is, the ductility and the opacity are not. And if this example seems fanciful, there are more mundane cases with which to support it. A piece of putty rests on a wire mesh, rather than falling through, due to the distribution of its molecular parts. Imagine now that the putty is heated, that the distribution of parts changes and that the putty does slip through the mesh. Two distinct higher-order changes are supervenient on the change in the distribution of molecular parts: one, a change in the shape of the putty, specifically a change to a shape that lets it through the mesh; and two, a change in size, in particular an expansionary change due to the effects of heating. These higher-order changes are not each relevant, let alone equally relevant, to the putty's falling through the mesh, though the story under discussion would imply that they are. The putty falls through the mesh because of the change in shape but despite the change in size.

The supervenience story fails to resolve our problem and we must press on in search of an alternative. The clearest case where we have states of different orders simultaneously invoked as causally relevant is that in which the higher-order state is picked out by existential quantification: that is, by a suitable use of the word 'some'. A clock makes a noise and the state responsible is the looseness of certain parts. Not knowing which parts and which state are responsible, however, we trace the noise to a state of the clock as a whole: that some of its parts are loose. Again, a piece of uranium emits radiation and the state responsible is the decay of certain constituent atoms. Inevitably, we will be unaware of which atoms and which state are responsible but, undaunted, we will trace the radiation to a state of the uranium as a whole: that some of its atoms are decaying. The some-state in each case is higher-order relative to the state of the loose or decaying constituents—the realiser state—in a manner that clearly parallels the role state. The problem is how it can be causally relevant at the same time as the realiser state: how it can relate to the realiser state in such a way that it has a distinctive relevance of its own.

The beauty of the cases involving some-states is that they make that problem relatively tractable; they point us towards a solution. It is more or less obvious in each of the examples given how the some-state can be causally relevant at the same time as the unidentified realiser state. The some-state is causally relevant in a way that presupposes the causal relevance of the realiser state: if that realiser state is not relevant then neither is the some-state. More specifically, the some-state is relevant so far as its realisation more or less ensures the presence and effectiveness of such a realiser state. Consider, perhaps subject to some restrictions, the various ways in which the higher-order state of the clock or the higher-order state of the uranium may be brought about. Under all or most of these ways of being realised, the realiser state is such as to bring about the noise or the radiation. And the actual realiser state does bring about the effect. Thus we can say that the higher-order state is causally relevant to the effect. Where the realiser state is relevant through producing the effect, as we may say, the higher-order state is causally relevant through ensuring, or more or less ensuring, the presence of that sort of productive state; it is relevant through raising the probability—non-causally raising the probability—of the presence of that sort of state.

How are we to describe the relationship between the higher-order state and the realiser state in cases where they are each causally relevant to a certain effect? A useful metaphor is suggested by the computer, in particular by the way in which an abstract computer program ensures that certain things will happen, through ensuring that things get organised in a certain way at the mechanical order. We can say, by analogy with this sort of case, that the higher-order type of state involves the realisation of a lower-order state sufficient to produce the type of effect that ensues; and we can say that in virtue of non-causally ensuring the presence of a suitable lower-order producer, the higher-order type of state programs for that type of effect. That some of the uranium atoms are decaying programs for the emission of radiation, even if the emission is produced by the decay of precisely these atoms or those, for the fact that some of the uranium atoms are decaying ensures or more or less ensures the decay of such atoms as will produce radiation.

The general idea in the program model, as we may now call it, is that a higher-order property is causally relevant to something when its instantiation ensures or at least probabilifies, in a non-causal way, that there are lower-order properties present which produce it. There are a number of separate elements in the model and it is worth spelling these out. A higher-order property programs for a certain result—say, the occurrence of an event E—just in case three conditions are fulfilled.

1. Any instantiation of the higher-order property non-causally involves the instantiation of certain properties—maybe these, maybe those—at a lower order.
2. The lower-order properties associated with instantiations of the higher-order, or at least most of them, are such as generally to produce an E-type event in the given circumstances.
3. The lower-order properties associated with the actual instantiation of the higher-order property do in fact produce E.

I hope that the notion of programming summed up in these clauses is intuitively clear. The factor that programs for an effect does not get to produce it, at least not in the sense in which the so-called producer state does so. It non-causally arranges things (it means that things are arranged) so that there will be such a producer state—maybe this, maybe that—available to do the work. For a given programming state there may be a hierarchy of progressively lower-order states that are causally relevant to the effect in question and it will arrange at the appropriate level for the presence of each of these. That some parts of the clock are loose arranges for the presence of these or those loose parts—the ones that cause the knocking—but it arranges also for the presence of such molecules as will make a noise in the course of the clock's operation.

Can we provide any argument for thinking that the programming relationship is crucial to the capacity of something like a some-state to be causally relevant, in particular to be causally relevant equally with the realiser state? Can we show that this is what is important rather than the fact, for example, that the some-state is supervenient on the realiser state? We can. Imagine a case where the supervenience remains but the programming disappears. John is giving a public lecture. Some members of the audience laugh very enthusiastically at one of his jokes—his family—and this embarrasses him. That some members laugh enthusiastically is supervenient on his family doing so but it is not causally relevant to his embarrassment; what is relevant is only that his family laughs in that way, for he would not be bothered if others did so. Thus the supervenience of a some-state does not ensure its equal causal relevance with the realiser state. What appears to be necessary is the programming relationship, for this is what is saliently missing in this example. There are many ways of realising the some-state—many ways of its being the case that some members of the audience laugh enthusiastically at John's joke—but not all of the realiser states will produce embarrassment in John. Thus the some-state in this case does not program for the embarrassment and that, plausibly, is why it does not have causal relevance side by side with the realiser state.

With the notion of a programming relationship in place, we can offer an account of how intentional states and the states that realise them can both be causally relevant to a certain effect: say, the performance of a certain action. The intentional state may program for that type of action, through arranging for a neural or electronic realiser state sufficient to produce such an action. Suppose the intentional state is the belief that p and that against the background of the agent's other intentional states this is held to explain the agent's A-ing. Under the account envisaged, the belief will program for the A-ing in the sense that no matter how it is neurally or electronically realised in the agent—and we may suppose that different modes of realisation are possible—the realising state will tend to be such as to produce an action of type A. The belief that p will not produce the action in the same sense as the realising state and if they are both causally relevant, they will be relevant in related ways. The intentional state will be relevant through arranging for the production, at the order of the realising state, of the action. The realising state will be relevant through producing the action at that order.

Should we accept the program account of how intentional states are causally relevant to the events they explain, in particular of how they are causally relevant consistently with the causal relevance of their realising states? The best argument for accepting it is that it would make sense of the causal relevance of intentional states, improving considerably on the most influential, current alternative: the story that gives them relevance through construing them as identical with electronic or neural states.

That story had the defect that it would make a state like the belief that p causally relevant but relevant in virtue of a property other than that of being the belief that p: relevant in virtue of being such and such a neural or electronic state. No defect of this kind detracts from the program account. Consider the case where the belief that p happens to have the same realiser state as the belief that q. Both the belief that p and the belief that q will have the same neural or electronic character but the program account explains how the belief that p may be causally relevant to an action of A-ing in a way in which the belief that q is quite irrelevant (Macdonald and Macdonald 1986). It explains how the belief that p may be relevant *qua* the belief that p. The belief that p will be relevant to the extent that if we look at the different ways in which that belief may be realised, the realiser will tend in each case to play a suitable role in producing the behaviour required for A-ing. The belief that q will not be relevant in this way. The realiser of the belief in the actual world—the one that happens also to be a realiser of the belief that p—plays a suitable role; but its other possible realisers would not do so.

But there are other considerations favouring the program account of how intentional states are causally relevant, apart from the fact that the account would resolve our original problem. Perhaps the most persuasive consideration is that the account would put intentional states on a par with a whole variety of causal antecedents; it would make them unexceptional sorts of causal factors. Not only are some-states causally programmatic. So too are a great variety of states that we invoke to explanatory effect in everyday reasoning.

An elastic eraser is pressed at the ends and it bends. Why did it bend? First answer: because of its elasticity. Second answer: because of its molecular structure. How can both of these states be causally relevant? Clearly, because the dispositional state, the elasticity, ensures the realisation of a lower-order state—a molecular structure—sufficient to produce bending under the pressure exerted. The dispositional state programs for the bending.

I try and fail to fit a square peg in a round hole of diameter equal to the side of the square. Why was it blocked? First answer: because of the squareness of the peg. Second answer: because of the impenetrability of this overlapping part of the peg. How can the squareness of the peg as a whole and the impenetrability of this particular part both be causally relevant? Again, the obvious answer would seem to be, because the squareness of the peg ensures the presence of an impenetrable part at a suitable point. The impenetrability of the part—now of this part, now of that—produces the blocking, and the squareness programs for it.

For a final example, consider a closed container that cracks under the pressure of the boiling water it contains: this, rather than the differential heating of the flask. Why did the container crack? First answer: because of the boiling of the water. Second (rough) answer: because of the momentum of this particular water molecule, or group of molecules, which shattered the first bond in the surface of the container. How can both states be causally relevant to the cracking? Easy. The boiling state of the water ensures a distribution of moving molecules such that there is almost bound to be some molecule that has a momentum and position sufficient to crack a molecular bond in the surface. The realising distribution does not produce the cracking but part of it does: this particular molecule. The boiling state is causally relevant so far as it more or less ensures the presence of such a productive factor and thereby programs for the cracking.

The squareness and the boiling-water explanations do not invoke states that are higher-order in the exact sense that a role state, or indeed a some-state, is higher-order. But in each case we find the crucial features that distinguish the case at the centre of our concerns. A basic state is agreed to be causally relevant—the impenetrability of the overlapping part, the momentum of the responsible molecule—and there is a question about how something less basic can be relevant too: in particular, something that does not collaborate with the basic factor either as a necessary part of a single causal whole or as a necessary part of a single causal chain.

We have made two points in favour of the program model of how intentional states have causal relevance: that the model would resolve our original
problem and that it would represent intentional states as causally relevant
factors of a common-or-garden kind. A third point worth adding immediately
is that it would also represent intentional states as significant causal factors.
Not all causally programmatic factors are significant. In particular, some such
factors are insignificant in this way: the knowledge of what lower-order factors
are operative is sufficient under normal assumptions to yield the knowledge
that they are operative too. A dispositional causal factor like the elasticity of the
eraser is insignificant in that manner. The knowledge that such and such a
molecular structure accounts for the bending of the eraser under such and such
pressure is sufficient to yield the knowledge that its elasticity is causally
relevant. But under the program model intentional states would not be insignificant causal factors of that kind in relation to the actions and other adjustments that they explain. The knowledge that such and such a neural or
electronic configuration was responsible in context for such and such an action
is not in itself sufficient for knowledge that a particular sort of belief was
causally relevant. The molecular structure of an eraser cannot produce bending, at least in the normal way, without the eraser counting as relevantly elastic.
But a neural or electronic configuration can produce a piece of behaviour
without the agent counting as having a relevant belief. The configuration
might occur, for example, without the overall neural or electronic organisation
that is necessary for the agent to be an intentional system.

Not only does the program model make intentional states into unexceptional causal factors, and into causal factors that are suitably significant. It also
has the virtue of being a plausible model. Whenever we take a program view of
how causal factors relate from one order to another, we must draw on certain
background assumptions. The assumptions that support the program construal
of the elasticity-molecular structure relationship are supplied by the very
definition of what it is to be elastic. The elasticity programs for a structure
sufficient to produce the bending because to be elastic is, by definition, to have
precisely such a structure. The assumptions that support the program view of
the squareness-impenetrable part relationship are supplied by elementary
geometry. And the assumptions that favour the program model of the tempera-
ture-moving molecule relationship are provided by a simple scientific understanding of what it is for a liquid to have such and such a temperature, in
particular to be at boiling point. It turns out that the assumptions required to
support the program view of the intentional state–realiser state relationship are
equally compelling.

The assumptions required are suitable design assumptions. If we are to think
that in a situation where it has suitable other intentional states, an agent's
believing that p programs for its doing A, then we have to assume that the
agent is designed so that in such a case it will be in a naturalistic state sufficient
to produce the response that constitutes A-ing. More generally, if we are to
think that an agent's intentional profile at any moment programs for the sort of

action that is intentionally required—say, that is rational in the light of that profile—then we have to assume that the agent is designed so that under normal functioning it will always be in a naturalistic configuration sufficient to produce such an action (Nozick 1981, chap. 4). But it is not unreasonable to make this sort of design assumption. We do so readily with the piece of artificial intelligence, as when we assume that the artificial chess-player will not normally let an opportunity pass to take an opponent's queen. We assume, in effect, that the designer is bound to have made the chess-player rational to that extent. In parallel to the artificial case, we may readily assume that any natural intelligence is going to have been designed, under evolutionary and perhaps cultural pressures, to meet suitable design specifications. We may assume that it will be designed so as normally to realise the sort of naturalistic state that produces any action—or other adjustment—that its intentional states require.

One proviso. We may reasonably make this design assumption provided that the action or adjustment is required in the sense of being rationalised: it is rational in the light of the intentional states involved. Mother nature has presumably provided agents with biologically appropriate desires—that is, desires whose pursuit tends to maximise inclusive fitness—and with a biologically appropriate sense of the desirable. But if this bequest is to work, she must also have equipped them in ordinary conditions to form beliefs that will not systematically thwart those desires—that is, rational, truth-seeking beliefs; and she must have enabled them generally to act rationally—that is, to act in a way that serves the desires well, according to the beliefs. In short, mother nature must have fashioned at least minimally rational agents (Cherniak 1986, but also see Stich 1990). If agents tend to be rational in the forming of beliefs and in the choice of actions then, assuming that their desires are biologically appropriate, this increases the probability that they will maximise their inclusive fitness (Dennett 1979, essay 1; Elster 1979, chap. 1). Up to now I have been careful not to identify intentional regularities with rational regularities, although I have used the latter to illustrate the former. But henceforth I shall assume that central to the intentional regularities that are satisfied by any intentional agent will be regularities characteristic of theoretical and practical rationality. Thus the sort of action or adjustment that can be said to be required by certain intentional states will generally be the sort of action or adjustment that they rationalise. Identifying those states as the cause will 'normalise' the response, showing how it falls under a certain rational norm (Pettit 1986a, forthcoming c).

I have made four points in favour of the program view of intentional states: that it solves our original problem, that it makes intentional states unexceptional causal factors and significant causal factors, and that it is a plausible model of those states, requiring only plausible background assumptions. There is one further point that remains to be made.

I mentioned earlier that any intentional system that displays intentional states with familiar contents—contents that bear on familiar entities and that can be articulated in ordinary language—is likely to display at the same time intentional states that have contents that can only be articulated in scientific psychology: contents that bear, for example, on subliminal perceptual cues. The idea is that intentional states may be nested, so that a system can act at once on the familiar object-centred belief that that tree is a eucalypt and on the unfamiliar cue-centred belief that some pattern or other is perceptually present (Jackson and Pettit 1992a). I ignored the nested picture earlier, opting to concentrate only on familiar intentional states. The reason was that earlier we lacked the resources to make sense of how both sorts of belief could be simultaneously relevant to the same action.

The last point to be made in favour of the program model is that as it makes sense of how intentional states and neural or electronic states can both be causally relevant to the same response, so it makes sense of how familiar and unfamiliar intentional states can both be relevant too. If the belief that that tree is a eucalypt depends for its causal relevance on the relevance of a belief with a cue-centred content, as both beliefs will depend for their causal relevance on the relevance of a certain neural or electronic state, then it is programmatically relevant in relation to that less familiar intentional state. The familiar sort of belief is such that its presence more or less ensures the presence of a productively suitable intentional state—maybe this, maybe that—in the less familiar category; its presence ensures that the system will be engaged, as we might say, with an appropriate cueing feature.

Many philosophers will resist the idea that the unfamiliar intentional states I have countenanced as a possibility really are intentional states. They will prefer to tie the category of intentional states to the familiar paradigms. Invoking a common term in this area, they may say that the unfamiliar states countenanced are sub-personal in a way that ordinary intentional states are not. I have some sympathy with this line but I think that we can only understand the motivation behind it, and we can only see the proper way to satisfy that motivation, when we put thought in the picture. Personal and sub-personal states can both count as intentional states in the generous sense allowed within the perspective of this chapter. But there may yet be a significant distinction between them. As will become clear in the next chapter, personal states can be taken as those intentional states that come under the control of thought, sub-personal states as those intentional states that do not.

In view of all the considerations rehearsed, I believe that we should endorse the program view of intentional states, using the program model to make sense of how such higher-order, role states can be causally relevant to the performance of actions and the making of other intentional adjustments. We shall have recourse to the program model again. We return to it in Part II, in defending the overall view of causal architecture that I describe as causal fundamentalism. Again we return to it in Part III, when we come to consider how the factors invoked in social theory can be causally relevant to the events they are supposed to explain. Reliance on the model is one of the distinctive features of this book.

Theories of Intentionality

The terrain that we have covered in this chapter has been intensively explored and charted in philosophical discussions of the past twenty years or so. It has been a major concern of traditional philosophers of mind and it has been the focus of attention for philosophers interested in plotting the likely fortunes of cognitive science and artificial intelligence. I believe that it might have been better had philosophers divided their attention more evenly between this terrain and the territory covered in the next chapter. Be that as it may, however, the fact that intentionality has been so thoroughly investigated puts me under an obligation, which I hope to discharge briefly in this final section. The obligation is to situate the sort of position emerging from our discussions within the constellation of positions that appear in the philosophical literature (see Sterelny 1990 for a survey).

The best place to start in situating the position we have adopted is with a formulation of what it is to believe something which we offered in the last section. A subject believes that p, we said, if and only if there is a context-free naturalistic state in that subject which ensures that the subject displays belief-that-p behaviour and which is embedded in the actual world, specifically in a history of suitable commerce with any items to which 'p' directly refers: it is a suitably anchored state, as we may put it. But this is just to say that to believe that p is to have in one a state that can play a certain causal role—that of producing belief-that-p behaviour—in particular a state of this kind that happens to be suitably anchored. As will be clear to any professional in the area, this formulation shows that the position adopted is a sort of functional-ism, or at least a close relative of functionalism. Functionalists say of any mental state to which they apply their theory that to instantiate that state is to house a realiser state which exercises a certain distinctive function: specifically, to house a state that plays a certain causal role. The position we have taken up makes us functionalists about intentional states, since we say of the belief that p, and by extension of other sorts of intentional states, that an agent instantiates an intentional state of that kind just in case it has in it a suitably anchored realiser state that is sensitive to potential inputs in such a way that the agent is bound to display belief-that-p behaviour.[6] The break with functionalism, if there is one, comes in the anchoring requirement. But that requirement is not foreign to the functionalist picture. It will be fulfilled from the moment that the functionally organised agent is actually exposed to any of the perceptual inputs that are relevant to the belief that p; exposure to such inputs will undoubtedly make for commerce with the objects to which 'p' directly refers.

But our functionalism about intentional states is not just any old sort of functionalism. The most striking thing about it to many eyes will be that it is a species of broad as distinct from narrow functionalism. Suppose that the proposition 'p' refers to something in the actual environment: that eucalypt, that parrot, or whatever. It will not be enough then for the belief that p that the agent is such, within the narrow confines of its surface, as to display belief-that-p behaviour in a suitable environment, where that condition can be satisfied in the absence of any actual commerce with the tree or the bird in question: where it can be satisfied by an atom-for-atom replica of the believer, which has been exposed to duplicates of the eucalypt and the parrot. The agent will have to satisfy a broad condition that requires both the internal and external worlds to be a certain way, and to have connected in a certain manner, not a narrow condition that puts a constraint only on the internal forum. The agent must be attuned to the external world in such a way that in its interaction with that world it mimics the belief function for the relevant proposition.

Broad functionalism implies, as it is often put, that beliefs and other intentional states are not always strictly in the head of the agent: that beliefs with contents that refer to things outside the agent's surface are not themselves contained within that surface, are not themselves hidebound. It entails a doctrine that is sometimes described as externalism (McGinn 1989). Two subjects that are internally identical may differ in some of the things they believe. Certainly they may differ in the things they believe, if they are located at different worlds: the worlds may not provide the same objects of reference. This is to say that our doctrine involves weak externalism. But the internally identical subjects may also believe different things at the same world: while the world provides the same objects of reference, the subjects may not have had suitable contact with the same objects. And that is to say that our doctrine involves strong externalism.[7]

I believe that that parrot is a rosella. A counterfactual self that had not confronted that parrot—even an atom-for-atom replica that had confronted an indiscernible parrot—would not believe in my present position that *that* parrot is a rosella, at least not prior to making contact with the actual parrot; he—counterfactually, I—would believe, at best, that the duplicate parrot is a rosella. The point relates to the anchoring requirement mentioned and an analogy may help to relieve any mystery that seems to attach to it. I am envious of Tom but a counterfactual self that had not come across Tom, even a counterfactual self that was an atom-for-atom replica, would not count in my present position as suffering the same envy. Not having come across Tom, but rather some duplicate of Tom, that self would not be targetted on the appropriate object.[8]

Externalism, even strong externalism, is ambiguous between two claims. The doctrine may be that in the event of appearing in my present position my counterfactual self would manifest the very state I actually display, but that that state would not have the property of being a belief that that parrot is a rosella. Or, more dramatically, the thesis may be that my counterfactual self would not even manifest that state, only a replacement. It is worth noting that on the higher-order, role state view of intentional states that we have adopted, this more dramatic thesis is supported by our observations. If my belief that p is the higher-order state of having a naturalistic context-free state within me, which satisfies a condition essentially involving the actual world, then that state cannot exist in my counterfactual self; that self lacks the required relation to the actual world. Notice on the other hand that if we identified the belief that p, not with the role state, but rather with the realiser state within the subject, then only the less dramatic doctrine would follow. My counterfactual self and I would share that very realiser state—the very state that is the belief—but in me it would have the property of being the belief that that parrot is a rosella, in him it would not (Jackson and Pettit 1988).

So much for the characterisation of the position developed in this chapter. Though we can do so only in brisk strokes, it remains to show how such broad functionalism relates to the other major positions adopted in the literature on intentionality.

All theories of intentionality, in the sense in which we are concerned with the phenomenon, try to relate intentionality to the naturalistic world in which it is manifested, in particular to the naturalistic make-up, neural or electronic, of intentional agents. The main distinction between such theories is whether they are compatibilist or incompatibilist: whether they regard intentionality as compatible or incompatible with how the naturalistic world is assumed in science to be. Our theory is clearly compatibilist and the first sort of alternative with which we should contrast it is the incompatibilist approach to intentionality. Unsurprisingly, this mainly takes the form of an eliminativist claim that science is right about the naturalistic world and that it shows our common-sense or folk psychology to be mistaken in postulating intentionality (Patricia Churchland 1986, Paul Churchland 1979, Stich 1983).

What motivates such an eliminativist theory? Mainly the thought that what neurophysiology is beginning to show us about higher mammals like ourselves—or what we are scientifically driven, from whatever sources, to think about our make-up—reveals that we cannot after all be intentional subjects. The idea is that when we look inside our skins we fail to find any belief or desire structures, or any structures of the kind that are required for the having of beliefs and desires. Is this idea persuasive? It should be clear from the discussion here that I do not think it is. As I have characterised intentionality, there is no reason to assume that intentional subjects must have any particular internal make-up. All that is required is the satisfaction of certain intentional regularities in the interaction of the system with its perceptual environment.

We might be disturbed by certain discoveries about our make-up. Thus the discovery that a subject is controlled by radio signals from Mars would presumably lead us to think that it is not an autonomous intentional system: only the larger system that includes the controller is intentional (Peacocke 1983, chap. 8). Equally, the discovery that a subject is rigged case-by-case to conform to intentional regularities in all actual and likely circumstances might lead us to deny that it is an intentional system (Block 1981). Certainly the discovery will lead us to that denial if the circumstances in which the rigging guarantees conformity to the regularities fall short of all the circumstances that are intuitively relevant; in that case the subject will conform accidentally to the intentional regularities, rather than responding under their control.[9] But short of such extreme possibilities, we can be sure that no discovery in neurophysiology, and more generally no scientific discovery about the make-up of apparently intentional subjects, is going to lead us to deny the intentionality of familiar subjects like ourselves. To be persuaded by neurophysiology that we are not intentional systems would be as mistaken as coming to be persuaded by atomic physics that this desk-top is not solid: that given there are parts of atoms unoccupied by smaller particles, it is really full of holes.

There are two sorts of premises required to defeat eliminativism in general, one analytical, the other empirical. The analytical premise is that a system is intentional just in case such and such a condition is fulfilled; the empirical premise is that that condition is satisfied by certain subjects. Our broad functionalism supplies us with a deflationary analytical premise, for it shows that what is required for being an intentional system is just the satisfaction of certain regularities in interaction with an environment. This premise is deflationary, so far as it breaks any connection between intentionality and internal make-up. It thereby makes it easy to defend the empirical premise that there are beings—in particular, human beings—that meet the condition involved and count as intentional subjects. We should have no hesitation in defending such an empirical proposition, given our success in mapping one another's responses onto intentional regularities. And so we should have no hesitation in rejecting eliminativism about intentionality. (For a fuller account of this defence see Jackson and Pettit 1990a, 1990c.)

If broad functionalism contrasts on the compatibilist-incompatibilist dimension with eliminativism, it contrasts on further dimensions with other compatibilist theories. One sort of compatibilist theory would make the intentional ontologically independent of the naturalistic—it would involve a kind of dualism—but we need only be concerned with alternatives that share with our approach the thesis that the intentional is supervenient on the naturalistic. We can divide the most influential of those alternatives, very crudely, into three kinds. (For finer distinctions, see Fodor 1985.) Each kind can be generated by starting from the broad functionalist position and then imposing a distinctive strategy; the strategies can be combined in different ways, yielding mixed alternatives, but I will only address pure options. The strategies available are: to narrow broad functionalism; to restrict broad functionalism in other ways; or to supplement broad functionalism.

To narrow broad functionalism, to go to what we can describe as narrow functionalism, is to say that believing that p, even if 'p' refers to things in the external world, requires nothing more than being internally equipped to produce belief-that-p behaviour; the world may lack the objects—the referents of the sentence 'p'—that such behaviour requires or the subject may have had no contact with those objects. It is to say not only that the belief that p is the context-free naturalistic state that produces belief-that-p behaviour in suitable contexts, but also that the presence of that state is sufficient to make it true that the agent believes that p, regardless of how the surrounding world is. Take the agent in this world that believes that that parrot outside my window is squawking. Imagine that that agent had occupied a world where there was no such parrot, only the illusion of one, but where this made no difference to the internal make-up of the subject. Or imagine that the agent is in the actual world but has never made contact with that parrot. The narrow functionalist says that even in such cases the agent would still believe that that parrot—a parrot, notice, with which it would have no connection of any naturalistic kind—is squawking outside my window.

This view is Pickwickian, and is not adopted in pure form by many philosophers, because it leads to absurd results. Consider this joke. Do you play the piano? I don't know; I've never tried. Narrow functionalism leaves room for similar results in the area of intentional states, for it allows that I may have beliefs about things that do not exist in my world or that exist unknown to me. Do you believe that the Mona Lisa is beautiful? I don't know; I haven't ever heard of it. Do you believe that two plus two equals four? I don't know; I'm not familiar with arithmetic. (For the makings of a sophisticated narrow functionalism which might escape these jibes—though for other reasons the author does not describe himself as a functionalist—see Dennett 1987, essay 5.)

Why would anyone want to narrow broad functionalism in the manner characterised? Few go for wholesale narrow functionalism, as I mentioned; few claim that narrow functionalism tells the whole story about intentional states. But many think that narrow functionalism gives the important truths about intentional states, in particular the truths that fix them as psychological determinants, and that any truths it neglects—say, to do with how an actual object can figure in a belief content—can be covered by suitable, independent additions; the restrictive functionalism to be discussed may be presented as a view of that kind (see McDowell and Pettit 1986). So why does narrow functionalism have such an attraction? More generally, why is it attractive to think of intentional states as being, at least in their essentials, narrow or context-free?

A variety of motivations have been suggested (McGinn 1989). I suspect that a primary one is the thought that beliefs and desires are causally relevant to the options they explain; that causal relevance cannot come from relational properties; and that the broad functionalist must be wrong therefore to take properties like that of being a belief that p as relational, specifically as contextually relational (Pettit 1986a). But it will be clear from our program account of the causal relevance of intentional states why this should not be regarded as a good reason for going narrow.

It may be that any properties that make their bearers causally productive in an underived sense have to be non-relational: I am thinking of properties—perhaps like mass or charge—that may make their bearers causally relevant to effects in a way that does not depend on the relevance of anything else. But there is no reason why properties that make their bearers causally relevant in the sense of being causally programmatic should not be relational, even relational to context. Thus there is no reason why the contextually relational property of being a belief that p should not make its bearer—the belief in question—causally relevant to various actions and other adjustments. An intentional agent may be designed so that under normal circumstances the satisfaction of that relational property by one of its intentional states—that is, its coming to believe that p—ensures in the presence of suitable other beliefs and desires that it is in a naturalistic state sufficient to produce an action or adjustment that the intentional states together rationalise.

Notice that we can devise an *argumentum ad hominem* against the narrow functionalist—wholesale or not—who rejects this line. The content properties of beliefs and desires are relational, even for such a functionalist; they require the naturalistic realiser of the belief or desire to be disposed to interact appropriately with other internal states: specifically, interact in such a way that the agent is equipped in an appropriate context to display belief-that-p behaviour. Thus if we could not explain how contextually relational properties are causally relevant, then apparently neither could our opponent do so. We could challenge that opponent with the fact that even on his approach content properties have a relational character, albeit they are not contextually relational.

The other two strategies for amending broad functionalism are more congenial than that of narrowing the doctrine in the manner just described. My only thought about them is that they are unnecessary; they are designed to cope with problems that are not genuine difficulties. I will try to be brief in dealing with them (see too Godfrey-Smith 1991).

The first of the other two strategies is that of supplementing or adding to broad functionalism. This strategy is usually motivated by the allegation that broad functionalism leaves content so indeterminate that the imputation of error is always avoidable. Suppose that a cat pounces at a doll in a spot where it has often found birds. It appears that we can interpret its behaviour—and represent it as rational—in either of two salient ways. We can say that it has a standing desire to catch birds and wrongly believed that the doll was a bird. Or we can say that its only standing desire is to pounce on birdlike things and that it acted on the correct belief that the doll was indeed a birdlike thing. Is there any way of resolving this sort of interpretative question? It may seem that there ought to be, on the grounds that if we can save the truth of the cat's belief by moving in this case to the second ascription, then we will always be able to interpret such an animal's behaviour so as to ascribe only true beliefs. If we are always able to do that, then a question arises in turn as to whether animals like the cat—or, by similar reasoning, robots—really do have beliefs and desires. After all, we know about beliefs and desires in the first place from our own case, and it seems to be distinctive of any believer like you and me that we can get things wrong.

There are well-known arguments to the effect that the interpretative slack alleged here is not as great as I have suggested (Dretske 1988, Fodor 1987, 1990). Those who espouse the additive or supplementary strategy think that there is the slack alleged and so they hold that broad functional behaviour does not determine the intentional states of a natural agent on its own; the contents of those states are fixed only with the help of other factors. The usual move is to introduce evolutionary considerations side by side with facts about broad— or perhaps narrow—functioning as co-determinants of the intentional states exhibited by any natural intelligence (Millikan 1984, Papineau 1987). Thus it will be said that the cat believed that the doll was a bird, not that it was a birdlike thing, on the grounds, roughly, that it was the development of a capacity for having beliefs about birds that was adaptive in the evolution of the species. The naturalistic state within the cat that plays the role of producing suitable behaviour is supposed to sensitise it to birds; the presence in cats of such a state was selected for the property of ensuring such sensitivity, not for the property of sensitising it to birdlike things.

The second of the other strategies for amending broad functionalism is to restrict it rather than to supplement it: in particular, to restrict it in a less dramatic fashion than is involved in going wholly narrow. Consider the move explored by Jerry Fodor (1987) of arguing that it is sufficient for believing that p that two conditions are fulfilled. The first condition is that the agent has within it a context-free naturalistic state, which plays in relation to other context-free states a role characteristic of belief as distinct from desire. And the second is that that state co-varies in such a way with the state of affairs that p—or its elements co-vary in such a way with the constituents of that state of affairs— that we can see it as indicating that p: we can see it as carrying the information that p. If an agent meets the conditions laid down by the broad functionalist for believing that p, then it will almost certainly meet these two conditions. But it will also meet other conditions besides. That is why the move to these two conditions can be seen as a way of restricting broad functionalism. Fodor's way of restricting broad functionalism is one among many. Another is explored by Fred Dretske (1988) when he argues that it is sufficient for believing that p that there is a context-free state in the agent which indicates that p—thus it meets Fodor's second condition—and which is recruited as a cause of suitable behaviour, in the learning history of the agent, *because* it indicates that p.

What is the motivation for restricting broad functionalism in this fashion? As I see it, the core motivation is an assumption about how to resolve the same sort of indeterminacy that motivates the additive strategy. Unlike additive theorists, restrictive functionalists hold that the relations postulated in a broad functionalist account of what it is to believe that p are the only matters relevant to determining the content of that belief. They argue, however, that it will not do to take all of those relations to be equally relevant to the determination of each content; they hold that if we go for such a non-restrictive ascription of relevance, then we will be unable to account for a certain possibility of error.

Restrictive functionalists assume not only that error is possible, but that it can be partial. Thus they assume that a subject may differ from the right-minded in believing that not q, while remaining in tune on other matters. Suppose then that there are two subjects who each believe that p, but who differ as to whether or not q. If the subjects converge in the belief that p, then there must be a restricted set of broad functional relations—one shared by such diverging subjects—that can account for their each believing that p. There must be some restricted set of relations to ensure that any subject that enjoys them believes that p: the set will suffice, independently of surrounding relations, to determine this belief content (Dretske 1988, pp. 154–55; Fodor 1987, chap. 3).

What are we to say of the additive and restrictive amendments to broad functionalism? My response is to deny the assumption they share and to adopt a relaxed attitude on whether either of the claims they advance can be upheld.

The assumption they share is that intentional subjects—all intentional sub-jects—enjoy a determinacy of content sufficient to make a certain sort of error possible. I argue in the next chapter that human beings, and more generally any subjects that think, must enjoy such a determinacy of content. But I see no need to ascribe the requisite determinacy of content to non-thinking inten-tional systems: for example, to cows and robots. I also argue in the next chapter that the determinacy that thinkers enjoy can be explained by the features that make thought possible. Thus I see no need to identify resources sufficient to explain determinacy in the features of non-thinking subjects. If there are suitable resources available, as additive and restrictive theorists each claim, then well and good; that is a sort of gain. But if there are not, that is no great problem.

Our account of intentionality suggests, as it happens, that the resources of non-thinking subjects will not be sufficient to ensure that every intentional or rational system will be indictable with error (cf. Dennett 1987, essay 4). According to the line developed, it is necessary and sufficient for counting as an intentional subject that a system is governed by intentional regularities in its interaction with a perceptual environment, at least under normal or ideal circumstances. But we have said nothing on how such favourable circum-stances are fixed. Thus, for all that we have said, the following may well be possible: that we can represent a given intentional system as having such and such intentional states relative to a certain specification of favourable circum-stances; that we can represent it as having different intentional states relative to another specification; and that for any piece of behaviour there is some way of playing with the specification of favourable circumstances that allows us to see the system as not being in error about anything.

We may wonder whether the cat has beliefs involving a bird or a birdlike thing: whether, as we may put it, the concept it deploys is that of a bird or birdlike thing. Let us suppose that the concept correctly applies to anything the cat is disposed to apply it to, under normal or ideal circumstances. But what circumstances are suitably favourable? Those in which all and only birds trigger the concept? In that case, we will ascribe a bird-concept. Or those in which all and only birdlike things do so? In that case we will ascribe the concept of a birdlike thing. There seems to be no reason in principle why we cannot in every case go to the error-diminishing specification of favourable circumstances (Fodor 1985). If we take this line, we will probably think of the creature as having attitudes with contents that are vague in the way in which a property like baldness is vague. In particular, we will think of it as having attitudes with contents so vague that it is never indictable with error.

It may be unsatisfying to be able only to ascribe such vaguely featured attitudes to a creature but the practice of ascribing them need not be useless. The attitudes will enable us to make intentional sense of the creature's responses—in ascribing them, of course, we will always cast the contents under an arbitrarily precise form—and they will aid us in predicting those responses. It is just that there will be no responses made by the subject that are intention-ally intelligible, but only in the light of false beliefs. If the creature ever emits responses that would only be intelligible in the light of beliefs that are certainly false, then the responses are not intentionally intelligible at all: they are the products of reflex or breakdown, and are not programmed for by intentional states. All of the creature's intentionally intelligible responses are intelligible in the light of beliefs that have contents vague enough to be represented as true.

But though it is possible to think of an intentional system as having states with an error-eliminating indeterminacy of content, it will almost always be methodologically desirable to treat an intentional system as having relatively determinate contents. It will almost always be much simpler and more illuminating—among other things, more illuminating about biological function and evolutionary history—to specify normal and ideal circumstances, and therefore to characterise intentional states, in such a way that we often ascribe error. Consider a machine that serves pretty well as an adding machine but that goes a bit awry with numbers between 1,000 and 1,010. It will be much simpler and more illuminating—for example, more illuminating about its likely structure and history—to represent it as an adding machine for which circumstances are unsuitable in a particular range of numbers, namely it malfunctions there, than to represent it as a machine that works equally well under all the circumstances in question, though not with the effect of adding. Similarly it will usually be more attractive to ascribe to animals and robots intentional states that mean that those subjects sometimes fall into error. But this sort of methodological consideration does not mean, either in the adding machine case or in the case of the intentional system, that there has to be an objective fact of the matter fixing the error-permitting characterisation as the right one.

Is the admission of indeterminacy of content compatible with the claim that intentional states are causally relevant to the actions and other adjustments of intentional subjects? Not perhaps on some accounts of causal relevance, and that may explain the resistance to allowing indeterminacy. But there is no problem with the admission of indeterminacy on our program model of causal relevance. Where we cite the belief that p, in combination with other states, X, as causally relevant, indeterminacy teaches that strictly we should relativise the claim to such and such a scheme of normalisation or idealisation. Suppose then that there is another scheme relative to which we ascribe the belief that q, combined with Y states, rather than the belief that p, combined with X. There will be an empirical equivalence in such a case between the belief that p, plus X, under the one scheme, and the belief that q, plus Y, under the other: there will be nothing in actual or possible experience to distinguish between them (Quine 1970, chap. 2). But given this equivalence, there is no problem with saying that they each program for the sort of response in question, and that in that sense they are each causally relevant to the response. Since the conditions are always going to be realised together, there is no problem in saying that the realisation of each more or less ensures the presence of a factor sufficient to produce the response.

I hope that this discussion of alternatives will help to put in perspective the sort of broad functionalism that we developed, though not under that name, in the first two sections of the chapter. Our commitments force us to reject the eliminativist theory of intentionality as well as the theory that would narrow the functionalism involved. But we can be relaxed about whether it is possible to defend either of the claims made by those I have described respectively as additive and restrictive functionalists. Neither the position that we have developed in this chapter, nor the position that we shall be developing in the next, forces us to make a judgment on these issues. On the contrary, they combine to make it possible for us to avoid commitment. That is a great benefit, as it enables us to side-step some of the most debated questions in contemporary philosophy of mind.

Notes

1. Both option-desires and property-desires can be represented as desires for corresponding propositions, as in the tradition of Jeffrey 1983. The corresponding proposition for the ordinary sort of property can be taken, roughly, as the existential claim that the property is realised. But unlike the option-desire, the property-desire is not a desire for the set of worlds—that set of worlds, 'warts and all'—at which that proposition holds. Rather it is the disposition, relative to any world under consideration, to desire that the proposition should be true there; if the proposition is not true at the given world, the desire amounts to a preference for the nearest world (or worlds) where it is; if the proposition is true at the given world, the desire amounts to a preference for that world over the nearest world (or worlds) where it is not true. Pettit 1991a, p. 512, fails to make this matter clear. Frank Jackson has drawn my attention to the fact that the distinction connects loosely with the distinction between having a desire for a proposition conditional on its realisation, as one has a desire for an option, and having a desire for the image of a world on a proposition: on this matter, see D. Lewis 1986, p. 318. I am grateful for an exchange on related matters with John Broome.
2. Some cognitivists defend the view, not that desires are cognitive states, but that while desires are non-cognitive, they need not be involved in intentional action: the only states necessary may be beliefs (Platts 1979, chap. 10; See Pettit and Smith 1990). Here I ignore this less compelling variety of cognitivism, since I am prepared to assume that desires are always present in intentional action.
3. There is a reason why someone might say 'no', apart from that which I go on to discuss. Someone might think that there are particulars involved in the intentional attitudes of human subjects that are not available to be objects of belief just because the *naturalistic* particulars are unchanged: I have in mind the sort of particulars that the Platonist about mathematics postulates. I ignore that Platonist possibility here.
4. Notice that we cannot turn the thesis into a necessary thesis by putting the Cartesian proviso in the supervenience base, as it were: by stipulating that any two possible worlds that are naturalistically identical and that are alike in obeying the actual naturalistic laws and in not containing any non-actual stuff are going to be intentionally identical. That may be a necessary truth but it is not a necessary *supervenience* truth; the absence of non-actual stuff is not part of what we can plausibly consider as a supervenience base.
5. In taking this line I do not have to commit myself to a controversial thesis about trans-world identity, as it it called. Even a theory like that of David Lewis (1973) will presumably have to mark a distinction between an actual object existing in another

world—in his language, having a counterpart there—and that object having a duplicate in that world, a duplicate that is numerically distinct from it. In taking this line I should also mention that I do not mean to suggest that there are not stronger supervenience theses available: theses invoking indiscernibility rather than numerical identity.

6. Functionalism about intentional states has come to be associated with some more specific commitments than those mentioned here but I abstract from these (see Block 1980; Dennett 1979, Introduction; Fodor 1985). I do not believe that the more specific associations have served the cause of clarity.

7. In the weak sense of externalism a state of mind is external in character if a counterfactual self, even an exact replica, in another possible world would not necessarily display it. In the strong sense a state is external if such a counterfactual self would not necessarily display it, even in the event of appearing in the actual world, in my present position; what is also required is that that self should have had some commerce with that world, in particular a form of commerce that I enjoy. The distinction is drawn in these terms in Jackson and Pettit 1992a; it coincides with a distinction drawn in McGinn 1982, pp. 6–9.

8. I phrase the point in terms of a counterfactual self, rather than in terms of a *doppelgänger*, in order to connect with the earlier discussion of supervenience; the break with the more common ways of doing things is not of great significance.

9. What to say if, as indeed Block intends, the rigging is such that the circumstances in which conformity to the regularities is produced do not fall short of the intuitively relevant ones? The question is dealt with in Jackson 1992. A thought that links with Jackson's is that in Block's system an intentional state like the standing belief that Canberra is in Australia would lack the intuitive unity we expect when we say that to believe such a thing is to have in one a context-free state that plays—and that has actually played—a certain causal role; its operation now would not depend on its operation earlier.

2

Thought

We now know what intentionality involves. To be an intentional agent is to interact with a perceptual environment under the control of intentional regularities, at least in favourable circumstances: in particular, as we have come to see, it is to interact with that environment in accordance with minimally rational regularities of a theoretical and practical kind. Any intentional system, in particular any active and deliberative system, will realise a range of intentional states, from beliefs to habits of inference to desires, where desires may involve what we have called property-desires or option-desires. Those intentional states are higher-order physical states: that is, states that each require the agent not to be in any particular physical configuration, but to be physically such that it displays an appropriate pattern of behaviour—say, belief-that-p behaviour—in its actual environment. Such states will be causally relevant to certain adjustments and actions on the part of the agents and may be invoked to explain them. They do not produce the adjustments and actions in the fashion of neural antecedents but they program for those responses: however the intentional states are neurally realised in any instance, the realising configuration is more or less bound in normal circumstances to produce the sort of response in question.

There are two things that we should now notice about intentional agents. First, there are certain conditions that it is in any agent's interest that its intentional states should fulfill. Second, consistently with being intentional an agent may or may not act intentionally in order to see that it fulfills those conditions. It is important to notice these things, because they tie up with the distinction between the thinking and the non-thinking, the thoughtful and the thoughtless, intentional subject. The thinking subject does things of an intentional kind in order to see that it satisfies the conditions in question. The non-thinking subject does not perform such actions; it lacks the capacities required.

Given our characterisation of the intentional agent in the last chapter, there are a number of conditions that it is clearly in the agent's interest that its intentional states should fulfill. I shall identify four, in a spirit of illustration. I do not say that this list is exhaustive. Two of the conditions bear on beliefs, two on desires.

What conditions is it in an agent's interest that its beliefs should fulfill? The most important ones can be summed up in the requirement that its beliefs should be true, having true contents: that for any belief that p which the subject holds, it should indeed be the case that p. We express this as the requirement of truth, in view of the canonical schema for truth: '"p" is true if and only if p'; that a subject's beliefs are true means that things are as the beliefs represent them to be. Why is it desirable for any agent that its beliefs should be true? It is desirable for any agent that it should secure results in action that it finds desirable: results, as they should also be, that serve its desires. But no agent is in a position to select actions that will serve its desires for certain; all it can do is to select actions that will serve its desires according to its beliefs. And so it had better be the case that things are as the beliefs represent them: it had better be the case that the beliefs are true. Otherwise it will be a fluke if the agent's actions serve its desires.[1]

But how are the agent's beliefs going to have the best chance of being true? The question directs us to more proximal conditions that it is in the agent's interest for its beliefs, and more generally its habits of inference, to meet. The agent's beliefs should be rational: they should be such that, under accepted criteria of evidence, they stand a better chance of being true than available alternatives. More specifically, they should be well supported inductively or deductively. They should be inductively rational in the light of observation or they should be inductively or deductively rational in the light of rational beliefs. If truth is the ultimate desideratum, inductive and deductive rationality is the more proximal one.

What now of desires? It is in any agent's interest that its property-desires should be founded in true beliefs: that the properties for the desire of which it acts are the desirable properties, the properties that it ought to desire. Equally, it is in any agent's interest that the option-desires it forms should be truly founded in a corresponding sense: that any option desired should be the desirable option; it should be the option that promises to do better than alternatives, in itself and in its probable outcomes, by the relevant desirable properties. No agent will know which desirability beliefs are true, only which hold out the best promise of truth: which it is most rational to believe, under accepted criteria of evidence. And so if it is ultimately in an agent's interest that its desires be truly founded, it will be more proximally in the agent's interest that they be rationally founded: that they should correspond to the desirability beliefs it is most rational to hold, under prevailing criteria. The overall condition is one of practical rationality. In the case of property-desires, practical rationality comes down to what we might call evaluative rationality; in the case of option-desires, it comes down to what is usually described as deliberative rationality.

We have identified four sorts of conditions that it is proximally in an intentional agent's interest that its intentional states should fulfill: conditions of evaluative and deliberative rationality on the practical side; conditions of inductive and deductive rationality on the theoretical. These conditions will be differently understood in different cultures and traditions, as different criteria of evidence come into play, but for our purposes here we may ignore such variations. We may speak of its being in the interest of an agent that its desires and beliefs should be rational, while allowing that there are different accounts available as to what rationality involves.

Apart from noticing that there are such conditions we should also notice that, for all that has been said so far, intentional agents may not act intentionally to promote the fulfilment of the conditions. They may not perform any actions with the goal of thereby furthering those conditions within themselves. We suggested in the last chapter that there are design reasons why we should expect any natural intentional subject to be minimally rational in relation to biologically appropriate desires. But an intentional agent might tend to be rational in that way without doing anything intentionally to promote its rationality. And equally an intentional agent might tend to be rational in the richer sense intended here and not do anything intentionally in order to further such rationality.

The point becomes obvious once we think about what is involved in doing something intentionally. Intentional actions clearly involve desire: an agent brings about a certain state of affairs intentionally, after all, only if it desired to bring it about.[2] We drew attention in the last chapter to the fact that not all of the rational responses that an intentional system makes are intentional actions in this sense. A system believes that if p, then q and receives evidence that induces the further belief that p. The adjustment whereby it comes then to believe that q should go through more or less autonomously, if the system is functioning properly, and it will not count as an intentional action. It will not be programmed for by any desire on the part of the agent.

The inferential response is the product of beliefs—the beliefs in the premises—and of the relevant inferential habit. The inferential habit involved is a goal-seeking state, as desire is, but it does not operate in a belief-channelled manner: it may lead to the adjustment without the agent's believing that the adjustment is likely to realise the goal sought. When desire plays a role, and when the response is an intentional action, there are a number of strategies that the agent is capable of adopting by way of realising the goal in question: when I act on a desire to drink that coffee, there are a number of ways that I may pick up the cup. If the desire leads to one strategy rather than another, that is because the agent believes that that strategy is likely to realise the goal. Unlike the inferential habit, the desire connects via a belief with the response it generates (see pp. 18–19 above).

The fact that intentional action requires desire makes it clear why an intentional system may not do anything intentionally to promote the satisfaction of our four conditions of rationality by its intentional states. The system will not do anything intentionally to that end, if it lacks the capacity to form the belief that by doing such and such, by adopting such and such a strategy, it is likely to further one or another condition. And it may lack the capacity to form such a belief. It may be insensitive to the property of truth, or evidential support, and may be incapable of having beliefs involving such properties in their contents. Those properties are sophisticated features that belong to the contents of intentional states in the first place—that is, to propositions—and to the states themselves in the second. It is clearly possible for a system that counts as intentional on our definition not to have any beliefs in respect of such properties; it is possible for a system to interact with its perceptual environment under the control of intentional regularities without instantiating intentional states of that particular kind. Thus it is possible for an intentional system not to be able to do anything intentionally to promote its satisfaction of conditions of rationality.

But while this is possible, equally it appears to be possible that an intentional system might be able to act intentionally with a view to having—or at least increasing the probability that it will have—intentional states that do satisfy conditions like those of evaluative and deliberative, inductive and deductive, rationality. Certainly there is nothing obviously incoherent in the notion of an intentional system with such an extended range of intentional activity. And equally there is nothing to suggest that an intentional system with that extended range of activity would be less efficient; on the contrary, the extended system could presumably be more efficient, having an extra capacity to promote the satisfaction of conditions that are important to its performance.

We have made two points: that there are a number of conditions that it is in the interest of any intentional system that its intentional states should fulfill; and that for all that has been said so far, an intentional system may or may not be capable of intentionally acting so as to promote its satisfaction of those conditions. These points made, we can introduce our distinction between thinking and non-thinking, thoughtful and thoughtless, intentional subjects. An intentional system is a thinking or thoughtful subject just in case it has the capacity to act intentionally with a view to promoting in itself conditions such as those of evaluative and deliberative, inductive and deductive, rationality. Otherwise the system is non-thinking or thoughtless.

A thinking intentional system need not have the capacity, with regard to all the intentional states that it instantiates, to act intentionally with a view to promoting the rationality of those states. We may countenance intentional states with contents that only scientific psychologists can have a go at articulating—states that bear on subliminal perceptual cues, and the like—and if we do, then we must admit that no ordinary intentional system will be able to bring such states under the control of thought; no system will be able to control states whose contents it cannot articulate. The thinking intentional system is required to be able to pursue rationality only in relation to certain intentional states, not necessarily in relation to all. As mentioned in the preceding chapter, the states with which it lacks the capacity to seek to be rational, the states it cannot bring under the control of thought, can be identified with what are often described as sub-personal states. Our focus on thinking systems means that henceforth we shall be entirely concerned with personal intentional states: in practice, with intentional states that have contents that can be readily articulated in ordinary language.

I assume that human beings like you and me are thinking subjects. The fact that I am labouring over the writing of this, or that you are bothering to read, is a vindication of the assumption. We are not just creatures who form beliefs and desires—more or less rationally—and proceed under favourable conditions to act in the way that they rationalise. We are also creatures who understand what it is for those states to be rational—albeit our understanding is marked by the images and resources of a particular culture and tradition— and who have a care for the attainment of that rationality. Thus we at least occasionally take time to try and check that the properties we cherish are the ones that we consider valuable, that the options we choose cohere with the values we cherish, that the empirically grounded propositions we endorse are genuinely supported by the evidence, and that the deductively derived propositions we believe do actually follow from the alleged premises. In taking time over such checks, we indulge in an intentional, desideratively grounded form of activity. We indulge in the activity of thinking.

But if we humans are thinkers in this sense, then which are the non-thinking intentional subjects? My inclination is to say that robotic intelligences of the kind generally imagined, and most animal intelligences, are non-thinking. The computer and the cow may interact with their environment under the control of intentional regularities. But neither shows any signs of thought, any evidence of intentionally seeking after the sorts of rationality we have mentioned. Thus, although each may revise existing beliefs in the light of new evidence, neither ever exhibits surprise at such evidence (Davidson 1985). We would expect such surprise, in some cases at least, from thinking subjects. Previously believing that p, the thinking subject will often have believed that the evidence was against the truth of 'not p'; and so the appearance of evidence that causes a shift to the belief that not p should induce surprise.

Before leaving the distinction between thinking and non-thinking subjects, there is one matter I should emphasise: the activity of thinking, as I conceive of it, is subject to important naturalistic limitations; it is not an unconstrained activity of pure spirit. Thinking involves the intentional application of checks to the connections that sustain certain beliefs and desires. But the intentionality of thinking is subject to limitations on at least two fronts. First of all, the checks that thinking deploys cannot all be intentionally checked themselves; and second, the connections to which the checks are applied cannot all be intentional responses.

The first limitation will be familiar to anyone who remembers Lewis Carroll's lesson, that every argument involves a rule of inference, which cannot itself be expressed as a premise. Take the argument 'If p, then q; p; therefore q'. Anyone who finds that argument valid must endorse the *modus ponens* rule that allows us to detach the consequent of a conditional, given the antecedent. So should we put that rule explicitly into the premises? Well, if we put a statement of that rule into the premises, then anyone who finds the resulting argument valid must endorse the further rule that allows us to detach the consequent, once given the antecedent and the *modus ponens* rule. And so we are no better off than before; there is still an undischarged assumption at work. The generalised lesson is that for any argument that is found to be valid, there must be a background rule of inference endorsed, a rule of inference that is not itself formulated in a premise. This lesson means that as an intentional subject thinks about what follows from what, and what it ought to believe or desire, then it inevitably deploys rules of inference that languish unexamined in the background. It cannot run checks the whole way up.

The second limitation can also be illustrated with respect to inference. Suppose that a thinking system checks potentially relevant evidence with a view to seeing whether it should believe that q. Imagine that it discovers that p and, given a firm belief that if p, then q, it infers that q. Will the inference that q be itself an intentional action? Surely not: it is an involuntary act of belief-formation. Just as thinking cannot run checks the whole way up, so thinking does not exercise intentional control the whole way down. The checking activity is intentional and thoughtful but actually coming to believe that q, actually drawing that conclusion, is a more or less autonomic response. Indeed, coming to believe that q had better be a response of that kind: it would hardly augur well for the prospect of truth if an agent came to believe that q because that engaged with its desiderative attitudes.

Although it is autonomic on the agent's part that it comes to believe that q in the case described, there is still an important contrast between its coming to believe that q after intentionally checking the evidence and its coming un-thinkingly to believe that q. We may mark that contrast in a terminologically natural way by saying that in the case where thinking occurs the agent makes a judgment: makes a judgment as distinct from just forming a belief. We shall follow this practice henceforth.

The second limitation, as we have described it, is ambiguous. If we say that an act of A-ing is not intentional, on the grounds that it does not connect with the agent's desiderative attitudes, there are two things that we may have in mind. We may mean that the act connects under no description with those attitudes: that it is non-intentional in the manner of a knee-jerk. Or we may mean that it does not connect with those attitudes under the A-description, though it does under some others: that it is non-intentional in the way my knocking over the milk is non-intentional, where my reaching for the butter—the very same action, we may suppose—is intentional. Is judging that p in the sort of case mentioned non-intentional period, or just non-intentional under the description of judging that p?

It should be clear that in the sort of case in question, where the formation of the belief that p is preceded by intentional checking of evidence, the act is only non-intentional under the judgment-that-p description. Given the checking of evidence that precedes it, the act may also be described as judging in accordance with the evidence. And though the agent does not judge that p intentionally, it is intentional on the agent's part that it judges in accordance with the evidence. The agent judges in accordance with the evidence because—or at least in part because—it desires to judge in accordance with the evidence; judging in accordance with the evidence is a goal that it seeks in a belief-channelled way. The agent does not judge that p, even in part, because it desires to judge that p; it is not a goal of the agent's that it should judge that p. The desire on which the agent acts would lead to the judgment that q, if the evidence made it likely that q. And all of this holds, notwithstanding the fact that judging in accordance with the evidence is in this case one and the same action as judging that p.

If it is possible for me to judge intentionally in accordance with the evidence, why does it seem impossible to judge intentionally that p? Why, as it is sometimes put, does it seem impossible to decide to believe that p? (Bennett 1990; Williams 1973, essay 9). Here is a suggestion. If an act of judging that p were intentional, that would mean that a suitable desire on the part of the agent is sufficient in the circumstances—given its beliefs, given a suitable context, and so on—for it to come to believe that p. But on a functionalist analysis a subject will believe that p only if it is suitably sensitive to evidence that not p: only if, at least in normal circumstances, it is disposed to give up that belief in the face of such evidence. Thus it is necessary for a subject, if it is to come to believe that p, that it is not faced with contrary, incontestable evidence. And in that case it cannot be sufficient for believing that p that the subject has a suitable desire. Hence, special circumstances apart, the subject cannot intentionally come to believe that p, cannot intentionally judge that p.

The second limitation on the intentional reach of thought should not really be surprising. You are intentionally reading this book but doing that involves submitting yourself to procedures that will evoke non-intentional, more or less autonomic responses: the responses of word-recognition, sentence-analysis, understanding, and the like. There are certain crucial responses required for the intentional activity of reading that you cannot directly will yourself; you can only wait on them to happen. And so it is with thought. There are crucial responses required for the success of thinking that you do not intentionally control—in general, acts of judgment—and in order to get on with the thinking enterprise you must trust in your nature to ensure that those responses are forthcoming (see Ricoeur 1966).

Two Requirements of Thought

So much for what it means in general for an intentional system to be a thinking subject. We turn now to the identification of the major requirements an intentional system must meet if it is to be capable of thought. There are two salient candidates: first, a requirement of intentional ascent, as I shall call it; and second, a requirement of rule-following.

Intentional ascent is the requirement that an intentional system should not only have intentional states with contents—states like the belief that p, the desire that q, and so on—it should also have intentional states that are about such contents. It should not only have attitudes that bear contents, it should also have attitudes towards contents. I describe the requirement as intentional ascent, by parallel with what W. V. O. Quine (1970, pp. 271-76) has called 'semantic ascent'. In semantic ascent a speaker talks about the words used to describe things, as well as talking about things themselves. In intentional ascent an agent comes to have beliefs about the contents of belief, as well as having beliefs about things more basic than contents.

The requirement of intentional ascent is probably intuitively intelligible but it may be useful to say something by way of clarification, before looking at the argument for the requirement. We should be clear, first of all, on what exactly contents or propositions are, in my usage of those terms. Up to now I have been taking the notions as intuitively given but it will resolve important ambiguities if I stipulate how I understand them. The content of a belief, or of any other intentional state, is picked out by the sentence used in the that-clause of a perspicuous ascription of the state; the content of what we ascribe as the belief that p is picked out by the sentence 'p', as it is used in that ascription. What is picked out can be equally well described as a proposition, since a content is a proposition that serves in a suitable role vis-à-vis an intentional state. All of this is more or less uncontroversial.

But can I use another sentence 'q' or 'r' to pick out the same proposition or content expressed by 'p'? And if so, under what conditions? What is it that individuates a proposition and makes one sentence appropriate to express it, another not? That is the tricky question in this area (Salmon and Soames 1988). I will sketch a line on the problem, though in a stipulative rather than an argumentative spirit. It is important to specify a line, because otherwise our discussion will be dogged by ambiguity. But my hope is that the points I make in this chapter are not necessarily tied to acceptance of the particular line I sketch; I believe that the points can be reformulated to fit with a variety of approaches.

By all accounts, and this is my first stipulation, sentences that pick out the same content or proposition must have the same truth-condition, the same truth-value at different possible worlds. But that is not enough, as I see things, since '1 + 1 = 2' and '1 + 2 = 3' have the same truth-condition—they are each true at all possible worlds, being necessary truths—and yet the belief that $1 + 1 = 2$ does not have the same content as the belief that $1 + 2 = 3$ (but see Stalnaker 1984). As a minimal extra, I stipulate that the sentences must also satisfy the second condition of having corresponding components with corresponding semantic values. Assume that for every singular term there is an object to which it refers and that for every predicative expression there is a property that it ascribes; a property in this context need amount to nothing more than the set of all actual and possible things that satisfy the predicate. The stipulation means that if two sentences express the same proposition then they have corresponding singular terms with corresponding referents, corresponding predicative expressions with corresponding properties predicated, and so on. The arithmetical sentences lack that sort of correspondence, though they have the same truth-condition.

If these are the only two conditions that contents must satisfy, then a content or proposition is a construction out of objects, properties, and the like such as has been described, under various refinements, as an abstract situation-type (Barwise and Perry 1983). But there is a problem even with identifying contents and situations. Consider the belief ascriptions 'John believes that this building is the library' and 'John believes that the Menzies building is the library'. The beliefs ascribed do not have the same content on the functionalist account of belief adopted in the preceding chapter: the first belief would interact with the desire to go to the library and cause John to enter the building; the second might not, for John might not know which building is the Menzies building. That means that contents or propositions cannot be straightforwardly identified with situations, for though the contents involved are different, the same situation is picked out by 'this building is the library' and 'the Menzies building is the library'.

I add a third condition that has to be satisfied by sentences that pick out the same content or proposition. This condition requires that the sentences not only pick out the same situation, but pick it out under the same mode of presentation: in a phrase, it requires that the sentences pick out the same *scenario*. The sentences in our example must not only pick out the situation involving a particular building and the property of being a library, they must present the building—and of course the property—in the same mode. What sort of difference in presentation makes for a difference in mode? Any difference, we can say, that makes for the sort of functional difference we find between the two library beliefs. The comment may not be satisfactory as a general gloss, but it will do for our purposes.

To sum up the conditions mentioned, then: in order to express the same content or proposition, sentences must not only have the same truth-condition, and must not only have corresponding parts with corresponding semantic values; their parts must relate to the semantic values in the same way: the sentences must present the situation involved in the same mode. A content or proposition, in brief, is a scenario: a situation under a mode of presentation, a situation presented in a certain way.

I hope that this serves to clarify the notions of content and proposition. The requirement of intentional ascent is that a thinking agent must be able to have attitudes towards contents as well as attitudes with contents, in particular that it must be able to have beliefs about contents as well as beliefs with contents, beliefs about scenarios as well as beliefs in scenarios. But if we are to understand the requirement fully, it is necessary to say something more, not just about contents, but also about aboutness. What is it for a belief to be about a content or proposition?

If I believe that that telephone is bugged, then the content of my belief is the proposition expressed by the sentence 'that telephone is bugged'. But so far as that proposition is the content of my belief, the belief is related in a certain way to the main constituents of the proposition: the particular telephone, on the one hand, the property of being bugged on the other. How are we to describe that relation? In the case of the telephone, we naturally say that the belief is about the telephone or, perhaps less naturally, that the telephone is an object of the belief. I propose that we should say the same thing about the property of being bugged and about any other constituent of the proposition or scenario: that my belief is also about that property, that that property is also an object of my belief. About the telephone, and about the property of being bugged, I believe that that telephone has that property.

The requirement of intentional ascent is that the thinking subject has to have beliefs about propositions or contents, in this sense of aboutness. The subject has to have propositions as objects of belief, not just as contents of belief. It not only believes and desires that p, that q, that r, and so on. It also has beliefs about the propositions expressed by 'p', 'q', and 'r': say, that they are true, that such and such evidence supports them, and so on. Otherwise put, the thinking subject believes propositions that contain other propositions as constituents. It is not sufficient for it to have beliefs with contents that have only non-propositional constituents: constituents like concrete things and their properties. It must also have beliefs with contents that nest other potential contents, other propositions, within them. It must have beliefs in scenarios that incorporate other scenarios as constituents.

With these points of clarification made, we can now turn to the argument for the requirement of intentional ascent. Why must a thinking subject, a subject that acts intentionally with a view to satisfying conditions of rationality, practise intentional ascent? Why does it have to form beliefs about contents as well as beliefs—and desires—with contents? The reason is fairly straightforward. If an agent was not able to form beliefs about contents or propositions, then it would be unable to believe that a proposition was true or false, or was supported by evidence or not, or was perhaps a desirable or undesirable proposition to realise. If it was unable to believe things of this sort, then equally it would be unable to form corresponding beliefs about its own beliefs or desires. It would be unable to believe that its beliefs were true or well supported by evidence, for example, given that a belief only has such properties in virtue of the proposition believed—the content of the belief—having them. But if an agent was unable to form beliefs of this kind, then it would not know what it was for the properties in question to be instantiated. And in that case it could not form desires for the properties in question and could not therefore act intentionally with a view to having its intentional states display the properties. In brief, it could not think.

I hope that this is enough to make clear both what the requirement of intentional ascent is and why any thinking agent must satisfy it. Before leaving the requirement, however, it may be useful to mention that if it is fulfilled, then there ceases to be a problem about how a variety of other conditions, in particular conditions also apparently required for thinking, can be fulfilled; thus I will not discuss those other requirements separately. One such condition, implicitly recognised already, is that the agent should be sensitive to properties of propositions like truth and evidential support. I assume that if propositions are objects of belief for an agent, then there is no problem as to how it can become sensitive to any property of those objects.

A second condition is that the agent comes to have beliefs, not just about belief-contents, but about states of belief: that the agent, as it is sometimes said, comes to have the concept of belief (cf. Davidson 1984, essay 11, 1985). This condition will be fulfilled so far as the subject becomes sensitive to whether a content has the relational property of being believed by this or that person. Its fulfilment in respect of the subject's own beliefs will mean that the subject believes it has certain beliefs and thus, in one sense, that it is a conscious believer (Bennett 1976, Mellor 1977).

A third condition is that the agent is able to go back to a content to form more and more beliefs about it: beliefs, for example, concerning its evidential relations to other contents believed and to observations. If the subject has a belief-content presented as an object, then there is no obvious block in the way of its being able to access the object reliably in this manner. We said that fulfilment of the second condition will mean that the subject is conscious in a certain way of its beliefs: it is state-conscious, conscious of the state of believing. Fulfilment of the present condition will mean that it is conscious in another way of its beliefs: it is, as we may say, content-conscious. But the comment will bear further elaboration.

Content-consciousness of belief parallels a certain consciousness of perception. Consider someone who has only blind sight in a region of their visual field (Humphrey 1986, pp. 59–60; Weiskrantz et al. 1974). Such a person appears to form basic beliefs about the positions of objects in the relevant part of the visual field, at least to the extent of being able to point to the objects or avoid hitting into them. But her impairment means that she is not able to go back to the object, as it were, to determine the answers to an indefinite range of questions about shape, size, character and the like. To that extent her sight is blind or unconscious. Normal perception differs from blind perception in allowing the subject reliable access to the object perceived for purposes of resolving a variety of questions and forming corresponding beliefs. Normal perception is content-conscious, as we may put it. I say, in parallel to the perceptual case, that if the content of belief is presented as an object to the believer, then normal belief may be content-conscious too. The believer may be able to go back again and again to the proposition believed to check on its different properties.

So much for the first requirement of thought and the associated conditions. The second requirement is that not only does the thinking subject have to practise intentional ascent, it also has to practise rule-following. Unlike the three further conditions just mentioned in connection with the first requirement, this requirement raises a problem even when the requirement of intentional ascent is fulfilled. I see it indeed as the only requirement apart from intentional ascent that raises a major independent problem. Thus I believe not only that the two requirements are necessary for thought, but that their joint satisfaction makes it reasonably unproblematic that thinking should occur. It makes it unproblematic in the case of thought that all the individually necessary conditions—and therefore the minimal set of collectively sufficient conditions—should be fulfilled.

As with intentional ascent, it may be useful to clarify what rule-following involves, before offering an argument for why thinking requires it. There are two questions that need to be answered. First, what are rules? And second, what does it mean to follow rules? In setting out the answers, and later in arguing that rule-following is a requirement of thought, I follow Pettit (1990a, 1990b). (See also Summerfield 1990.)

What are rules? As I invoke the notion, and as it is commonly invoked, rules are normative constraints that determine that one member—or perhaps one subset—of a set of options is more appropriate in some way than alternatives. The option may be the most polite, as with a rule of etiquette; the most becoming, as with a rule of fashion; the just option, as with a rule of fairness; the proper verdict, as with a rule of evidence; the right thing to say, as with a rule of truth-telling; or whatever. The rule in every such case is the constraint in virtue of which the option is appropriate.

The rules with which we shall be concerned are relevant in an indefinitely large variety of situations—for an indefinitely large variety of option-sets—and it will be useful if we build the assumption of such open-ended relevance into our conception of rules. That a normative constraint is relevant in an indefinitely large number of situation-types means that there is no effective, mechanical way of specifying the option-sets to which it applies. If there were, then the constraint could be replaced by a mechanical procedure: the procedure would identify each type effectively and then identify the appropriate option. Most of the rules with which we are familiar, and not just those that concern us here, involve normative constraints that are relevant in an indefinite variety of situations. There is usually no mechanical way of specifying even the different situations where a rule of etiquette applies; such rules, as it is often put, are open-textured: there seems to be no limit on the mechanically specifiable types of situations where they dictate the appropriate response.

What sort of an entity is going to represent a normative constraint over an indefinite variety of situations? One type of entity that could serve in that role would be an indefinitely large set of pairs, one for every situation-type to which the rule is relevant: the first member in the pair would be the type of situation involved, the other the appropriate type of option. We might refer to such an indefinitely extended object as the *rule-in-extension*. Another sort of entity that could serve in the role required would be the abstract object which is conceived as having the property of identifying the appropriate option for every relevant situation to which it is applied; the entity would be an abstract function that, given a situation-type as input, delivers the appropriate type of option as output. We might call this the *rule-in-intension*.

It may seem strange to think of such more or less objective entities as rules. After all, someone may say, aren't rules things that we address to ourselves, like injunctions. Aren't they recipes for action or standards of success that we employ in guiding our conduct? Here it is important to distinguish between normative constraints and the linguistic or quasi-linguistic formulations of those constraints. The normative constraints are rules, in the sense in which I am concerned with rules, their formulations are not. I admit that we often refer to those formulations—say, the injunctions in a manual of etiquette—as rules. But I regard this as a secondary usage. Consider an analogy. It is common to refer both to reliable regularities, and to the formulations of such regularities—

the formulas found in scientific texts—as laws of nature; we already commented on this in the first chapter. If the analogy with rules is imposed, it is the regularities, not the formulas, that are laws in the proper sense of the term.

So much for what rules are. What then does it mean to follow a rule? That is the second question posed above. Here the important distinction to draw is between conforming to a rule and following it. To conform to a rule is to fix upon one option among a set of alternatives that the rule identifies as appropriate. This may mean forming a belief that the rule deems correct, forming an intention that the rule deems suitable, performing an action that the rule deems appropriate, and so on through the gamut of possible agent-responses. Conforming to a rule in this sense does not require a knowledge of what the rule is, or even a knowledge that conformity has been achieved. All the more obviously it does not require a desire on the agent's part to conform, or an attempt to satisfy that desire. That a subject conforms to a certain rule is a relational fact about its response and it tells us nothing, in itself, about the agent's state of mind.

To follow a rule, however, as distinct from just conforming to it, is to exhibit a certain state of mind; it is not just to satisfy a response-specification. To follow a rule is to conform to the rule intentionally, to conform to the rule through trying to conform to it, to conform to the rule because of acting on the basis of a desire to achieve conformity. An agent follows a rule in the responses it displays, be those responses beliefs, desires, or actions, just in case its responding in accordance with that rule is intentional. Thus it follows a rule only if a rationalising set of beliefs and desires holds out that pattern of response as a goal, and programs for its appearance. This notion of following a rule, it should be noticed, is a success concept: to follow is to try to conform and to succeed in conforming. There is a weaker notion of following a rule in common usage as well, a notion that requires just trying to conform to the rule, whether successfully or not. In place of the weaker notion I shall always speak of the agent trying to conform to the rule or trying to follow it.

There may seem to be a problem with the notion of an agent intentionally forming beliefs that conform to a rule or, more generally, intentionally displaying more or less autonomic responses that conform to a rule. We know that it cannot be intentional on an agent's part that it forms the belief that p or that q. So how can it be intentional that it forms a belief in accordance with a certain rule? Here we should recall the point made in the introductory discussion: that though an agent does not intentionally judge that p, it may intentionally judge in accordance with the evidence, where so judging is just judging that p. It should be clear by parallel with that case that an agent may not intentionally form the belief that p, though it does intentionally form the p-belief in accordance with a certain rule. That it forms the belief that p is not due, even in part, to any desire that it has; that it forms its belief in conformity to a rule is due, at least in part, to such a desire. In the context of its other beliefs the desire to conform to the rule programs for the subject's forming its beliefs in conformity to the rule but the desire does not program for the subject's forming the belief that p; after all, the desire would tend to promote the belief that q, did the rule make that belief the appropriate one to form.

So much for what rules are, and for what following rules amounts to. This is enough by way of clarifying what I have identified as the second requirement of thought, side by side with the requirement of intentional ascent. It remains now to present the argument for why this requirement is necessary. Why does every thinking subject have to follow rules?

To think is to act intentionally with a view to having intentional states that satisfy certain conditions: for example, the conditions of evaluative and deliberative, inductive and deductive, rationality and, ultimately, a condition involving truth. Whether an intentional state satisfies such a condition is determined in any situation by whether its content makes it an appropriate state to form or hold there. Whether a belief is true depends on whether the content is true; and whether a desire has a true foundation depends on whether the content of the corresponding desirability-belief is true. Whether a belief is inductively or deductively rational depends on whether the content is suitably supported by the evidence at hand and other propositions believed. Whether a property-desire is evaluatively rational is determined by whether the property is presumptively desirable. And whether an option-desire is deliberatively rational is determined by whether the option is presumptively desirable. In sum: a belief is true if and only if its content is true, rational if and only if its content is well supported; and a desire is truly founded if and only if the corresponding desirability-content is true, rational if and only if the corresponding content is well supported.

The point of these observations is to show that if a subject's thinking is to be successful, then what the subject has to try and ensure is that the intentional states it forms or holds satisfy the normative constraints represented by associated propositions. Suppose I believe that p and I am anxious that my beliefs should be true. What I have to try and ensure, in order to satisfy my anxiety, is that I believe that p only if the actual world is one at which it is the case that p. What I have to try and ensure, more proximally, is that I believe that p only if the actual world seems, by accepted criteria of evidence, to be one at which it is the case that p. Suppose on the other hand that I desire a property or option and am anxious that my desire has a true foundation in desirability. What I have to try and ensure is, ultimately, that I desire the object only if the actual world is one at which the property or option is desirable; and, proximally, that I desire the object only if the actual world seems, by accepted criteria of evidence, to be one at which the property or option is desirable.

The picture emerging is that propositions are rules of thought. They are canons that dictate, for a given way the world may be or seem to be, whether or not it is right for a subject to believe them under that eventuality: at that possible world. A proposition is sometimes said to be a function that takes possible worlds as inputs or arguments—in the way a function for determining primes would take numbers as arguments—and which gives true or false as output or value, in the way in which the number function would give prime or non-prime as value. This picture of propositions needs some amending (see Lewis 1983b, chap. 10) but, even as it stands, it may help us to see how propositions can serve as rules of thought. The function for primes would dictate how I should go in determining for a given number, whether it is a prime or not. And the proposition, being a function for possible worlds, dictates how I should go in determining for a given possible world, whether the possible world is one at which to believe the proposition. But if the proposition dictates how I should go at any possible world, it will dictate how I should go at the actual world. It will dictate whether or not to believe the proposition at the actual world, depending on which-way-that-things-may-be—which possible world—is actual or appears to be actual.

All of this said, the connection between thought and rule-following must be manifest. To think is to try to conform to the rules which certain propositions represent: with beliefs, to try to conform to the rules represented by their contents; with desires, to try to conform to the rules represented by the associated desirability-propositions. Thinking is the intentional attempt to submit oneself, ultimately to the regimen of truth, proximally to the discipline of rationality. To be successful in thought is not just to come to believe something correctly, for example, not just to come to believe in a way that conforms to the requirements of the content; it is to follow the requirements of the content in the belief adopted: it is intentionally to come to believe that proposition correctly.

The upshot is clear. Thinking not only involves the intentional subject in practising intentional ascent, coming to have beliefs about propositions as well as beliefs—and desires—with propositional contents. It also involves the subject in following the rules for belief- and desire-formation that such propositions represent. It involves the subject in trying to be faithful to the requirements of the propositions it contemplates in the beliefs and desires that it forms and holds. The propositions are not just objects of belief for such a subject. They give the subject its bearings as it tries to ensure that its beliefs—and desires—are suitably rational. They are the landmarks by which it must orientate if it is to succeed in the enterprise of thinking.

With the two major requirements of thinking isolated, we turn now to the question of how they can in principle be fulfilled. In the next section we deal with how intentional systems can practise intentional ascent, in the section after that with how they can practise rule-following. In each case, we ask the question of how the requirement can be fulfilled, against a background of naturalistic assumptions: assumptions that the world involves no entities of a kind undreamt of in our sciences, even if the entities it does involve are organised in a fashion that the sciences do not comprehensively chart (Pettit 1992). I assume, as I mentioned earlier, that the two requirements are the only independently problematic conditions for thought and that if we can explain how they can be fulfilled, then we will have adequately explained how thought is possible.

A note on methodology. I have spoken in each case of explaining—naturalistically explaining—how the requirements can be fulfilled. The explanation sought is not an evolutionary explanation, not an account of how the requirements might have come to be fulfilled by our ancestors under certain selectional pressures. Nor is the explanation sought an explanation that would direct us to the pressures that produce intentional ascent and rule-following in species or individual. What we want to do is, taking certain plausible naturalistic capacities and connections as given, show that they are sufficient to ensure that agents can practise intentional ascent and rule-following. We want to provide a constitutive explanation, as it were, rather than a causal explanation: an explanation that shows that certain naturalistic materials serve to constitute intentional ascent and rule-following. The enterprise is continuous with our attempt to make sense of intentionality. We tried to show that it is sufficient for a system to count as intentional that it should display a certain pattern of interaction with an enviroment, in particular a pattern of interaction that is naturalistically unproblematic. We shall now try to show that equally it is sufficient for a system to be a thinking system, it is sufficient for a system to practise intentional ascent and rule-following, that it should display certain other naturalistically intelligible capacities and connections.

The First Requirement: Intentional Ascent

The requirement of intentional ascent comes essentially to this: that the thinking subject must be able to have the sort of relation to propositions that any intentional subject has to the constituents of propositions that it believes; that it must be able to connect with scenarios in the way in which any intentional subject connects with the constituents of the scenarios in which it believes. I believe that the telephone is bugged and thereby have a certain relationship to the particular telephone in question and the property ascribed to it in my belief: this is the relationship, as we described it earlier, of having a belief about that object and about that property. If I am to be a thinking subject, then I must have the same sort of relationship to propositions or scenarios themselves.

What enables an intentional subject to have a belief about an entity like a concrete object or the property of a concrete object? What makes it the case that one of the beliefs it has is indeed about that object or property? In approaching that question we must assume that normal and perhaps ideal conditions have been fixed in some way, so that problems of indeterminacy do not arise; this assumption will be vindicated in the next section, where it becomes clear how such conditions can be non-arbitrarily fixed for a thinking subject. Assuming that conditions are normal, and even perhaps ideal, we can say that one of a subject's beliefs will be about a particular object or property if and only if the belief is sensitive to precisely that object or property.

The sensitivity required can be spelled out as follows, on lines already foreshadowed in the preceding chapter. Fixing the property, the subject holds the same belief independently of how, over an indefinite range of variations, the object is constituted or presented; move the telephone or present it in a different light and I retain the belief that it is bugged. And fixing the object, it holds the same belief about the property independently of how precisely the property is realised or evidenced; change the way it is bugged, or alter my evidence concerning the fact, and I continue to believe that the telephone is bugged. If we are to explain how the requirement of intentional ascent is fulfilled for thinking subjects, then we have to show how they might become sensitive to propositions or scenarios in the way in which I am sensitive to the telephone and the property of being bugged.

Consider a case that is like the telephone one but that may be more helpful as a point of departure. Imagine that a trained chimp learns that food is always to be found in a box with a triangular lid, no matter what the size or shape or colour of the triangle. Such a chimp is sensitive to the property of triangularity. We may imagine that it forms a belief about any triangular box presented— that the box is likely to contain food—a belief that retains the same content over all sorts of possible differences in the nature of the box. The task before us is to explain how thinking subjects might become sensitive to propositions in the way in which the chimp is sensitive to triangularity.

The chimp is always presented with a particular triangle and the secret of its sensitivity is that the particularities of the triangle hardly register or matter. The chimp ignores everything about the particular triangle on hand other than what it shares with other instances: namely, the property of triangularity. The triangle on hand is not just an example of triangularity for the chimp, but an exemplar. The particular instance exemplifies the property, in the sense that it has no role to play in the economy of the chimp's responses that could not be equally well played by any other triangle (Goodman 1968). The chimp looks through the concrete instance to the property of triangularity. The instance is a manifestation or presentation of that property.

Is there anything to exemplify and present a proposition or scenario to a thinking system in the way in which the property of triangularity is presented to the chimp? That is the question we face in this section. I will consider three answers in turn. I see serious problems with each of the first two and so I reject them. The third answer I adopt.

The first answer is the one that our chimp analogy most readily prompts. The chimp is presented with the property of triangularity through its various instances, instances that serve for the chimp as exemplars. We might suggest, in parallel, that a thinking subject could be presented with a proposition through something that exemplifies that proposition, that it could be presented with a scenario through experience of a situation that realises it. The obvious suggestion is that a proposition, say that the cat is on the mat, might be exemplified for a thinking subject by its truth-maker at the actual world: by the concrete situation involving the cat on the mat at the appropriate time. A subject that is capable of believing that the cat is on the mat will be sensitive to that situation—this actual one or any possible one—being disposed on observing its realisation to form that belief for the appropriate time. Moreover it will be sensitive to the observation in a way that is invariant over an indefinite range of differences in how the situation may be, consistently with being a truth-maker for the proposition, consistently with being a realiser of the scenario. Thus there seems to be no difficulty in the idea that as a concrete instance may exemplify triangularity for the chimp, so the concrete truth-maker might exemplify a proposition for such a subject. It might enable the subject to look through the details of the situation presented to see the proposition or scenario that it makes true.

Natural though it is, this first answer to our question runs into problems. The main difficulty is that if the only exemplar that is ever available for a proposition is the truth-maker at the actual world, then there is no way for a false proposition, an unrealised scenario, to be exemplified to a thinking subject. Thinking subjects have to be concerned with propositions independently of whether they are true or false. Their aim is to isolate propositions that are probably true, and propositions that it is rational to believe or desire, among a more comprehensive set, and neither the comprehensive nor the isolated set is going to coincide with the set of true propositions. Thus thinking subjects have to be concerned with propositions that have no actual truth-makers and that cannot therefore be exemplified by actual truth-makers. And so the first answer to our question collapses. There may be ways of patching up the answer but the prospects are not promising and I leave it be.

The second answer to our question is also prompted by the chimp analogy. We might think that though propositions or scenarios cannot be directly exemplified by their actual truth-makers, still there is room for another, indirect sort of exemplification. Every proposition or scenario has constituents and we might think that if we can explain how constituents are exemplified to thinking subjects, then we will be able to explain how the propositions they compose are also exemplified. The idea is this. I am sensitive to the cat, to the mat, to the present time, and to the relational property whereby one thing rests on another. Perhaps those individual sensitivities are sufficient materials out of which to build a sensitivity to the proposition—even the false proposition, as we may suppose—that the cat is on the mat now. Perhaps the combination of constituent-exemplars can make that proposition or scenario present to me as something to have beliefs about.

This second answer is not an easy one to make vivid or appealing; we have to imagine the thinking subject assembling particular presentations of objects and things in order to present itself with a proposition or scenario in which they figure. But in any case the answer runs into problems of the same order as the first. The difficulty here is not that the imagined exemplars exemplify too narrow a class—true propositions rather than propositions generally—but that they exemplify a class that is too broad. Take the items that are to exemplify our proposition: presentations, respectively, of the mat, the cat, the present time, and the on-relation. Why should that set of presentations exemplify the proposition or scenario that the cat is on the mat at the present time rather than simply the collection of corresponding entities: the set consisting of the cat, the mat, the present time, and the on-relation? Even if thinking subjects could assemble the individual exemplars required, there seems to be no reason to think that the assemblies would give them contact with the appropriate entities: with propositions rather than collections, with scenarios rather than sets of props.

It seems clear that the most obvious ways in which propositions or scenarios might be presented—the exemplifying ways that most closely parallel the way in which triangularity is presented to the chimp—are not going to work.[3] We must look for some relationship other than exemplification to show how it is possible for propositions to be presented to thinking subjects. The third answer to the question of how a proposition might be presented to a thinking subject invokes a relationship that is closely analogous to that of exemplification, a relationship involving proxies rather than exemplars. The best way to introduce the proxy-relationship is by an example of how it might work.

But a word of warning about the example. The idea to be illustrated is that propositions might be presented via proxies to thinking subjects. Now proxies are stand-ins or signs and, as is often noticed, signifying comes in many varieties. At one extreme we have natural signs—my blushing is a natural sign of my embarrassment—and at the other we have signs that are produced with the overt intention to communicate: say, my rolling my eyes in your direction, where you are a third party, in order to indicate my embarrassment (Sperber and Wilson 1986). In the example that follows, I introduce proxies or signs that are not communicative in any strict sense but that are at least produced by the parties to whom the propositions are presented. This should not mislead anyone into thinking, however, that under the third answer propositions can be objects of belief only for communicating, or only for sign-producing, creatures. It just happens that the example that I find most helpful in conveying the third answer to the question about intentional ascent involves creatures of that kind. We will return to the point later.

Imagine creatures, say a family of mice, that behave in such a way that we can ascribe to them the belief at different times that the cat is on the mat. Imagine now that nature or culture has been kind enough to the mice to have encouraged an association between that situation and a certain sound-sequence. Any mouse that sees the cat on the mat at a particular time not only responds by forming that belief and behaving in the normal hiding or fleeing way; it also gives off this sequence of sound. And any mouse that hears the sequence of sound responds at that time as it would normally respond to seeing the cat on the mat: it comes to believe that the cat is on the mat and to behave appropriately. The sequence of sound serves as a proxy for the cat-on-mat-now situation.

This piece of imagining is not naturalistically implausible, since it is a feature of some non-human animals that they utter and respond to warning calls in just this manner. Vervet monkeys are said to utter one sort of sound when a snake appears on the ground, another when an eagle appears in the sky, the sound serving in each case to arouse the response associated with observation of the appropriate situation (see Dennett 1987, essay 7). In order to bring our imagined mice into line with such models, we might think of them as having a variety of calls, each bearing on the position of the cat or on the position of some other dangerous creature. We might even think of these calls as being articulated, with one element figuring whenever the cat figures, a different element figuring whenever the relation of being on something is involved, and so forth.

The relationship the third answer proposes in place of exemplification is the proxy-relationship that a warning call—say, that the cat is on the mat—has to the corresponding situation. Clearly an instance of that call cannot exemplify the situation, for it is not an instance of the situation. But equally clearly it serves a function in the economy of a mouse's responses that is akin to the function served by a particular triangle in the economy of our chimp's responses. The mouse that hears the call does not concern itself with the particularities of the actual sound-sequence produced; all that matters to it, all that is relevant to how it responds, is the character that the sequence shares with other such calls, the character that distinguishes it from all-clear calls, from calls bearing on the dog, and so on. The mouse, like the chimp in our earlier example, neglects the details of the call in favour of something that is more abstract. The chimp neglects the particularities of different triangles, being sensitive only to the triangularity they all instantiate. The mouse neglects the particularities of different ways in which the cat-on-mat-now call may be sounded, being sensitive only to the scenario for which they all go proxy. As the chimp looks through the details of this or that triangle to the property it exemplifies, the mouse looks through the idiosyncracies of this or that call to the scenario for which it goes proxy.

The similarity between the chimp and mouse examples makes it plausible that, as the concrete triangle makes the property of triangularity present to the chimp, so the warning call makes the cat-on-mat-now scenario present to the mouse. As the chimp connects with the property, responding to the concrete triangle only on the basis of its property-exemplifying features, so by responding only to the features of the warning call that are relevant to its proxy-role, the mouse connects with the scenario for which the call goes proxy. As the property is there for the chimp to form beliefs about, so the scenario is there for the mouse to form beliefs about. The call makes the scenario into a possible object of belief for the mouse; it means that the mouse can have a belief that is directed towards the scenario as well as a belief with the scenario as content.

It is clear what beliefs the chimp forms about the property of triangularity: that this or that lid has it. But what beliefs is the mouse liable to form about the cat-on-mat-now scenario? Its normal hiding or fleeing response requires only the belief that the cat is on the mat. Thus the presenting of the scenario by the call does not necessarily lead to the formation of a belief about the scenario. But we can easily imagine how such beliefs might also figure.

Suppose that many of the calls made are false alarms, perhaps because of sincere error on the part of observers. It is going to be in the interest of the mice to be able to form the belief that such error has occurred and to inhibit formation of the belief that the cat is on the mat. If a mouse is to form a belief that error has occurred, however, then it is going to have a belief about the scenario presented by the call. The belief formed will be: about that scenario, it doesn't hold; about that scenario, it isn't true. The possibility of error introduces the need for mice to have beliefs, not just about the truth of propositions, but also about the support they command. If error is always a possibility, it is going to be in the interest of the mice to be able to form the belief that error is likely or unlikely with a particular call. The belief formed will then be: about that scenario, it is well supported; about that scenario, it is probably true.

Apart from sincere error, the sort of thing that is liable to disrupt life among our articulate mice is deception: the case where a mouse, aware that the cat is not in sight, nevertheless makes the cat-on-mat call and perhaps thereby gives itself an advantage in the search for food. It is going to be in the interest of the mice to identify such cases of deception, distinguishing them from cases of sincere error, since each mouse's welfare may depend on the punishment or isolation of deceivers. If the mice are to identify those cases, however, they are going to have to be able to form another sort of belief about the scenario presented by the call. They will have to be able to form a belief of the kind: about that scenario, the calling mouse doesn't believe it; about that scenario, the calling mouse is not sincerely asserting it.

How are we to depict the content of beliefs about scenarios or propositions, whether beliefs about their truth, their support, or their acceptance? The most obvious way links up with an analysis of corresponding linguistic contexts by Donald Davidson (1984, essay 7). Take the belief about the scenario, that it is not true. The obvious way of representing the mouse's belief-content would be by using the sound that presents the scenario or by using an equivalent expression in our language—the sentence 'the cat is on the mat now'—and then adding that that (i.e., that scenario just presented) is not true: 'The cat is on the mat now. That is not true'. This representation easily goes over into the ordinary expression, of which it can then be taken as a more perspicuous version: 'It is not true that the cat is on the mat now'. And a similar story applies for the other contents mentioned: 'The cat is on the mat now. That is probably true'; 'The cat is on the mat now. The calling mouse does not believe that'.

Although I have stuck close to the fanciful example of the mice, I hope that the general outline of the third answer to our original question has become fairly clear. How is intentional ascent possible? In particular, how can propositions or scenarios be presented to thinking subjects? They cannot be presented by exemplars such as truth-making situations or the constituents of such situations. But they can be presented by physical elements, such as the sound-sequences of our example, which go proxy for the scenarios rather than exemplifying them. They go proxy for the scenarios in the sense that in appropriate contexts they evoke the same responses as those scenarios themselves: they are functionally equivalent in their intentional effect. The relationship of going proxy is clearly very close to that of exemplifying. The proxy serves exactly like an exemplar in its effects on the relevant intentional system and that is why it can make present, as a potential object of belief, that which it goes proxy for. The only difference between the proxy and the exemplar is that it is not itself an instance of that which it makes present; it is, as the name suggests, a substitute or sign.

The connection between the proxy-relationship and the exemplar-relationship is worth stressing in order to bring out the continuity between how scenarios are presented, on the story developed here, and how other more or less abstract entities may be presented to a simple creature like a chimpanzee. But it should also be mentioned that the proxy-relationship clearly may look forward to more sophisticated developments, as it looks back to simple exemplification. Thus John McDowell (1980), to whom I am indebted for the proxy-idea, has argued that the relationship is an elementary form of the relationship whereby an assertion means that something is the case (see also Pettit 1987b). When the call goes proxy for the cat-on-mat-now scenario, that is sufficient for it to mean that the cat is on the mat now. If my assertion 'The cat is on the mat now' has the same meaning, under this account, that is by virtue of its continuity with the proxy-call case. My assertion may not inevitably elicit the same response among my fellows as the observation of the cat on the mat: they must allow, like the mice in the extension of our story, for possibilities like error and deception. But it retains enough continuity for it to mean that the cat is on the mat now in the same way that the call among the mice has that meaning.

I see no major problems with the third answer to the question of how intentional ascent is possible. Like the exemplification story with a property such as triangularity, it offers a plausible, naturalistic account of the presenting of propositions as objects of belief. Propositions can be presented, in a naturalistically unproblematic way, via proxies or signs. The story does not imply that only systems that produce artificial proxies for situations, only systems that have the capacity to signify, are capable of having propositions as objects of belief. Once we see that propositions can in principle be presented as objects of belief by such proxies, then we should recognise that the proxies that serve that purpose do not have to be signs employed by the agents themselves. For all that we have said, the function of presenting propositions as things about which to have beliefs may be adequately fulfilled by signs that originate elsewhere: signs that fail, not only to be communicative, but even to be the product of the believers. The possibility may need to be motivated but the important point for us is that nothing I have said here rules it out.

But though the third answer to the question about intentional ascent is compatible with this sort of possibility, it remains the case that according to that answer propositions are presented by signs, whether or not these signs are produced by the parties involved, and whether or not they are produced communicatively. The presentation of propositions as objects of belief is inherently sign-involving. In this respect the account may call to mind the old tradition that thought requires language (Wells 1987). But if we countenance that association, we must make the following two matters absolutely clear. First, we have argued not that language is required to explain the ability to form beliefs and other intentional attitudes, only that it is necessary for explaining the ability to form beliefs about propositions: beliefs that have propositions as their objects, not just their contents. And second, we have denied that language in the sense required has to be communicative, or has to be produced by the agents that form the beliefs; it may come to nothing more than a language of natural signs.

Even if the argument developed here makes a connection between thought and language, I should mention that it does not entail the language of thought hypothesis, as it is called (Fodor 1975). According to that hypothesis, the physical realisers of beliefs and other intentional states involve syntactically structured and semantically significant entities like sentences. As a subject asserts something through uttering a sentence, so on this hypothesis a subject believes something through appropriately instantiating a sentencelike state. Obviously the sort of language with which our argument connects thought involves public signs, not language in the bare sense of this hypothesis. The sentences of a language of thought, if there is such a language, are not items of which the believer is aware, let alone items that might present propositions as objects to have beliefs about (see McGinn 1984, chap. 4).

I have argued that the sign-involving account of intentional ascent is satisfactory, and satisfactory in a way in which the other alternatives are not. The account is the best explanation I can see of how propositions are presented and, since it does not seem to be particularly implausible in any of the assumptions it forces upon us, I am prepared to endorse it. We might go into greater detail in defending the account but that hardly seems to be necessary. While it is essential to understand how propositions might be presented to naturalistic subjects—while it is essential to demystify the process involved—nothing much hangs in the argument that follows on how exactly the presentation is achieved. It would be a waste of energy to devote much more time to a defence of the picture sketched.

The Second Requirement: Rule-Following

Preliminaries

It should be clear that an intentional system might have beliefs about propositions or scenarios—that it might practise intentional ascent—without being in our sense a thinking subject. It might have such beliefs, even beliefs that certain propositions are true or false, well or ill supported, without doing anything to try to ensure that its own beliefs are true or well supported. It might have beliefs about propositions, in short, without intentionally taking those propositions as rules to dictate what it should or should not believe—or, of course, desire. The requirement of rule-following, therefore, is an additional requirement of thought.

Is there any difficulty in the idea that an intentional system for which propositions are objects of belief—a subject that has achieved intentional ascent—should intentionally treat those propositions as rules for belief-formation? If a subject intentionally forms beliefs in accordance with rules then, as we use the term, it forms the beliefs by making judgments. Again, if a subject intentionally forms desires in accordance with suitable rules—the associated desirability-propositions—then it forms them by making corresponding judgments in regard to those propositions. Hence the question before us, in a phrase, is whether there is any difficulty in the idea that a subject that has achieved intentional ascent should intentionally treat propositions as rules for judgment: rules for determining when a judgment is true or at least rational.

The answer to the question is that yes, there is a difficulty that threatens this prospect. Consider the difference between what is required just for conforming to a rule and what is required for following it. A system need not target or address a rule in order to conform to it—the rule need not become an object of belief and attention for the subject—since the conformity may not be intentional. Moreover the set of rules with which such a system conforms may be so indeterminate as never to require us to ascribe infidelity; the point at which we fail to find a rule with which its behaviour conforms may be a point at which we see it as no longer in rule-conforming mode: it may be a point at which the system breaks down, for example. But a system must target or address a rule in order to follow it or to try to follow it, for following is necessarily intentional. And, moreover, the set of rules such a system tries to follow must be fixed determinately enough to allow for infidelity; if they were not, then it would make no sense to think of the system setting out, fallibly, to try and conform. The problem is to explain how the rule-following subject, in particular how the thinking subject, can target or address determinate rules of this kind: in the thinking case, determinate propositions.

The difficulty connects up with our discussion of determinacy of content at the end of the preceding chapter. I argued that with a non-thinking intentional subject there may well be an open-ended indeterminacy about what precisely are the propositions or scenarios it believes and desires; that, depending on which circumstances are stipulated as normal or ideal, the subject can be represented as having intentional states with any of a variety of contents; and that, at the limit, it may even be possible always to represent the contents of the states so that the subject cannot be charged with error. We defined intentional systems as subjects that are governed by intentional regularities in their interaction with a perceptual environment and under one specification of normal or ideal conditions, a system may obey one set of intentional regularities, under a second specification it may obey a different set, and so on. Contents will be determinate, or relatively determinate, if one set of normal or ideal conditions is fixed as objectively right. But while there will be methodological considerations that make some sets more convenient than others, there may be nothing to select a single set or family of sets as objectively right.

We can reasonably continue this relaxed attitude on indeterminacy when we deal with intentional systems that have achieved intentional ascent, developing attitudes towards contents as well as attitudes with contents. Consider the mice discussed in the previous section. We assumed in our discussion, as we said at the beginning, that favourable conditions were such that the mice have beliefs about the cat. But we can be open to the thought that it may be objectively indeterminate whether they have beliefs that are strictly about the cat—beliefs that are false if occasioned by a look-alike—or beliefs about the-cat-or-anything-like-the-cat. We can be open to the thought that they have beliefs, as we may come to think of the matter, about inherently vague contents as well as having beliefs with inherently vague contents. I have attitudes towards vague entities like heaps and baldness, so why should there be any problem with something's having attitudes towards vague contents?

The reason we can live with indeterminacy for non-thinking intentional systems is that while they display states that it is appropriate or inappropriate to hold in the light of certain propositions, while they display states for which certain propositions constitute rules of truth and rationality, they do not try to conform to those rules; the conformity may be reliable but it is, nonetheless, non-intentional. The rule-conforming system does not have to target or address a rule with which it conforms, since conformity will not be intentional. And the rules with which it conforms may be fixed so indeterminately that there is no room for ascribing infidelity.

But if we can live with such indeterminacy for non-thinking intentional systems, even intentional systems that practise intentional ascent, we cannot do so with intentional systems that think. The thinking subject does not just conform to certain rules; it tries to conform to them. Specifically, the thinking subject tries to make judgments in a way that accords with their contents, that is, the propositions or scenarios endorsed; it tries to make judgments that are true or at least rational, given those contents. This means that for every such judgment, the content of that judgment must be fixed on some basis other than that the judgment is true or rational; otherwise there would be nothing for the subject to try to fit its judgment to. The subject can try to respect the requirements of a proposition, after all, only if the proposition targeted is fixed independently of the success of the enterprise. It follows then that the thinking subject's intentional states—or at least the states that come under the control of thought—must have contents that are determinate enough to leave open the possibility of error in any of the judgments that it can try to get right. If we are to vindicate the notion of the thinking subject, we must be able to explain how it can identify suitably determinate propositions as rules of judgment.

Suitable determinacy does not mean absolute determinacy. It will be sufficient if we can explain how thinking subjects can target propositions that are determinate enough to make it possible for them to go wrong in their beliefs: determinate enough, in particular, to make it possible for them to go wrong in any situations where they try to get their beliefs right. What is ruled out is that the contents should be so vague that even in situations where the subjects try to get their beliefs right, there is no way they can go that is certainly wrong: no way they can go that means, indisputably, that their assault on truth is unsuccessful. Such determinacy allows that a proposition targetted by thinking subjects may be vague at margins where their concern for forming true judgments about that type of proposition runs out. An example will help to make the point vivid.

Suppose that we ascribe to certain subjects the desire to get their beliefs right about a certain type of proposition: say, the proposition that a person at a later time is the same person as someone at an earlier. Suppose further that we ourselves distinguish this proposition into some more specific claims: say, the claim that the later person enjoys a unique psychological continuity with the earlier, the claim that the later enjoys a unique bodily continuity, and the claim that the later enjoys both. Suppose, finally, that we imagine a situation where psychological and bodily continuity come apart—two people have their brains exchanged—and that we try this case on our subjects to see which precise claim they were targetting in asking about personal identity over time.

One response that they might make is to evince no interest in that kind of case. If they do make that response, then we must say that the proposition they aspire to have true beliefs about is vague at the relevant margin. Such vagueness will not matter, since it will be consistent with their actual enterprise of thinking, though not with the enterprise of making a judgment on the imagined case.

The position, in summary, is this. Non-thinking intentional systems merely conform to rules and no problem arises as to how they target rules that are determinately enough fixed to make infidelity possible. Thinking systems try to conform to rules and so there is a problem of this kind. The problem is to explain how a thinking subject can identify a determinate proposition with a view to trying to conform to the rule it represents: with a view to trying, given how the world is or seems to be, to make true or well-supported judgments involving that proposition. The task before us is to see if we can provide an answer. In this section we will approach that task in two stages: first, drawing on Saul Kripke's (1982) discussion of Wittgenstein on rule-following, we will show how difficult and even impossible the task seems to be; and then we will set out to accomplish the task, finding a way around the various difficulties involved. But before we go to those discussions, there is one further preliminary matter to get out of the way.

While propositions are rules of judgment for a thinking subject, rules for what it ought to believe at any possible world, they are what we might describe as non-basic rules. A non-basic rule—strictly, a non-basic rule for a given subject—is a rule such that the subject can intentionally conform to it, the subject can follow it, only through intentionally conforming to some other rule or rules. A basic rule is one of which this is not true: a rule such that the subject can intentionally conform to it directly. The subject will conform to a basic rule by doing other things, things involving the contraction of muscles or whatever, but it will not conform to the rule by doing anything else intentionally. At the intentional level, the rule will speak directly to the subject, guiding it about what is required. Notice that the divide between the basic and the non-basic is not just an on-off matter. Without being absolutely basic, one rule can be more basic than another: it will be more basic than the other if the subject follows the other by following it. (See Hornsby 1980, chap. 6, on basic actions; also Danto 1973.)

Propositions or scenarios are compound entities that are true or well supported in virtue of their component parts' meeting corresponding conditions. Thus the true scenario must be existentially satisfied in the sense that the entities it involves or presupposes actually exist; this condition is trivially realised with properties, at least under many accounts, but not with particulars: if that parrot outside my window is an illusion then there is no scenario or proposition expressed by the sentence, 'That parrot outside my window is squawking'. Again the true scenario must be attributively satisfied in the sense that the particulars and properties it involves must be related in the manner ascribed in the proposition: that parrot, if there is one, must actually be squawking, and so on. In order to try to conform to such a proposition in making true judgments, any thinking subject like you or me will distinguish the requirements of existential and attributive satisfaction, and will try in the first place to conform to them; it will try to see that its judgments are existentially and attributively satisfied, keeping track of the particulars required and checking that they possess the properties ascribed to them. Those requirements may not represent basic rules for the subject to respect—they may be followed by the following of yet simpler rules—but they are certainly more basic than the rules represented by propositions.[4]

The fact that propositions are non-basic rules means that the problem with how thinking subjects can follow propositions in their judgments reduces to the problem of how they can follow the more basic rules involved. Thus we should look at the rule-following problem with a view to more basic rules than propositions, and ideally with a view to rules that are likely to be absolutely basic. I propose to focus primarily on the more basic rule-following that a subject conducts as it tries to ensure that a judgment is attributively satisfied; in particular I propose to focus on that sort of rule-following in the simplest sort of case, where a single property is ascribed to a particular object. There ought to be no difficulty about how the lessons drawn generalise to other cases and to other rules. Thus I assume that it will be sufficient for explaining how the second requirement of thought can be fulfilled, if we explain how it can be fulfilled in this case.

As it tries to ensure the attributive satisfaction of its judgments in the sort of case I envisage, a thinking subject will be looking to the attribute or property in question to see that it really belongs to a given object: the property will be predicated of such an object, as the property of squawking is predicated of that parrot outside my window. Having fixed on the object in question, the thinking subject will be trying to conform in the simplest case just to the requirements of the property in forming its judgment: it will be trying to ensure that it ascribes the property only to a member of its extension, only to something that actually has the property. It will be trying to ascribe the property of squawking only to squawking things, the property of being a box only to boxes, the property of being triangular only to triangular things. The rule that any such property represents is a rule that the subject will have to try and follow on any of an indefinite number of occasions, since the subject will have to try to satisfy the rule whenever it tries to make a true judgment involving the property.[5] With some properties—those that are defined in terms of other properties—the rule will be non-basic: it will be followed by following more elementary property-rules. But unlike the rules represented by propositions, at least some property-rules will have to count as basic. Thus they will be a suitable focus for our discussion.[6]

In the preceding section I argued that a subject for which a proposition is an object of belief must rely on signs to represent that proposition. In the most plausible case, and certainly with creatures like us, the sign for a simple proposition involving a particular and a property will be articulated into components, with one element corresponding to each of these constituents of the proposition. Thus the subject that tries to follow the property-rule in the judgments it makes may be trying, at the same time, to determine when the property-sign is appropriate. If it is a sign-producer, it may be trying to determine when it is appropriate to produce the sign, uttering the word 'squawking' or 'box' or whatever. It may be useful to keep this in mind in the discussion that follows but I shall not emphasise that aspect of things. Were I to do so, the rule-following involved in thought might seem to be a matter of correctly applying words, and the rules followed in thought might begin to seem like linguistic rules: like standards of correct speech or even recipes for speaking correctly. This would be very unfortunate, for the rule I try to follow in forming beliefs involving the property of squawking or the property of being a box is not a rule in the sense of a formula for using a word. It is nothing less than the constraint represented by the property itself: that is what I try to remain faithful to in my judgments, and indeed in my speech too.

For related reasons, I will try to avoid talk of concepts. Such talk can also lead us to think of the rule-following problem involved in thought as one of how to apply wordlike entities to things. It may be philosophically unobjectionable to say that when a subject comes to form the judgment that a parrot is squawking, it applies the concept of squawking to the parrot. But it may still be misleading to introduce this mode of talk. Talk of applying concepts may suggest that forming judgments involves surveying an inner language—the language of concepts—and trying to satisfy the right standards in applying that language to things, as if the only rules around were formulae governing the use of words. When I speak of rule-following, as I have already stressed, I mean the enterprise of remaining faithful to certain objective constraints: in the case of thought, to the sorts of constraints represented by propositions and, more proximally, by particulars and properties. This focus would be jeopardised by casting the problem in terms of concepts, or in terms of public language, and so I shall avoid such modes of talk in this discussion. It will prove convenient to introduce talk of concepts later, in Chapter 4, but for the moment we do well to eschew it.

The Problem

Rules, as we have seen, are normative constraints that apply in an indefinite variety of situation-types. Thus the sort of property-rule on which we will focus is a normative constraint that applies in all those judgments where the issue arises as to whether the property is instantiated or not. Abstracting from possibilities of marginal vagueness, such a normative constraint might be taken as a rule-in-extension or a rule-in-intension. The rule-in-extension would consist of the set of pairs that each involve, first, a different situation where the question of how to apply the rule arises and, second, the correct response for that situation. The rule-in-intension would consist of an abstract function that delivers the correct response as output for every situation presented as input. With the property-rule, the rule-in-extension would be fixed by the infinitely large set of its instances in actual and possible worlds and the rule-in-intension would consist in a function that told us of every suitable entity whether or not it was an instance of the property.

The problem of rule-following is how a subject like one of us can be put in a position to try and conform to a determinate or relatively determinate entity of such a kind: an entity like a rule-in-extension or a rule-in-intension. What sort of thing could be normative over an indefinite variety of cases and yet engage in the manner required with our intentional projects? Putting the question the other way around, among the things that do engage in a suitable manner with our intentional projects, what items could serve as indefinitely normative constraints? There are two conditions that must be fulfilled by any satisfactory candidate: first, that it meet the objective condition of being or fixing a normative constraint that applies in an indefinite variety of cases; and second, that it meet the subjective condition of engaging appropriately with our intentional projects: of being something to which a creature like one of us can try to conform.

The subjective condition breaks down into at least three distinct sub-conditions and it will be useful to distinguish these, if we are to get the measure of the problem on hand. The first sub-condition is that the rule must be determinable or identifiable by a finite subject, in particular that it must be determinable or identifiable independently of any particular application. If the subject is to try to conform, then there must be something presented to it to which it can address its efforts. And if the subject is to be in a position to try to conform in any instance, then the rule to which it is to try and conform must be presented independently of how the rule applies in that instance. Allow that the rule is partly identified as requiring such an such an option in this or that situation, and it makes no sense to think of the subject trying to be faithful to it in that situation.

The second sub-condition that a rule must satisfy if it is to engage with the intentional projects of a creature like one of us is that it should not only be identifiable as a target of conformity for a finite subject, it should also be capable of instructing the subject, so to speak, on what it requires in the different instances where the subject tries to conform. This means that the rule must be directly readable, in the sense that the finite subject can tell straight off what it seems to require—this is the case with basic rules—or can tell what it requires by the use of rules whose apparent requirements it can ultimately tell straight off: this is the case with non-basic rules. Unless a finite subject can read off the requirements of a rule in this way, then it is not in a position to try to conform.

The third sub-condition complements the second. The notion of trying to conform presupposes the possibility of failure, for it means that the rule that is targeted by the subject is fixed on some basis other than what the subject actually does; it means that the identity of the rule is fixed in a way that is consistent with the subject's not succeeding in its plan of conforming. This possibility of failure means in turn that while a rule that is to engage with our intentional projects must be readable by a finite subject, it can only be fallibly readable. No matter how directly the rule speaks to the subject, no matter how quickly the subject can tell what the rule seems to require in a given situation, that fact alone cannot provide a guarantee that it has got the requirement of the rule right. The subject must be a fallible authority, in at least one sense of that phrase. It may be in a position to know what a rule requires in a given situation. It may even be in a position to know that it got the rule right in that situation. Whether these claims are allowed will depend on how precisely the limits of knowledge are drawn. But no matter how knowledge is understood, the subject cannot be in a position to rule out altogether the possibility that it might get a rule wrong. The rule must be fallibly readable, as we shall say. This condition, it turns out, is one of a family of relevant fallibility constraints; later we shall come across another two.

To return then to the problem of rule-following, the challenge is to identify something that can simultaneously satisfy the objective condition of being a normative constraint that is relevant in an indefinite variety of situations and the subjective conditions of being independently identifiable, directly readable, and fallibly readable. I say that the challenge is to identify something that can satisfy these conditions, not something that does necessarily satisfy them. The point is important. We will have dealt successfully with the problem before us if we can show that the apparent tension between the objective and subjective conditions is not a block to the reality of rule-following. The challenge we are concerned to meet says that necessarily rule-following is illusory, on the grounds that there is nothing for rule-followers to know; there is nothing that can fit the objective and subjective conditions we have distinguished (Kripke 1982, p. 38). This is a strong form of scepticism about rules and we will meet it if we can argue that possibly rule-following is a reality: that, at the least, the tension between the objective and subjective conditions does not show that it is an illusion.

There are two ways in which we might think of meeting the challenge before us: by taking something we know to satisfy the objective condition and by then showing how it can also satisfy the subjective constraints; or by taking something that certainly satisfies the subjective constraints and by then showing how it can also satisfy the objective condition. But both paths look to be blocked and that is the essence of the rule-following problem. The rule-following problem was first formulated in the work of Ludwig Wittgenstein (1958, 1978). In setting out the problem I follow the version in Saul Kripke (1982), adopting roughly the same presentation as in Pettit (1990a). (See also Fogelin 1987 and Wright 1980; and, for an overview, Boghossian 1989.)

Take the sorts of entities we know to satisfy the objective condition: the rule-in-extension and the rule-in-intension. The rule-in-extension is not capable of satisfying the subjective conditions, because it is liable to be an infinitely large set. There is no way that I could get in touch appropriately with such an infinite object. There is no way that I could get in touch with the infinite extension of a property across actual and possible worlds: say, the extension associated with boxes or triangles or games. And, to take the sort of rule discussed by Kripke, there is no way I could get in touch with the rule-in-extension associated with the plus-function: the rule determining what number is the referent of 'x + y', for any two numbers x and y. 'The infinitely many cases of the table are not in my mind for my future self to consult' (Kripke 1982, p. 22).

What of the rule-in-intension? What, for example, of the addition function, as Frege would conceive of it, which determines the correct option in any judgment about the sum of two numbers? What is there against the idea that this abstract object might be able to satisfy the subjective conditions, engaging a finite mind appropriately? Here the problem is to explain how a creature like one of us is able to get in contact with such an abstract object. It does not affect our senses like a physical object and so we are not causally connected with it in the ordinary way. So how then does it become present to such a creature? The obvious sort of answer is to say, like Frege, that it does so via an idea—or some such entity—that the subject can contemplate. But then the suggestion boils down to one that we consider in a moment, a suggestion that we will find wanting.

Moving from the entities that can clearly satisfy the objective condition on a rule to entities that look more likely to be able to satisfy the subjective conditions, the question here is whether such entities can be objectively satisfactory: whether they can serve as normative constraints over an indefinite variety of cases. Kripke mentions two main candidates for entities of this kind: first, actual or possible examples of the application of the rule in question, such as examples of a property or examples of addition; and second, introspectible states of consciousness, as for instance the sort of *quale* that might be thought to be associated with identifying a box or with adding numbers together. There is a special problem with the second candidate, which is that often no plausible *quale* is available. But, more important, there is an objection that applies equally to both candidates, and indeed to any finite object that is proposed for the role in question.

The objection, and this is clearly derived from Wittgensteinian materials, is that no finite object can unambiguously identify a constraint that is normative over an indefinite variety of cases. Consider a series of examples of addition: $1 + 1 = 2, 1 + 2 = 3, 2 + 2 = 4$, and the like. Or consider any set of examples of boxes or triangles or games. For all that any such finite object can determine, the right way to go with a novel case remains open. 'Plus', as we understand it, forces us to say that $68 + 57 = 125$ but the examples given do nothing to identify the plus-rule as distinct from, say, the quus-rule, where this says that the answer in the case of 68 and 57 is 5. 'Is a triangle', as we understand it, forces us to say that a square page, diagonally folded, is a triangle but the examples given, if they do not include this case, will be consistent with the folded page's not being a triangle; perhaps the rule illustrated outlaws paper triangles or perhaps it outlaws triangles made by folding. The fact is that any finite set of examples, mathematical or otherwise, can be extrapolated in an infinite number of ways; equivalently, any finite set of examples instantiates an infinite number of rules.

It appears then that a subject cannot be put in touch with a particular rule just on the basis of finite examples.

> When I respond in one way rather than another to such a problem as '68 + 57', I can have no justification for one response rather than another. Since the sceptic who supposes that I meant quus cannot be answered, there is no fact about me that distinguishes between my meaning plus and my meaning quus. Indeed, there is no fact about me that distinguishes between my meaning a definite function by 'plus' (which determines my responses in new cases) and my meaning nothing at all.
>
> (Kripke 1982, p. 21)

The problem raised extends to *qualia*. 'No internal impression, with a *quale*, could possibly tell me in itself how it is to be applied in future cases' (Kripke 1982, p. 43). If the impression has a bearing on future cases, say on the application of 'plus', it will be capable of being extrapolated in any of an infinite number of ways. How then am I supposed to grasp a particular rule in contemplating the impression? How is the impression supposed to make salient just one of the infinite number of rules that it might be held to illustrate? The question extends from qualitative impressions to all mental objects of contemplation, including the sort of idea postulated by Frege. 'The idea in my mind is a finite object: can it not be interpreted as determining a quus function, rather than a plus function?' (Kripke 1982, p. 54).

The upshot of these considerations is that it is extremely mysterious how there can be rules for creatures like you and me to follow. Such rules are required to satisfy two sorts of conditions, objective and subjective, which no familiar entity seems to be capable of simultaneously satisfying. Now a number of responses are possible at this point. One is to go sceptical and deny that there are rules: deny that there is anything there for rule-followers to know about, when they claim to know what a rule requires. A second is to go dogmatic and, insisting that of course there are rules, argue that they are *sui generis* (Goldfarb 1985). But such responses are not attractive and, ideally, we should look for some way around the challenge. We should look for an account of how finite creatures like you and me can follow rules.

Before leaving this discussion of the problem, we should mention a further response to the rule-following challenge, one that Kripke himself spends a lot of time in debunking. This response says nothing on what rules are but still insists that there is such a thing as rule-following. It identifies following a rule with displaying a disposition to go on after a certain pattern, say a pattern in applying the word 'plus' to new cases. I will not delay over this theory since, although it attracts a variety of criticisms from Kripke, the basic flaw is already crippling. The theory does nothing to explain how in following a rule I am directly but fallibly guided by something that determines the right response in advance. A disposition may determine what I do but it cannot provide this sort of guidance; it gives me nothing to take instructions from, nothing to be guided by. 'As a candidate for a "fact" that determines what I mean, it fails to satisfy the basic condition on such a candidate, . . . that it should *tell* me what I ought to do in each new instance. Ultimately, almost all objections to these dispositional accounts boil down to this one' (Kripke 1982, p. 24).

The Solution

Any non-sceptical response to the challenge about rules has to vindicate the idea that a rule-following subject tries to conform to entities that satisfy the objective condition: constraints that are normative over an indefinite variety of cases. Let us assume then that if a subject follows a rule, it is put in touch with an entity of this kind. We can think of it as a rule-in-extension or a rule-in-intension.

The question that arises under this assumption is how a rule-in-extension or rule-in-intension—henceforth I shall simply say, a rule—can satisfy the subjective conditions, being independently identifiable, directly readable, and fallibly readable. This is a question, at base, about how a rule can be suitably presented or represented to a subject like one of us, since there is no possibility of a rule communicating itself immediately: there is no possibility of the subject's 'mainlining' the rule. Let us concentrate then on this presentational or representational issue, looking in turn at how independent identifiability, direct readability, and fallible readability might be realised. In exploring the issue, we shall keep basic rules in mind: rules of a kind that can be identified and read without the use of other rules. If the issue can be solved for such basic rules, it can be solved also for non-basic ones. The examples I use will be property-rules of the kind that we have chosen as our focus and the rule for 'plus' on which Kripke concentrates. I shall assume for convenience that the examples used are all basic rules.

Independent identifiability

What material, material directly accessible to an intentional subject like one of us, could serve to identify a basic rule for us? The outstanding candidate is: examples of its application. If we assume that they are basic rules, the plus rule might be presented as the $(1, 1, 2) - (1, 2, 3) - (2, 2, 4)$ rule, the rule for box as the $(X)-(Y)-(Z)$ rule, where X, Y, and Z are all boxes, and so on. It appears however that this candidate has already been ruled out. Any finite set of examples instantiates an infinite number of rules, as Kripke stresses. And doesn't that mean that no set of examples can represent a determinate rule for an agent?

No, it does not necessarily mean this. Instantiation is a two-place relationship between a set of examples and a rule and it certainly has the feature of being a one-many relationship: one finite set of examples instantiates many rules. But the relationship that is of concern to us when we ask whether a finite set of examples can represent a determinate rule is not instantiation but exemplification. Exemplification is a three-place relationship, not a two-place one: it involves not just a set of examples and a rule but also a subject for which the examples are supposed to exemplify the rule. Although any finite set of examples instantiates an infinite number of rules, it may be that for a particular subject the set exemplifies just one rule. Nothing has been said to disallow this possibility.

We saw earlier how particular triangles may come to exemplify triangularity for a relatively simple creature like a chimpanzee, so that the chimp comes to be sensitive to that property. The chimp is always presented with particular triangles but the secret of its sensitivity is that the particularities do not matter. The chimp ignores everything about the particular triangle on hand other than what it shares with other instances: namely, the property of triangularity. The triangle on hand is not just an example of triangularity for the chimp, but an exemplar. The particular instance exemplifies the property, in the sense that it has no role to play in the economy of the chimp's responses that could not be equally well played by any other triangle (Goodman 1968).

Now the particular triangles with which the chimp is presented are not just instances of triangularity; they are instances of an infinite number of different properties. So why do we think that it is triangularity that is exemplified, when that is only one of an infinite variety of properties that are instantiated? The answer is, surely, that among the commonalities that might have caught the chimp's attention, and might have dictated whether some new presentation is going to be treated like the old ones, it is triangularity—give or take a certain indeterminacy—that engages with the creature. The chimp forms a disposition in response to the actual examples presented, a disposition to treat certain further presentations in the same way, and all others differently. And the other examples that get favoured under this disposition are just, so far as we can see, triangles.

We know that no finite set of examples can *instantiate* a determinate rule. But it does not follow, as we can now see, that no finite set of examples can *exemplify* a determinate rule for an agent. Suppose, as is generally allowed, that the agent forms an disposition or inclination to extrapolate in one direction rather than another from any set of putative exemplars of a rule. Given the inclination supposed, the examples presented will exemplify the corresponding way of going on, the corresponding rule of response, for the agent. They will make that rule salient. They will serve to direct the agent to that rule in the way that the particular triangles direct the chimp to the property of triangularity. Looking at particular triangles, all that the chimp registers, as it were, is the triangularity. It overlooks the specific details of the examples and it ignores any other commonalities between them. By analogy, the rule-following agent may look at examples like (1, 1, 2), (1, 2, 3), (2, 2, 4), and register only the addition way of going on; or look at examples of triangles or boxes or games and register only the ways of going on associated with those properties.

If this sort of story goes through, then there is no problem about how a rule-following agent can use examples to present a certain rule, a certain way of going on, to itself. Looking at the instances, it will be able to have the rule as a salient object of belief; it will be able to target or address the rule. The subject will think of it ostensively as *that* rule, the one exemplified by those instances. The addition rule will be presented as that rule, the one that goes (1, 1, 2),

(1, 2, 3), (2, 2, 4); the box rule as that rule, the one that goes X, Y, Z; and so on. Or at least the rules will be presented in this ostensive fashion if they are truly basic; otherwise they may be introduced by definition in terms of ostensively presented rules.

These reflections give us the naturalistic makings of a story about how a rule can be independently identifiable by an agent. The rule can be identified through being exemplified by a certain set of instances. And it can be independently identified, identified independently of any instance where it applies, to the extent that no particular instance figures essentially among the exemplars. The addition rule can be identified independently of any particular example, since no particular example is necessary for making the rule a salient object for the agent; the box rule can be identified independently of any particular box, since no single box is essential as a paradigm for making the property visible; and similarly in other cases.

We have been dealing with the issue of how a rule can be presented to an agent as something for it to try to follow. And it appears that resources of exemplification such as we human beings share with other creatures may be sufficient for this task. The story given will raise many questions but I propose to postpone consideration of them until later. To give attention to the problems at this stage would disturb the development of my account of rule-following. And in any case the problems to be considered are best dealt with, when all the elements of the account are in place.

Direct readability

So much by way of explaining in naturalistic terms how a rule might be independently identifiable to an intentional system. The next matter to explain is how an independently identifiable rule, how a rule presented in certain exemplars, might be directly readable by the agent. We assume that the rule is basic rather than non-basic. What we have to explain then is how it can speak directly to the agent, in particular how it can guide the agent, without the intermediacy of any interpretation, in new sorts of cases. Happily, this is the easiest matter of all to explain, under the approach taken here.

Under our proposal, a rule is made salient to an agent just to the extent that some examples elicit an inclination to extrapolate in a certain way to new cases. But if this is so, then just to think of a new case relative to the rule that is made salient is to be disposed, after garnering suitable information about the case, to read the rule in one or another way. There is no need for any intermediary between the contemplation of the rule and the judgment on the case presented. Reading the rule will be a matter of letting it lead, a matter of authorising the inclination in which it reveals what is required in the new case.

Consider an example. I have been introduced to the rule for forming box-beliefs: to the property in virtue of which something is rightly believed to be a box. The introduction to the rule will dispose me without any hesitation or reflection, without any need to go through maneouvres of interpretation, to rank the container in which my new shoes arrive as a box. Or if there is any hesitation or reflection, that will not bear on what the rule requires in the case described. Rather it will be needed for gathering suitable information about the case: it will involve feeling the box to see that it is a solid, turning the box over to see that it has a bottom, and the like.

Most stories about rule-following invoke an extrapolative disposition to explain how, given exposure to some examples, I can be prompted to go on in a certain way in sorting out other cases. But, for all this says, the inclination might present itself to me in the manner of a hypnotically induced compulsion: a brute and blind tendency to divide up the things to be sorted in a particular fashion. And if the inclination presents itself in this way then there is no hope of its identifying anything that meets what Kripke calls the basic condition on any candidate for a rule: 'that it should *tell* me what I ought to do in each new instance' (Kripke 1982, p. 24).

The story that I offer does point us towards something that can meet that condition. In virtue of my inclination, I can use certain instances to identify a rule for myself: a rule that I can directly ostend as *that* rule, the one featured saliently in such and such examples. But if I now look to new cases, and ask whether they fit the rule or not, the positive or negative inclination that they induce in me will present itself as a disposition to judge that yes, they do, or no, they don't. As I hold up the new cases to see how they fit with the rule, I have to wait and see whether they present themselves in a positive or negative light: whether they present themselves as of a kind with the original examples, fitting the rule exemplified in those instances. How the new cases present themselves will be fixed by the way my inclination goes. And so the firming up of the inclination, in a positive or negative direction, will amount to nothing more or less than a revelation as to whether or not the new cases fit the rule.

To sum up, then, my story agrees with most other accounts in positing an extrapolative disposition but it gives the disposition a distinctive role. First, it invokes the disposition to explain how some instances can present a rule to an agent: to explain how the instances can exemplify the rule. Second, it invokes the disposition to explain how the agent can be directly prompted in new cases: to explain how the rule is read in such cases. Third, it connects the presentational and the prompting aspects of the disposition in order to make sense of what it is for a rule to speak directly to a subject, telling the subject how to go on. Because the inclination serves the presentational role, its significance in the role of prompting the agent's responses in new cases is transformed from what it might otherwise be. The prompting does not involve an unintelligible leaning: an exogenously inspired urge to go this way rather than that. It appears as a leaning that the rule itself elicits in the agent, a response that it forces on any subject that understands it.

There is one further point to emphasise, before we leave the issue of how rules are directly read. This is that the proposal I am making does not require a rule-follower to have any awareness of the inclination generated by the examples that exemplify a rule, let alone to attend to that inclination in itself. I speak of the inclination making salient one of the rules instantiated by the examples: of the subject identifying the rule—via the examples—in virtue of the inclination. But none of this is meant to suggest that the rule-follower focusses on the inclination. The rule-follower will focus simply on the examples and—in them, as it were—on the rule they manifest. The inclination explains how the examples exemplify or manifest a particular rule but it does this without having to feature in consciousness.

Consider an analogy. When I look at a physical object, all that is directly presented to me is a sequence of profiles: now this profile, now that, as I move around the object. Yet in experiencing those profiles I see the object itself, in the perfectly ordinary sense of that verb. I see it, as we might say, *in* the profiles; indeed I scarcely notice the profiles, focussing as I do on the object they manifest to me. What explains how the profiles manifest *this* sort of object, conforming to the ordinary image of the middle-sized spatio-temporal continuant: *this* object, rather than any of the many ontological fancies—say, a sequence of unconnected scenes—that are strictly consistent with the set of profiles? Presumably something about my psychology. Presumably, in fact, a disposition that I share with others of my species, to neglect the detail of the profiles in favour of something that is taken to be constantly present in all of them.

The relevance of the analogy should be clear. As I see a particular sort of object in these profiles, so I may see a particular rule manifested in such and such examples. As the profiles efface themselves in my attention, yielding centre stage to the object, so the examples may command less attention than the rule they exemplify. And, to come to the point I want to emphasise, as the disposition that explains why I see a certain sort of object is something of which I may not be aware, so the inclination that explains why I am directed to a particular rule need not figure in my consciousness either. This analogy may be the best way of grasping the sort of proposal I am trying to develop.

Fallible readability

The story told so far may serve to explain the independent identifiability and the direct readability of certain rules, in particular certain basic rules, but it does not serve to make sense of the fallible readability of the rules. Fallible readability means that for all that the rule-follower knows, it cannot rule out the possibility of its getting the rule wrong in a particular instance. The subject may identify the box-rule in certain instances and go on without difficulty to construe the rule in new cases, ranking this object among boxes, that object outside. But no matter how unproblematic the exercise, the subject cannot eliminate the possibility of going wrong in this extrapolation. It cannot eliminate the possibility of counting a non-box as a box, a box as a non-box.

No room has yet been made for fallibility of this kind. Certain instances exemplify a rule for me in virtue of the inclination they generate. And that rule is extended to new cases through the operation of the inclination. I may get a case wrong, through not informing myself adequately about it: I may treat a hologram as a solid object, for example, and then take it as a box. More on this possibility later. But I cannot get the rule itself wrong, so it would seem. I cannot take the case as given, try to read how the rule applies and then, despite my efforts, misread the rule. For to try to read the rule in the new case—we assume the rule is basic—will be to let my inclination lead. And if the rule is identified in virtue of that inclination, how can the inclination mislead? The problem is probably immediately perceptible but, in any case, it can be motivated by reflecting on the connection between inclination and rule, under our account of rule-following.

Under the story we have developed, a certain sort of biconditional holds a priori: it holds in such a way that we do not have to employ ordinary empirical checks to establish its truth.[7] Suppose that we are dealing with a subject S and we are concerned with the rule, r, that is made salient to S by certain examples. It is going to be a priori on the story so far that a new case x is an instance of the rule r if and only if S is disposed, given full information about x, to treat it as similar in relevant respects to cases that it takes to exemplify r. In other words, according to our story, there is going to be an a priori connection between the rule r and the inclination in virtue of which certain cases exemplify the rule. The rule, r, is just that rule which corresponds case by case with the responses the inclination prompts the subject to make, in extrapolating from the original examples. Thus it is a priori knowable that as the inclination goes, so goes the rule.

If the rule is a priori connected in this way with the subject's inclination, then there is no room for fallibility. The subject is in a position to know that it cannot possibly be misled, as it allows its inclination to adjudicate on whether a new case is or is not of a kind with instances that it takes to exemplify a certain rule. If we assume it is fully informed about new cases, following the rule will be an enterprise in which it cannot fail, and cannot see itself as failing. The inclination will be definitive of how the rule should go in every new case. It will not represent an attempt to divine how the rule should go; it will be itself the determinant of how the rule should go.

If a rule is to be fallibly readable—if it is to be something that a subject truly tries to follow—then it cannot be a priori connected with the subject's inclination in this manner; it must be connected only in an a posteriori way, so that it is a matter of empirical discovery that the subject gets the rule right in a given situation. We must therefore revise the story told in explaining how a rule can be independently identified and directly read. We must revise it in a way that preserves the independent identifiability and the direct readability, while making room for fallible readability as well.

If an inclination is a priori connected with a rule, then it correlates with that rule which fits it exactly: the rule correctly read in the responses it prompts. If the inclination is to be a posteriori connected, then it must connect with a rule that is related to it in some other way, a rule that it may not exactly fit. What other way is there for a rule to relate to a subject's inclination, consistently with the general sort of story told?

Only one possibility seems to be open. The inclination must continue in the presentational and prompting role it is assigned in the story so far, if we are to be able to explain the independent identifiability and the direct readability of rules. But if such an inclination is to mislead the subject on some occasions, and not on others, then there must be a distinction between favourable and unfavourable circumstances of operation. And in that case the rule with which the inclination connects must be that rule which it fits, provided that favourable circumstances obtain. Under such an account the inclination in itself will not be a priori connected with the rule; it will be a priori connected with it in favourable circumstances only. And so there will be a possibility of securing fallibility.

In order to develop this idea, we need a theory as to how certain circumstances might get established as favourable. There are a number of features the circumstances must satisfy. First of all, they must not be identified as favourable by the fact that they lead to correct readings of the rule; they are supposed to help in the identification of that rule and they could not do this if they were themselves identified by reference to the rule. Second, favourable circumstances must not be identified by a procedure—say, by an inventory of favourable circumstances—which will enable the rule-follower sometimes to know that its circumstances are favourable and its inclination reliable; otherwise fallibility would be jeopardised, since in decidably favourable conditions the rule-follower could rule out the possibility of error. Third, favourable circumstances must be identified by a procedure that does not require the rule-follower to have the concept of favourable circumstances itself; this is because thinking subjects like you and me do not always have this concept: it is an invention of philosophical theorists, not a concept that is necessarily shared by the subjects of the theory.

Can we suggest a story then as to how certain circumstances might get established as favourable for a thinking subject, a story that would give favourable circumstances these features? Favourable circumstances divide into normal and ideal conditions: roughly, circumstances in which certain perturbing influences are absent, on the one hand; and circumstances, on the other, in which certain helpful factors—say, a wealth of information—are also present. I have a story to tell in each case.

The story I have to tell about normal circumstances may be motivated by a more or less conjectural account of how human beings get the concept of a colour—say, the concept of red—going. According to the account I envisage, people have red-sensations—things look red to them—as a matter of primitive experience. Given those sensations, they find English postboxes, ripe tomatoes, and heated metals similar in a salient respect. And this enables them to use such examples to indicate a certain property, namely the common colour. What colour? All they can say is, *that* colour, pointing at relevant examples. The colour is ostensively fixed for them.

Well to a certain extent anyhow. For it turns out that sometimes a ripe tomato looks different by their lights—and, no doubt, by the lights of theorists—from how it does at other times, and indeed that it looks different as between different people. This offends against a supposition that its colour is stable. The way people make sense of the variation, given the supposition of colour stability, is to find a feature of the occasions when it looks different, or of the individuals to whom it looks different, that marks them off as not counting. Thus the colour they identify by reference to certain examples as *that* colour is not whatever colour property the objects present, but whatever property they present under circumstances that can be allowed to count: whatever property they present to normal observers on normal occasions, as theorists might say.

My preferred characterisation of normal circumstances generalises a line suggested by this genealogy of colour-concepts. I believe that there are two assumptions that we spontaneously and systematically make as participants in an area of discourse when we form and discuss our beliefs. These are assumptions, respectively, of intrapersonal and interpersonal constancy. The intrapersonal assumption is that something is amiss if I find myself reliably inclined to make different judgments at different times—in particular, judgments different by my own lights—without any justifying difference in collateral beliefs or whatever. The interpersonal assumption is that something is amiss if you and I find that we are reliably inclined to make different judgments—again, judgments different by our lights—without any such justifying difference (see Craig 1982).

These assumptions mean that in any area of discourse theorists can identify normal circumstances as those that, lacking the perturbing factors that participants might identify, are not disposed to generate interpersonal or intrapersonal discrepancies of response. Normal circumstances are identified by reference to how the participants carry on; they are the circumstances that survive intrapersonal or interpersonal negotiation about how to handle discrepant responses. This account of normal circumstances contrasts sharply with more standard approaches; these usually resort to an inventory of normal circumstances or of the factors that must be absent in normal circumstances (see Wright 1988). We might describe the account as *ethocentric*, from the Greek word 'ethos', which can mean habit or practice. The account is ethocentric, because it identifies normal circumstances by reference, first, to the habits of response among subjects—say, their disposition to have certain sensations in the presence of red objects—and, second, to their practices of negotiation about discrepancies in those responses.

The ethocentric account of normal circumstances meets the conditions already mentioned. First, normal circumstances are non-circularly identified: they are not identified as circumstances that facilitate correct readings of the rule in question, only as circumstances that engage in a certain way with the practices of negotiation among participants. Second, normal circumstances are identified by a procedure such that fallibility is preserved: the rule-follower can never know that conditions are normal in such a way as to be able to rule out the possibility of error, for later negotiation may always identify them as abnormal. Third, normal circumstances are identified by a procedure such that the rule-follower does not have to have the concept of normal circumstances; the concept can remain the property of theorists.

But not only does the story give us a satisfactory account of normal circumstances, it also suggests a similar, ethocentric story for ideal circumstances. Imagine that we rule-followers find ourselves sometimes inclined to give a response that is different by our lights to an intuitively similar situation when a further feature of the situation—further information of a certain sort—comes into view. There is a discrepancy in our responses but, unlike the sort of case considered earlier, the discrepancy is explained by a difference in the information available. Imagine now that we always favour the response that is based on such fuller information: we discount earlier or other responses, taking the information to be relevant. Sometimes the information that is designated as relevant in this way will be such that there is always more of it available about any situation; sometimes it will be more obviously bounded. But in either case the theorist can introduce the notion of circumstances that are not just normal but ideal. These will be circumstances that not only lack what we rule-followers would put down as perturbing influences but circumstances where we have all the relevant information that could ever be available.[8]

The concept of ideal circumstances introduced in this way has all the features that recommend the parallel concept of normal conditions. Ideal as distinct from just normal circumstances are not identified by reference to the rule in question. Ideal as distinct from just normal circumstances are identified by a procedure such that, at least in the case where more and more information is available, the subject cannot rule out the possibility of error on the grounds of having all relevant information. And ideal as distinct from just normal circumstances are identified by a procedure such that the rule-follower does not have to have the concept of such conditions in its own repertoire.

With this ethocentric account of normal and ideal circumstances in hand, we can introduce favourable circumstances into the story told to make sense of the independent identifiability and the direct readability of rules. We know that if the story is to leave room for the fallible readability of rules, then it must represent the connection between the rule-follower's extrapolative inclination and the rule followed as a posteriori. We know, again, that the story can do this if it makes it a priori, not that the inclination gets the rule right in all circumstances, but that it gets it right under circumstances that are favourable. So let us now revise the story so that it has this effect.

It should be clear how the revised version should go. The subject takes certain examples to exemplify a particular rule, given the inclination that they generate. The subject reads off that rule in new cases—as always, we assume the rule is basic—letting the inclination speak for how it should go. But the subject does this without any basis for ruling out the possibility of error, for it allows that negotiation with itself across time, or negotiation with other subjects, may establish that the circumstances that prevail in any reading are not favourable and that its response under those circumstances is mistaken. However direct, therefore, the subject's reading of the rule is nonetheless fallible. The inclination will ensure that the examples exemplify the rule. But the rule exemplified will not be whatever rule corresponds in further cases to the actual working of the inclination. It will rather be the rule that corresponds to the working of the inclination under favourable conditions. Thus the rule exemplified will be one that the rule-follower may try to follow, without having any guarantee against going astray.

This presentation of the story is elliptical in one respect. The subject, we say, allows that negotiation may establish that certain circumstances are not favourable. But of course it is very important that the subject may do this, without itself having any concept of favourable circumstances. For that concept is one that may only belong to theorists, not to participants. The important point is that the subject gives a certain role to negotiation about discrepancies across time and persons. It must make appropriate assumptions about intertemporal and interpersonal constancy. It must be disturbed by the appearance of discrepancy across times or persons. And it must go in search of factors that enable it to discount one or other of a pair of discrepant responses in order to resolve the dissonance felt. If the subject does behave in this way, then the theorist will recognise that the rule exemplified for the subject by certain examples, the rule it tries to follow, is not whatever rule corresponds to the inclination engendered by those examples; rather it is the rule that corresponds to the inclination under circumstances that survive negotiation: under circumstances that are, in the ethocentric sense, normal or ideal.

The ethocentric account of normal and ideal circumstances enables us to revise our original story, then, so as to make sense of the fallible readability of rules. And it does this while allowing the story to continue to explain the independent identifiability and the direct readability of rules. But the ethocentric addition to the story also has other benefits. It enables us to make sense of two possibilities that we have hitherto ignored or passed over; these, like fallible readability, exemplify plausible conditions of human fallibility.

The first is the fallible identifiability, as distinct from readability, of a rule: the possibility that the examples designed to exemplify a rule for a subject may actually fail to do so, cueing the subject into the wrong rule. It should be clear, with the ethocentric twist in place, that we can explain this possibility by the fact that the examples are not presented under favourable conditions. As perturbing influences can lead me to misread a rule so, equally, they may lead me to misidentify a rule in the first place. Thus it may be that I am correct in the later reading of a rule, as it were, though the rule I follow is not that which I originally took relevant cases to exemplify.

The second possibility illuminated by the ethocentric twist has already been mentioned in passing: it is the possibility that a subject may not be fully informed about a given case so that, reading a rule correctly in the case apparently confronted, it fails to read it correctly in the case actually on hand. What is it for a subject to be to be fully informed about a given case? The answer may be formulated, once again, with the help of our notion of favourable conditions. Under the ethocentric story, subjects will assume that there can be no discrepancy across time or persons in how a given case is rightly taken. Suppose that if a discrepancy is traced to one party's having more information of one or another kind, then the response based on the extra information is the one that is allowed to count. We know then that to be fully informed about any case among subjects of this ilk is to have access to all information of those kinds. The notion of full information becomes unmysterious.

It is time to summarise. If a rule is to be fallibly readable, then it cannot be a priori connected with the inclination in virtue of which it is exemplified and under the influence of which it is read. It must be connected to that inclination in an a posteriori fashion, so that it is a matter for empirical checking that the inclination leads a subject correctly or incorrectly. What can the connection be? In particular, what can the connection be, given that we have to keep enough of our earlier story in place to explain the independent identifiability and the direct readability of rules? The answer to which we have been driven, and there is no obvious alternative, is that the rule is a priori connected to the inclination under circumstances that count, by the ethocentric criterion, as normal or ideal.

With this question answered, we have all the elements of our solution to the rule-following problem in place. The solution in general, and not just the account of favourable conditions it presupposes, may be described as ethocentric. The essential claim is that a proper appreciation of the role of habit and practice—for short, a proper appreciation of the role of ethos—makes sense of how intentional subjects can follow rules, in particular rules of thought. The habit of response—the disposition or inclination set up by certain examples—explains how a rule can be independently identified for a subject, and how it can be read directly by the subject. The practice of negotiation about discrepant responses, intertemporal or interpersonal, explains how the rule can be fallibly read, as well as explaining some other associated matters: the two possibilities discussed.

Given the solution to the rule-following problem, we can see how an intentional subject can get to be a thinking one. In order to think, the subject must be equipped to try and make judgments that are appropriate—true or at least rational—in the light of their contents: judgments that are faithful to the propositions or scenarios that constitute their contents. That means, as we analysed things, that in order to think the subject must satisfy two requirements: it must be able to make judgments about propositions or scenarios; and it must be able to treat those propositions or scenarios as rules. We saw in the last section how it can get to make judgments about propositions, finding proxies or signs to make propositions present. And now in this section we have seen how it can identify suitably determinate propositions—suitably determinate contents—and intentionally try to conform to them in the judgments it forms. Specifically, we have seen how it can identify rules through the following of which it can intentionally conform to the requirements of propositions: rules like that associated with the property of being a box or triangle or game, rules like the rule that governs addition. We assume that there are no problems in extending the solution to other relevant cases.

We should be getting on to discuss the plausibility of the ethocentric solution to the rule-following problem. But before we do so, there is one further point worth making. At the beginning of this discussion of rule-following I argued that while intentionality itself does not require an error-permitting determinacy of content, thinking does. I said that if we could establish non-arbitrarily what conditions were normal or ideal for a given intentional subject, then a corresponding determinacy would follow, but that there may be no non-arbitrary way of doing this with non-thinking subjects. It is worth mentioning now that the clue to the determinacy of content enjoyed by thinking subjects, under the ethocentric proposal about rule-following, is precisely that normal and ideal conditions are non-arbitrarily fixed by their own habits and practices.

The thinking subject aims to be faithful to a rule of judgment that is exemplified by certain examples and, if those examples fix a rule that is determinate enough to allow error, that is because normal and ideal conditions for reading off the rule in question are fixed by how things are with the subjects themselves, not merely by a methodological convention. They are fixed by the practices of negotiation that the subjects follow, intrapersonally and with one another. We may well think that it is indeterminate whether the cat has beliefs involving birds or birdlike things, for there may be nothing to fix the matter. But we cannot think that it is indeterminate whether rule-following human beings have one sort of belief or the other, whether they target or address one sort of object or the other. Suppose that two people conflict in their credal responses to a birdlike doll. We cannot think that they may both be right, with different circumstances being favourable in each case and with each party targetting a different content. For the parties will reject that possibility themselves and will negotiate to the point of identifying the conditions of one or other as comparatively abnormal or non-ideal. In pointing us towards certain conditions as the objectively normal and ideal set, the practices of negotiation will remove the possibility of an error-eliminating indeterminacy of content.

A Plausible Solution?

It is one thing to say, as indeed I do say, that the ethocentric solution looks to be the only available account of how rule-following takes place. It is another to argue that the account is plausible. The fact that the solution is the only way to explain the possibility of rule-following does constitute an argument in its favour but such an argument might be outweighed by other considerations. It might make assumptions, for example, that prove too implausible to be accepted. So what are we to say about our solution? Does it represent a plausible line?

I shall break the question down into three separate issues. First, does the solution invoke materials—say, human capacities—that are not realistic posits? Second, does it claim that more can be done with those materials than seems to be really possible? Third, does the solution have any consequences that make it inherently hard to accept?

First question

Does the ethocentric solution invoke materials that are unlikely to be available? As already stressed, it invokes two distinct posits, each of which goes with the word 'ethos': on the one hand, the habit of response that a set of examples is supposed to engender in an intentional system like one of us; and on the other, the practice of negotiation about discrepancies of response across subjects or times. I see no difficulty with either posit.

Take the habit of response first. Any intentional subject must be disposed, at least under certain conditions, to form suitable beliefs on being presented with certain situations: to form beliefs that are made true by those situations. But if on being presented with suitable situations, a subject is disposed to form corresponding beliefs involving addition, for example, or corresponding beliefs involving the property of being a box, a triangle, or a game, then that ensures that it has developed an inclination to go on from certain examples of addition or examples of the property in question to other examples. Thus any thinking subject will be equipped with the sort of extrapolative inclination, the sort of responsive habit, which our ethocentric story requires us to postulate.

What of the practice of negotiation? There is no way of arguing, as in the case of the responsive habit, that this is inevitably present among any intentional subjects. But there is evidence that the practice is actually present in subjects like you and me and that is sufficient to still any disquiet about postulating it. One of the most striking facts about the way we carry on is that we do not take a *laisser-penser* attitude in our dealings with one another or in our dealings with ourselves over time. Whenever we form discrepant beliefs in response to apparently the same situation, whenever we appear to disagree with ourselves or with one another, then we tend to look for a resolution of the divergence. We may countenance no-fault disagreements in some marginal cases (see Price 1983, p. 403; 1988, pp. 159–62); for example, we may think that there is no fault in your finding something amusing, and in my not doing so. But in general we think that disagreements call for explanation and, if possible, reconciliation.

Second question

The second question to be raised about the ethocentric solution is whether it claims to do more than is realistically feasible with the materials that it employs. Here the main challenge will be to explain how we can expect a finitely based disposition or inclination to play the role of identifying—or, strictly, of enabling certain instances to identify—a rule that applies in an indefinite variety of situations. We know that the disposition is bound to be vulnerable to all sorts of contingencies that would arise as the agent tests it across some relevant, if unusual, situations; we know that the disposition is bound to burn out or misfire under some relevant inputs, as the computer will break down under overload. So how can we think that such a shaky, vulnerable thing might help to identify a rule that is relevant in an indefinite variety of cases? How can we assume that a finite set of examples can direct subjects to something relevant across an indefinite variety of situations and, consequently, that the extrapolative disposition generated by those examples can hold out across all the relevant cases?

We do not have to assume that the actual disposition at work in making a rule salient to a subject could go on functioning properly across an infinite range of cases. We need only assume that there is some appropriate idealisation of the actual disposition such that under that idealisation it would function in the required way. Otherwise put, we need only assume that the disposition would function in the required way, *ceteris paribus*; it would enable instances to identify the rule for every case where other things are equal. But even this assumption may be challenged. How are we reliably to tell what would happen if other things were equal? How are we to tell what would happen, for example, if a subject were able to extend the disposition we associate with adding, to infinitely long numbers? The challenge is backed by Kripke (1982, p. 27): 'Surely such speculation should be left to science fiction writers and futurologists. We have no idea what the results of such experiments might be.'

This is not a telling assault. For it appears that the attack, if successful, would tell against every sort of theorising. We continually idealise in the fashion illustrated, we continually commit ourselves to other-things-being-equal projections, as we theorise about the world around us (Blackburn 1984b, pp. 289–91; Fodor 1990, pp. 94–95; Pettit and Price 1989). Consider the sort of commitment I make when I say that any factor is reliably connected with some other: for example, that the volume of a gas varies inversely with the pressure of the gas. I say that if we increase the pressure of a gas in the actual world then the volume decreases, other things being equal. And I say that if we were to increase the pressure of any gas, by any amount, then again the volume would decrease, other things being equal. With some magnitudes of pressure or volumes of gas, the requirement that other things are equal might mean, for all that I know, that the universe would come to an end (Fodor 1990, p. 94). But this does not matter. The ignorance as to what would happen in such cases does not invalidate my commitment to the relevant gas law. If this is standard scientific practice, then Kripke can hardly criticise the commitment ascribed to rule-following subjects in my story. They take certain examples to present a rule that is relevant across an indefinite range of cases and so, by my account, it must be the case that the disposition that makes the rule salient would hold up, other things being equal, across that range. But that assumption is, by relevant standards, a common-or-garden commitment. It is not something for which we can particularly fault them.

Still, this response may not be sufficient to quiet the concerns raised by Kripke about idealisation. Perhaps his suggestion is, not that it is generally inappropriate to make counterfactual commitments under idealisation—under the assumption that other things are equal—but only that it is inappropriate to do so when the sort of idealisation required is not appropriately fixed: that is, when there is nothing suitable to fix what sorts of things are required to be equal (Kripke 1982, p. 28). When I commit myself about what would happen to a gas, other things being equal, I may not be able to spell out what has to be equal but I have a general sense of what is required. In particular, I have a sufficiently good sense of what is required not to have to admit, circularly, that other things will be equal only if they ensure that the gas law holds. The suggestion may be that when we invoke how the disposition of the rule-followers would go, other things being equal, the only way we can spell out what is required for other things to be equal is to say that they are such as to allow the subjects to get the relevant rule right.

But this line of criticism does not tell against our ethocentric story either. As the story was developed to make sense of the fallible readability of rules, it directs us to the mode of idealisation appropriate to the claim that the disposition of participants can hold up, other things being equal. Other things will be equal for the operation of the disposition, we may say, just in case they allow a negotiated convergence of response, across times and persons, in the way in which things currently allow it. Things must not change so that convergence is achieved collusively or coincidentally. And things must not change so that the possibility of convergence fades altogether out of view. We cannot say exactly what may be involved in things' remaining the same in those regards. But in that respect we are no worse off than with the gas law. The important point is that we have a general sense of what is required that does not reduce to the assumption that other things will be equal only so long as they enable the disposition to pick out the right rule.

The objections we have been considering suggest that there is no possibility of rule-following subjects identifying an indefinitely relevant rule in virtue of the presence within them of a common-or-garden extrapolative inclination; the idea is that no finite, vulnerable inclination could serve this sort of role. As against those objections, I have urged that there is such a possibility: that a suitably idealised inclination, an inclination as it functions under favourable circumstances, may well pick out a rule of indefinite range. But now another, related objection may strike someone.

My argument may establish that there is indeed a possibility that an extrapolative inclination can play the role required—there is nothing incoherent in the idea—but it does not establish that every extrapolative inclination actually plays that role. It leaves it open that as two subjects rely on their inclinations over certain cases to identify a common rule, as they assume that their inclinations will converge in circumstances that survive negotiation, they are actually misled. The argument leaves open the possibility that their assumption is mistaken and that in certain cases they have not yet confronted, their inclinations will diverge wildly and non-negotiably. In short, it leaves open the possibility that the idealised inclination on which they rely will indeed break down. The account is designed, as we saw, to rebut the sceptical claim that necessarily rule-following fails: that there is nothing for rule-following possibly to consist in (Kripke 1982, p. 38). But it leaves in place the weaker, sceptical observation that it is not necessary that rule-following succeeds: would-be rule-followers may be deluded in thinking that there is a rule they are targetting, for they may be deluded in thinking that there is anything they will prove capable of converging on.

But now it may be objected that if the account leaves that possibility in place, then would-be rule-followers cannot know which rule they are trying to follow, since there may be no rule there to answer to their inclinations. Trying to follow a rule seems to require knowing which rule is to be followed, as trying to hit a target seems to require a knowledge of which target is to be hit. And so the possibility that the account leaves open is sufficient on its own to mean that there is no rule-following. Or so at least it may be said.

The answer to this objection becomes obvious once we recognise that it is possible to have fallible knowledge. My knowing that p may be inconsistent with its not being the case that p. But my knowing that p is consistent with its being logically possible, for all the evidence available to me, that not p. I may know things that are not strictly entailed by the evidence available to me; I may know things on the basis of less than logically fool-proof evidence. The possibility of such fallible knowledge explains how a subject may know which rule it is following, even though the evidence does not logically rule out the possibility that there is no rule there—that there is no rule adumbrated in relevant extrapolative inclinations—for the subject to follow. I may know which of my children is in the garden even though it is logically possible, for all that I can see through the slats of the blind, that there is no child in the garden and that I am fooled by a play of light in the trees. And equally, by more or less exact parallel, an intentional subject may know which rule it is following, even though it is logically possible, for all the evidence available, that there is no rule in its sights: that no rule corresponds even to the idealised version of its extrapolative inclination.

We have been dealing with the general question as to whether the ethocentric story is unrealistic, not in the posits it introduces—the habits and practices it invokes—but in the things that it claims can be done with such materials. I hope that enough has been said to show that the story is not unrealistic in this way. If it remains surprising that the habits and practices invoked in the ethocentric approach are sufficient to put a subject in touch with a rule of indefinite range, then I can think of only one further consideration to put in place. This, to recall a point made earlier, is that the rules which a subject targets need not be defined over every instance; they may be vague at certain margins. We have abstracted from vagueness in most of our discussion, assuming that the intentional subject aims at a precise rule-in-extension. But this abstraction is motivated by considerations of convenience only. It may be salutary to remember that what a subject aims at in many cases may be, not this exact rule-in-extension, and not that or the other, but rather something that is vague relative to the differences between them.

Suppose that a subject has been trained, as we have been trained, to think in terms of the property of redness: to target that rule and to try and form only true beliefs in regard to it. Now consider something that is as red things are, up to the point where it is looked at—up to the point where even a photon of light falls upon it—and which then comes to be as green things are?[9] There are three rules-in-extension that differ only in what they require in this sort of case: one fixes the object as red, a second as green, and a third as neither. So which rule does our subject target? The answer may be that the subject does not care about this sort of case and that the rule it targets is vague relative to the three possibilities distinguished. Having come across the case, the agent may decide, more or less arbitrarily, to track just one of the precise targets. But up to that time its target may be the vague rule, not one of the more specific ones. We should remember that something less than full determinacy is all that need figure in rule-following; we should recall the points made about determinacy earlier in this chapter. The phenomenon of rule-following is already problematic enough; we should avoid making it seem unnecessarily mysterious.

Third question

We have dealt now with two questions that bear on the plausibility of our ethocentric account of rule-following: the questions as to whether the account is realistic in the materials it invokes—the habits and practices it posits—and in the things that it claims can be done with those materials. We come, finally, to a third sort of question that can be raised about the account. The question is whether the account has any consequences that make it inherently hard to accept. There are a number of counterintuitive consequences that may seem to flow from the account and I shall look at three. Notice that these consequences are alleged to be genuinely counterintuitive, not just to be inconsistent with prevailing philosophical outlooks. We shall be looking later, in chapter 4, at how far the ethocentric story is philosophically uncongenial—in particular, uncongenial to realist tastes—as distinct from being downright counterintuitive.

It is often assumed that with some properties that are involved in our beliefs, we think of those properties in a way that refers us to favourable or standard conditions. Take the property of a remark that it is amusing, of a person that he or she is irritating, or of a chair that it is comfortable. To be amusing a remark must amuse or at least, as we will hasten to say, amuse appropriate people in appropriate situations. To be irritating a person must irritate but it will not do just to irritate the irritable; the person must irritate even normal people, as we will tend to say, in normal situations. To be comfortable a chair must offer comfort to someone who sits in it but, again, the sort of person it is required to seat comfortably is only the normal person, not someone who suffers from chronic back-pain, for example. A first counterintuitive consequence that may be associated with the ethocentric account of thought is that it represents us as thinking about all basic properties, and indeed all basic rules of thought, in the subject-relative fashion in which it is alleged that we think of the properties illustrated.

This consequence would be counterintuitive, for it should be clear that we do not relativise to particular subjects in the way in which we think of the property of being a box, or a triangle, or a game, let alone in the way in which we think of an operation like addition. So does the ethocentric account entail that we must always think in such terms of anything that serves as a basic rule of thought?

No, it does not. In developing the ethocentric account I stressed that rule-followers themselves may not have the concept of normal or ideal conditions, let alone think of certain rules of thought in terms of such conditions. Presented with suitable examples, they will think of a rule they intend to follow in an ostensive way, as *that* rule, the one exemplified in those examples. In reading the rule elsewhere—and perhaps in checking the rule exemplified—they will be disturbed by any intertemporal or interpersonal discrepancy of response, and will be disposed to negotiate about its sources. That means that theorists may introduce the notion of normal or ideal conditions, and may identify the rule targeted as the rule that the inclination associated with the original examples inclines the subjects to follow under such conditions. But nowhere in this story is there any suggestion that the subjects must themselves think of that rule in a subject-relative fashion.

In introducing a reference to normal and ideal conditions, the story does not tell us how the subjects think of the rule, how they conceptualise it. Rather it gives us an account of the conditions under which the rule can get to control their thought, being something that they try to follow. It is a story about the conditions for possessing intentional states that target that rule, not a story about the way in which the rule is represented within those states. The story is that subjects like you and me can identify a basic rule of thought if—and apparently only if—we are equipped with suitable habits of response, and the responses to which we are disposed are subject to certain practices of negotiation. And such a story may hold, without the elements it deploys appearing within the modes of thought of the subjects themselves.[10]

Now to a second counterintuitive consequence that may be thought to follow from our account of rule-following. The account says that something is a box just in case we normal subjects are disposed to see it as saliently similar to cases that they take to exemplify that property. Suppose then that we were differently disposed from how we actually are. Doesn't that mean, under the ethocentric story, that different sorts of things would be boxes? Generalising the possibility illustrated, doesn't the story have the global effect of making certain matters of fact—the fact that this is a box, the fact that that is a triangle, the fact that 68 plus 57 equals 125—sensitive in a highly counterintuitive way to possible shifts in our dispositions? Not necessarily. It is possible to avoid these consequences by taking a particular line on a question that our account has left open.

According to the account developed, the extrapolative inclination of subjects is a priori connected with the corresponding rule under favourable circumstances. Does that connection hold necessarily? Does it hold under every possible variation, so that even if the inclination had been different, it would have picked out the right rule under favourable conditions? Otherwise put, is the rule that rule, the one that corresponds in favourable conditions to the inclination in the actual world, so that at a possible world where the inclination fires differently it gets the rule wrong? Or is the rule whatever rule corresponds under favourable conditions to the inclination, so that in the possible world where the inclination fires differently it still gets the rule right? On the first, contingent view of the connection between inclination and rule, we keep one and the same rule in mind—the one relevant in the actual world—as we ask whether we would have got it wrong if the inclination had been different; thus we say that yes, we would have got it wrong. On the second, necessary view of the connection, we focus on a different rule as we ask about the situation where the inclination is different and so we say that no, the different inclination would not have got the rule wrong (see Davies and Humberstone 1981).

There is no conflict between either reading of the inclination-rule relationship and the requirement that rules be independently identifiable, directly readable, and fallibly readable. That is why I have left the question open about whether to take the connection as contingent or necessary. But the issue now raised about the ethocentric account makes it clear that we should prefer the contingent reading, at least in most cases. That reading blocks any counterintuitive sensitivity of matters of fact to possible shifts in our extrapolative dispositions. As the connection between inclination and rule is a posteriori under our account, so it is contingent as well.

Consider that possible world where our counterparts are led by a counterpart inclination to claim that 68 plus 57 equals 5; their extrapolative disposition from exemplars of adding takes them in a different direction on this sum. We certainly do not want to say that had we been in that world, had our extrapolative inclination been like those of our counterparts there, then the sum of those numbers would have been 5. The contingent reading of the inclination-rule connection enables us to avoid alleging any such sensitivity. Under the contingent reading, we would not have been faithful to the rule with which our inclination corresponds in the actual world—under favourable conditions—had we been disposed to extrapolate in the manner of our counterparts. And so we may say that had the inclination been different, the arithmetical matter of fact would have stayed the same and we would simply have got it wrong.

There is one last association with a counterintuitive consequence that I would like to consider in defending the ethocentric account of rule-following. The account does not mean that we have to say in the actual world that had our inclinations been different, then certain matters of fact would have been different too. This we have just argued. But doesn't it mean that in the actual world, for all we know at any time, our extrapolative inclinations may generate intertemporal or interpersonal discrepancies that prove incapable of negotiation? And doesn't it imply that if our inclinations suffer such misadventures, then there will be highly counterintuitive consequences?

Our ethocentric account certainly means that for all that theorists or participants know at any time, the inclinations in virtue of which people aspire to target a certain rule may be subject to later breakdowns, as would-be rule-followers find that their responses to apparently similar cases diverge wildly between different times or across different persons. We commented on this possibility in discussing the second question raised about the ethocentric approach. So what would happen, according to our account, if rule-followers were plagued by non-negotiable discrepancies? And would the consequences really be counterintuitive, as alleged in the objection we are considering?

If there were interpersonal divergence, then individuals would have to give up on the assumption of interpersonal constancy, at least in the relevant cases. But they could each still think that they were targetting an elusive rule, a rule they could read only in a fallible manner. What then of the more extreme possibility? What would happen if there were intertemporal divergence, so that individuals found, severally or collectively, that they were disposed without apparent reason to treat similar cases differently at different times? There are two cases to distinguish. One is where the population is struck down at a particular time with a once-for-all change in disposition or dispositions. The other is where the population is struck down with randomness, so that there is a continuing shifting of dispositions across time. In the first case, our account would predict that the individuals would have to think of their earlier selves as targetting different rules from those they now try to follow. In the second, it would predict that individuals would have to give up on the aspiration to follow rules; they would have to despair of trying to be faithful to independent guidelines.

I do not think that any of these predictions can be used to cast doubt on the ethocentric story. The scenarios envisaged are fairly outlandish and we do not have an intuitive grip on what would happen if they came about. But the ethocentric predictions seem, in any case, to be perfectly reasonable. We can certainly envisage coming to the conclusion that we are following different rules of thought from others or that we are now following different rules from those we followed previously, though it may not be easy to work out the view that we would then take of our compatriots or of our earlier selves. It may be less easy still to envisage coming to the conclusion that there is nothing objective to guide us in thought, since coming to that conclusion itself would seem to require that there is at least one matter of objective fact. But I do not think that we can lay the problems of imagination that we encounter here at the door of the ethocentric story. The problems are due to the extreme scenarios envisaged, not to the theory that generates predictions as to what would happen in those scenarios.

This completes my discussion of questions that may be raised to cast doubt on the ethocentric solution of the rule-following problem. The questions considered do not exhaust the issues that might be raised. But I hope that our treatment of them is sufficient to indicate the resources of the ethocentric story and to establish it as a plausible solution to the rule-following problem.

As the first chapter developed and vindicated the notion of the intentional subject, so in this chapter we have developed and vindicated the notion of the thinking subject. The first chapter is not uncontroversial, as it follows a particular line through familiar territory in recent philosophical and psychological discussion. But this chapter is controversial in a very different way. It breaks new ground in marking the distinction between non-thinking and thinking subjects, although that distinction has been implicit in some recent literature (Cummins 1983, chap. 3). And it breaks new ground in offering an account of what is required for thinking, and of how those requirements can be fulfilled by naturalistic subjects. It was necessary for us to break this ground, as human beings are clearly thinking systems, not just intentional ones. We would be ill-equipped to tackle the questions raised in the remainder of this book, if we approached them with anything less than a full appreciation of the sort of thinking thing—the sort of *res cogitans*—that we are.

There is much more to say in elaboration of how thinking subjects operate. We have yet to see in any detail how a subject can get to address propositions as objects and to submit itself to the normative discipline of trying to respect the requirements of the propositions in the beliefs it forms. But that task must be postponed for now. It involves discussing the social performance of intentional subjects and we turn to that discussion in Part II—specifically, in

Chapter 4—where we consider mind and society. The task of detailing the workings of thought also involves discussing the ways in which intentional subjects get to understand and interpret one another and we turn to that task in Part III—specifically, in Chapter 5—where we consider mind, society, and theory. This has been an austere, analytical discussion of the possibility of thought. The later elaborations of the basic story should add flesh and muscle to the bones.

The most striking theme that emerges from our consideration of the thinking subject is that the capacity for thought requires the enjoyment of a certain sort of interaction. The thinking subject must interact with other subjects or, at the least, it must interact with its past selves. Only by investing such other subjects or selves with a certain authority on the reading of the rules it addresses, does it manage to target rules about which it may go wrong: does it manage to target rules that represent an external constraint on the success of its enterprise. Thus the purely solipsistic subject, the subject isolated from the society of past and present, would be incapable of thought. This theme will be taken up again in our later discussions and will be given a considerable sharpening. For we shall be arguing that if the thinking subject is to satisfy a certain plausible constraint of scrutability, then not only must it enjoy society with itself across time, it must also enjoy society with other subjects. The thinking and scrutable subject cannot be a purely solitary individual.

Notes

1. Stich (1990, Chapter 5) rejects this claim but under assumptions that are quite different from those endorsed here. Where I assume that beliefs come with contents given—I try to vindicate the determinacy of content later in this chapter—Stich takes beliefs as uninterpreted brain-states. Thus Stich thinks that there is an issue as to whether I should want my beliefs to be reliably true under the ordinary practice of content-ascription—true, period—or true under this or that deviant interpretation: true*, true**, or whatever. Stich argues that if I intrinsically value truth—truth, period—that can only represent a quaint preference that certain culturally neutral objects, my beliefs, should have a culturally parochial feature, truth-relative-to-the-local-interpretation. I would counter that under my way of taking beliefs it represents a preference that certain cultural objects, my beliefs-with-contents (as given under the local interpretation function), should have a culturally neutral property, truth. Again, Stich argues that if I instrumentally value truth, that disposition needs to be supported by a demonstration that truth will do better than truth* or truth** or whatever in the service of my desires: in the procurement of what I find desirable. I would respond that if my desirability-beliefs and my desires come with contents given, in particular with contents given under the local interpretation function, then it would be a major coincidence if they were better served by true* beliefs rather than true ones (see Harman 1991, p. 198).

2. Notice that I avoid the problem of giving sufficient as well as necessary conditions for a response to be an intentional action. The response will have to be rationalised by the agent's beliefs and desires and it will have to be caused by those beliefs and desires.

To avoid the problems of deviant causation, however, a further condition will also have to be satisfied: the response will have to be caused by those states, because it is rationalised by them, and not because of some fluke. But how to spell out that condition? For an up-to-date survey and treatment, see Bishop 1989. I believe that the program model can help us in dealing with the problem. The idea it suggests is that a non-deviant response will be counterfactually resilient under variations in how the triggering states are realised, whereas the deviant response will be forthcoming only when a certain flukey connection is present. That idea should be distinguished from the more common suggestion that the non-deviant response will be counterfactually sensitive to certain variations, whereas the deviant response will be insensitive. But I cannot explore the merits of the idea here.

3. If I am wrong about this, then what follows in this section is unnecessary but no serious harm is done to the overall position defended in the chapter. The first or second answer would also be compatible with that position.

4. The considerations mustered here give grounds for the view that the thinking subject will have to satisfy a constraint 'that no one can have a belief with a particular content . . . without grasping the constituent concepts of that content' (Davies 1990, p. 45). It is an open question, however, whether there is a sense in which this constraint has to be satisfied by the non-thinking, intentional subject.

5. There is a contrast here with the rule represented by a proposition. With a proposition there may only be one act of conformity that any subject makes, though it is an act that the subject may revise: this is the judgment in favour or against that proposition. The proposition is a normative constraint that is relevant over an indefinite variety of situations, in the sense that for an indefinite variety of possible worlds it dictates whether it is appropriate or not—whether it is correct or rational, for example—to judge in favour of the proposition there. But a thinking subject only ever confronts one possible world—the actual one—and so, it need only judge in accordance with the proposition on one occasion. This makes it awkward to raise the rule-following problem in relation to propositions and it provides another reason for focussing on the rule represented by something like a property.

6. There is no inconsistency between the assumption that a thinking subject follows a propositional rule by following the property-rules and the argument in the last section that a proposition cannot be presented as an object of thought just through presentation of its constituents: specifically, the argument that such a mode of presentation would fail to distinguish the proposition from a mere collection of constituents. That argument established that if a proposition is to figure as an object of belief for a subject, and if it is to figure therefore as a rule the subject can follow, then it must be presented other than via its constituents; it must be presented in something like the proxy manner described. The fact that this is so leaves it possible, as I claim now, that nevertheless the subject follows the rule represented by the proposition through following rules associated with its constituents. Notice that if we could explain propositional rule-following, and thought, without having recourse to the story given about intentional ascent—if we could explain it with just the constituents account of intentional ascent—this would not disturb the broad flow of my argument; it would only mean that that story was not an essential part of my narrative.

7. Here and henceforth the notion of the a priori is introduced without a commitment to any particular theory. As I see things, the notion may even be understood in a Quinean spirit. Quine (1974) admits a distinction, after all, between truths that are admitted, or that are derivable from truths that are admitted, by anyone who learns a language, and truths of which that is not so.

8. I do not mean to suggest that the only concept of ideal conditions involves an ideal of information. My intention is mainly to illustrate how we might get such a concept going.

9. The example is Mark Johnston's.

10. In a congenial discussion of concepts, Christopher Peacocke (1989, 1990, 1992) introduces the idea of a possession condition for a concept, in particular the idea of a condition that requires two things: that the agent find a certain response primitively compelling and that it find the response compelling because of certain features of the situation in question. Our account of rule-following identifies a possession condition of this kind. An agent will have the concept of a box—or whatever—only if it finds it primitively compelling to group other boxes with certain paradigms and only if it finds this compelling because of features of those paradigms: the compulsion is not driven from within. More on these matters in Chapter 4.

II
MIND AND SOCIETY

Preview

The upshot of the first part of the book is a distinctive picture of the mind, in particular the human mind. With this picture in hand, I move on now to consider the main issues raised by social ontology, the main issues concerning the impact of society on human psychology. As the first part of the book is marked by the more or less novel way of distinguishing between thinking and non-thinking subjects, so the second is marked by what I think is an original distinction between two issues in social ontology. I believe that the issues have been systematically confused over a long period and that no progress can be made in this area until they are carefully kept apart.

The first issue divides what I choose to describe as individualism and collectivism. This is the question as to whether society involves the presence of any regularities or forces that compromise the picture of human beings as intentional agents: the picture endorsed in our image of ourselves, and charted in the first part of the book. Many social scientists, and many philosophers too, have suggested that did we have a full understanding of the factors at work in social life, then we would realise that the common-or-garden, intentional image of human beings is radically mistaken. We would realise that we are deluded about the extent to which we are rational in the beliefs and desires we form, and rational in the actions we select in service to those beliefs and desires. We would realise that the intentional image of ourselves as more or less autonomous subjects—as autarchical agents—is a conceit that lacks foundation in reality. The thinkers who maintain this view are collectivists, in my terminology, while those who reject it, those who deny that social forces or regularities are inimical in this way to intentional autonomy or autarchy, are individualists.

This first issue in social ontology has to do with how far individuals are compromised from on high, by aggregate or structural factors. It is a vertical issue, as we may put it. The second issue, by contrast, is of a horizontal character. It is the issue between what I call atomism and holism. The question bears, not on the relation between high-level factors and individual human beings, but on the relation between the individuals themselves. It is the question as to how far people's social relationships with one another are of significance in their constitution as subjects and agents. Atomists occupy an extreme position, according to which it is possible for a human being to

111

develop all the capacities characteristic of our kind in total isolation from her fellows, if indeed she has any fellows. There is no incoherence, as it is often put, in the notion of the solitary individual. Holists deny this claim, arguing that one or another distinctive capacity—usually the capacity for thought—depends in a non-causal or constitutive way on the enjoyment of social relationships.

This part of the book, as the chapter titles indicate, offers a defence of individualism in the first debate, holism in the second. The emerging position, holistic individualism, is a novelty in the arena of theoretical debate, both among philosophers and among social thinkers more generally. It raises fascinating questions as to how we should pursue social explanation and as to how we should go about political evaluation, and these are the questions that the third and final part of the book addresses.

Collectivism is an exotic and extreme doctrine and much is said in the third chapter on its orgins—its motivating sources—and on other doctrines with which it is regularly confused. I draw distinctions between collectivism proper and a variety of such theses. I distinguish it, to begin with, from a number of entirely credible doctrines: first, as already mentioned, the holistic claim that people depend non-causally on their relations with one another for the possession of distinctive human capacities; second, the reciprocity doctrine that intentional agents are reciprocally affected by the social entities they help to constitute; and third, the revisability thesis that social science can force some revisions on our intentional psychology. I distinguish it also from the individual dispensability claim that, although people often effect big things on the social stage, things such as those accomplished by Napoleon, still the things effected would in most cases have come about by another hand, had the individuals involved not been around. And, finally, I distinguish it from the social inevitability claim that, although individuals are as we think of them in intentional psychology, their opportunities are so socially limited that what they end up doing is determined by social circumstances. Such theories may be of great interest to practising social scientists but they do not compromise our picture of intentional agency in the manner that I take to be distinctive of collectivism.

I argue that collectivism proper is motivated by the recognition, particularly in the work of social scientists like Durkheim, that there are certain social-structural regularities at work in social life, certain regularities that are discontinuous with intentional regularities. These regularities do not belong to the intentional realm, as the regularities of chemistry do not belong to the realm of physics, and, furthermore, they are not logically or superveniently fixed in place by intentional regularities: we can imagine keeping the intentional regularities and varying the structural ones. Yet they require that people display certain intentional responses. They are regularities such as that whereby an increase in unemployment is said to occasion a rise in crime, urbanisation to produce a decline in religious practice, the social benefit of social stratification its appearance or persistence, and so on. Collectivists fasten their attention on such alleged regularities, being impressed by the fact that they obtain reliably, and obtain discontinuously from the regularities that hold at the intentional level.

Their collectivism is born of the belief that in order for social-structural regularities to have this status, something like the following must hold. Either the structural regularities are in potential conflict with the intentional regularities and override them. Or things have been selected and rigged in the social arena so that the structural regularities are sustained reliably across the vagaries of our intentional attitudes; the structural regularities outflank the intentional regularities, as we may put it. Collectivists hold either by the overriding thesis or by the outflanking thesis. Under the overriding thesis, structural regularities represent more powerful forces than intentional regularities; under the outflanking thesis they represent a deeper order in the way intentional responses are organised.

My defence of individualism consists in an extended critique of each of those theses, together with a demonstration that we can countenance social-structural regularities without recourse to either extreme; we can countenance them without thinking that the intentional autarchy of human beings is compromised by the existence of overriding or outflanking effects. I argue that there is no problem for intentional psychology in recognising that a regularity such as that which relates unemployment and crime is at once reliable and discontinuous from intentional regularities.

That such a regularity is discontinuous from intentional regularities means two things. First, it belongs to a different realm from the intentional regularities: it is not a psychological regularity, for example, such as that which would relate the perception of unemployment to a rise in crime. Second, it does not obtain just on the basis of intentional regularities: we can imagine a world in which human beings have exactly the same sorts of intentional characters and dispositions as here—they obey the same psychological laws—but in which an increase in unemployment is not connected in the same way with a rise in crime. The regularity can certainly obtain without people being aware of the unemployment, or having any such psychological connections with it; that is to say, it can belong to a different realm. And equally the regularity can obtain without being fixed in place by the intentional regularities. If the regularity obtains in our world, that is partly in virtue of people's intentional make-up and partly in virtue of the fact that the unemployed are given greater motive and opportunity to commit crime. Thus we can imagine that the regularity fails in a world in which people have the same intentional make-up, because we can imagine that in that world the unemployed are not given extra motives or opportunities for crime: perhaps the climate is so benign that unemployment offers a prospect, mainly, of sunny days on the beach.

Once we see that social-structural regularities do not offer a challenge to intentional psychology, then the temptation to believe in the overriding thesis wanes. But my critique of that thesis serves to undermine it in any case. The thesis asserts either that there are no intentional regularities at all, so that structural regularities win by default, or that there are such regularities and that they are defeated in conflict with structural regularities. The first, more extreme version of the thesis is outlandish: our intentional psychology is perhaps one of the best-tested programs of understanding, being a program that we all employ in our everyday exchanges, and it is barely conceivable that any social-scientific results would wholly undermine it. The second, limited version of the thesis, on the other hand, runs into conflict with the most compelling picture available of the relation between higher levels of causality—say, the chemical, biological, psychological, and structural levels—and the level explored, at least ideally, in physics. On that picture, which I describe as causal fundamentalism, the causal regularities at higher levels are determined superveniently by the regularities and background conditions that obtain at the physical. If all higher regularities are fixed on such a common basis, then there is no room for the sort of conflict envisaged in the more limited version of the thesis.

So much for the overriding thesis. The other collectivist doctrine alleges, not that structural regularities override intentional, but that they outflank them. It holds that we would not be around if we were not the sorts of intentional creatures who are disposed to respond in such a way that the structural regularities obtain. Where the overriding thesis represents social-structural regularities as predetermining what we do, the outflanking thesis alleges that we are predestined to behave in the way required by those regularities. My critique of this thesis is that the sort of selectional process that would have to have operated for human beings to be compromised in this way is extremely unlikely to have been present in the development of our species or our societies. The doctrine is not quite as implausible as the overriding thesis but it is implausible in a comparable measure.

With the issue between atomists and holists, as with that between individualists and collectivists, there have been many statements as to what it involves. The holist maintains that individual human beings depend on the enjoyment of social relations for the appearance of certain distinctive human capacities. But depend how? It cannot be enough, though this is not always made clear, that they depend causally on the enjoyment of such relations. Whoever thought that people are causally uninfluenced by their social relations? I argue that the dependence thesis involved has to be of a constitutive rather than a causal character. It has to mean that, at least given certain contingent conditions that distinguish the human predicament, part of what is involved in the having of some characteristic human capacity is the enjoyment of social relations.

In the chapter on thought I argued that the ability to think presupposes a subject who enjoys a certain sort of interaction either with herself across time or with other subjects. I argue now that if the ability to think is required to meet a further condition, a condition that is certainly characteristic of the human situation, then it presupposes a subject who enjoys a certain sort of interaction with other people. In other words, it requires community, as the holist maintains. There is no prospect of the solitary thinker, no prospect of the sort of possibility that the atomist has to countenance.

The extra condition that enables me to run this argument is a requirement of publicity. It is the requirement that, whatever a subject thinks, whatever the content of her beliefs and desires, it is possible for another subject to know what that content is. It is possible for another knowledgeably to identify which rules the subject is following in her thought, and to identify them as rules that she can follow herself. The rules, and therefore the content, are commonable: they can be claimed as a common possession. I argue that this commonability condition can be realised only if people defer to one another, and not just to themselves at later times, in identifying the contents or propositions to which they try to be faithful. Their thinking must be the sort that involves interpersonal interaction.

The key idea in the argument is this. If the property I target in box-identifying beliefs—if the extension of my concept of box—is fixed just by my extrapolative inclination over time, then it is always possible that you are wrong about the extension in question; you may use your own extra-polative inclination as a prosthetic device for identifying the extension but you cannot know that it is appropriate to do so. On the other hand, if the extension is fixed by an extrapolative inclination that we are assumed to share—subject to non-perturbing conditions—then one of two things must hold. Either there is no such inclination shared between us—at some point we just go different ways—and I have nothing in mind with the concept of box; my thought fails of an object. Or there is such an inclination on offer—the proof is in the successful practice of convergence—and you can tell what I have in mind; you, after all, share in the determination of what that extension is. Thus, if I have anything in mind with the concept of box, then it is something to which you, in principle, have equal access.

It transpires that if we human beings are thinkers, in particular if we are thinkers who are able in principle to know which contents we are each targetting, then our thinking involves social interaction; it involves us in a project of mutual orientation and engagement. Having argued for this holistic doctrine, I face the charge that such a position goes naturally with relativism: with the claim that as communities differ in the practices they foster, so will they diverge in the concepts with which they furnish individuals and, it is even said, in the worlds which they put at the disposal of individuals. I face the charge that my holism commits me to a view that different societies construct different realities, so that the community of knowledge divides into groups that are hermetically sealed off from one another.

In the last section of Chapter 4 I confront this accusation and try to iden-tify the relevant consequences of the sort of holism espoused here. The holism espoused connects with a relativistic sort of stance to the extent that it makes it a priori that if the people in a community take something to fall under a basic concept, C, in conditions that come out as normal and ideal—in conditions identified after the pattern described in Chapter 2— then it is C. The holism involves a doctrine under which the responses of people are authorised—they become authoritative—on certain matters. For example, people's sensations of redness may become authoritative, under normal conditions, for whether something really is red or not.

I argue that the response-authorisation involved in our holism does not commit us to thinking, in a reductivist way, that people really address their own responses in their thought and talk: that colour-discourse is really about colour-sensations, for example. I also argue that it does not commit us to any form of intersubjective idealism, an idealism under which shared responses like colour-sensations would actually make things coloured. But I concede that the response-authorisation does involve a certain anthropocentrism, making human beings the measure of truth and falsity in certain matters. It may mean, for example, that normal observers in normal circumstances cannot be in ignorance or error about the colour of something that is presented to them.

But I argue that we can happily embrace this kind of anthropocentrism. In particular, we can embrace it consistently with holding by a realist view of our relation to the world: a view under which learning about the world is a matter of discovery, not invention. This realist aspect of our anthropocentrism comes out in the fact that we can continue to hold by two distinctive claims. We can continue to believe in epistemic servility: in the fact that in seeking to know about things we have to attune ourselves to an independent reality. And we can continue to believe in cultural openness: in the fact that there is nothing that the members of one culture can know that is inaccessible in principle to people outside.

3

For Individualism, against Collectivism

In the first two chapters we focussed on the make-up of individual, intentional subjects. Our concern in this chapter and the next will be with the life of individual subjects in interaction. We will be exploring social as distinct from psychological ontology: the theory of what goes to constitute the social world, the theory of what there is and how things stand in the social arena. This sort of theory belongs as much to social scientists as it does to philosophers but in these chapters I shall deal with the subject in a more or less philosophical manner. The emphasis is partially remedied in the final part of the book, where I turn to the significance for social and political theory of the picture developed in our earlier discussions.

Social ontology has traditionally been focussed on one general issue: that of how individual agents relate to society. Usually it is assumed that individuals relate to society as parts relate to the whole and then the question is whether the whole is something more than a sum of the parts or—the expressions are treated as equivalent—whether the parts are transformed through belonging to the whole. But the assumption that individuals are parts of a social whole, in any strict sense of parthood, is itself questionable (Ruben 1985, chap. 2). And in any case it is a mistake to think of social ontology as being centred on a single issue.

I believe that there are two quite distinct issues that have equally concerned social thinkers and that discussion in the area has been systematically vitiated by their being run together. The first issue has to do with whether individual agents are compromised in their agency by aggregate social regularities, whether a knowledge of how those regularities work would undermine our view of those agents as intentional and thinking subjects: for short, our intentional or folk psychology. What our intentional psychology comprises is explicated in the first two chapters: it amounts to the doctrine that in certain conditions identified as favourable we conform—and can conform intentionally, in the manner of thinking subjects—to intentional regularities, specifically to those regularities that we identify as rational and appropriate in our everyday accounting. The issue of whether aggregate social regularities undermine this intentional psychology is vertical in character. It bears on how far individual agents are affected, as it were, from above; in part-whole terms the issue is whether the whole is more than the sum of its parts.

The other issue in social ontology is of a horizontal character rather than a vertical, for it bears on how far individual agents are affected by one another, not affected from above; in part-whole terms it might be represented as the issue of whether the parts are transformed through jointly belonging to a single whole. The issue is whether individual agents non-causally depend on their social relations with one another for the possession of any of their distinctive capacities: say, for the possession of the capacity to think. Causal dependence is granted on all sides and the question is whether there is a more fundamental sort of connection between something like the capacity to think and participation in community. Does a thinking being like one of us have to belong to a society, have to enter into certain social relations with other human beings, in order to be able to think?

I shall be arguing in the next chapter that these issues are independent: that it is possible to adopt any combination of answers to them. But it should be clear even on first inspection that they address different matters and should not be treated together. It is something of a mystery why they should have been so systematically assimilated in discussions over the past couple of centuries. Perhaps one reason is that the part-whole terminology makes it easy to see them as a single issue: as already mentioned, the question of whether the whole is more than the sum of the parts is often treated as equivalent to the question of whether the parts are transformed by belonging to the whole. We shall be looking at other reasons for the confusion when we come to the argument for the independence of the issues.

Given that we distinguish these two issues within social ontology, we have to regiment the terminology that is traditionally used to describe the opposed ontological positions. Traditionally individualists are equated with atomists, collectivists with holists—the longer terms are Latin in origin, the shorter Greek—but I shall break those equations. I shall use the longer, Latin terms for opposed positions on the vertical issue and the shorter, Greek ones for opposed positions on the horizontal. Individualists deny and collectivists maintain that the status ascribed to individual agents in our intentional psychology is compromised by aggregate social regularities. Atomists deny and holists maintain that individual agents non-causally depend on their social relations with one another for some of their distinctive capacities.

In this chapter I discuss the vertical issue in social ontology and argue for individualism. In the next I discuss the horizontal issue and argue for holism. The upshot, holistic individualism, is a distinctive position to adopt and in the third part of the book, I go on to look at its significance for social and political theory. My discussion of the vertical issue in this chapter will be in three sections. In the first I look at the issue itself, distinguishing two different ways in which people's autonomy might be compromised by aggregate social regularities. In the second section I argue that it is not compromised in the first way and in the third I argue that neither is it compromised in the second; there is no reason to think, as I put it, that aggregate social regularities override or outflank the intentional regularities that characterise individual thinkers.

One further comment. The issue between individualism and collectivism appears recurrently in the history of social science, as social scientists take up the continuing if sometimes mooted refrain that they have things to say that give the lie to our ordinary psychological sense of ourselves. But many of the things that social scientists think of as compromising intentional psychology, and as supporting collectivism, do not do so in any strict sense. Interesting though they may be—and we will be looking at five such claims later—they are things that individualists can equally admit. The propositions that really do challenge intentional psychology—the overriding and outflanking theses discussed in sections two and three—are not the most popular claims in contemporary social science and are not very easy to motivate. My discussion of those doctrines, then, will not catch the eye of many social scientists and may be found tedious by some. But I think that nonetheless it is important to pursue that discussion; it represents the only way of laying the collectivist spectre to rest.

The Issue

Background

The issue between individualism and collectivism is motivated by the place of social laws in human life. In the first part of this book we have seen that it is part of being an intentional and thinking subject, and part therefore of being human, that one's interaction with the environment is governed, at least in favourable conditions, by intentional regularities. But it is a fact of experience that while we may be individually governed by intentional laws, we are collectively subject to a variety of social regularities. And so the question arises as to whether those social regularities compromise the rule of the intentional regularities in any way and thereby rob us of our agency. This is precisely the question between individualism and collectivism.

When I speak of regularities, as mentioned in the first chapter, I always have reliable or lawlike regularities in mind; I do not intend the term to cover accidental patterns. We defined intentional regularities as regularities that involve intentional properties and in practice we assumed that they always involve intentional properties in their consequents: they govern intentional responses like actions and attitudes. It is natural to take social regularities in a parallel manner as regularities that involve social properties, specifically as regularities that govern the realisation of social properties. But what is a social property? The question is controversial and I take a quick, stipulative line (Ruben 1985, chap. 3).

A property is social, I say, just in case its realisation requires that a number of individuals evince intentional responses: they display certain attitudes or perform certain actions, at the same or at different times. The definition is almost certainly too wide to fit all our intuitions but it will serve us reasonably well. It allows certain 'natural' predicates like 'goes to sleep at nightfall' to express social properties, when more than one person is involved. It allows 'institutional' predicates like 'is a judge' to express a social property, even when it applies only to one individual. All of that is as it should be. And of course it allows the significant predicates that we use of social entities like groups or practices or trends to express social properties. Indeed a social entity will count as social only so far as its essence is given by a social property. It is not of the essence of a judge—not of the essence of the person, as distinct from the role—that she has a social property. But it is of the essence of the court or the jury or the trial that it satisfy a certain social property: a property requiring the presence of certain intentional responses in a number of people.

If a social regularity governs the realisation of a social property, and if the realisation of such a property requires a number of individuals to exhibit certain intentional responses, then we can see why social regularities raise the question between individualism and collectivism. For it appears that social regularities come from outside the domain of intentional regularities and dictate the occurrence of certain patterns in intentional responses. It appears that social regularities invade the territory of intentional regularities and usurp their authority. And if that is so, then there is a question as to whether social regularities compromise our status as intentional subjects, our status as subjects who are governed by intentional laws.

If we conform to familiar, intentional regularities then there is a sense in which we enjoy a minimal sort of autonomy. We will be generally responsive to rational pressures in the formation of intentional states and we will be generally faithful to our intentional states—to our beliefs and desires—in the actions we perform. Where those pressures prescribe a response, we will display the response; and where the pressures allow any of a variety of responses, we will be able and liable to adopt any: we will not be forced in any particular direction. Or at least we will display such responsiveness under circumstances that count as favourable. Stanley Benn (1988) describes the minimal autonomy involved here as autarchy. The question raised by social regularities is whether they compromise the autarchy ascribed in intentional psychology. Do they show that we are not what we seem: that we are less our own men and women than we would like to believe?

This question is not a telling challenge with all social regularities. Perhaps the best thing to do in motivating it is to distinguish between different types of social regularities and to identify the sort of regularity that offers the most challenging threat. I shall distinguish between social regularities that are causally continuous with intentional; social regularities that are logically continuous with intentional; and social regularities that are discontinuous in each of these ways with intentional regularities. I describe the first two categories as social-intentional regularities and the last category as social-structural regularities. I shall argue that it is only social-structural regularities—in particular, the regularities of this category that are unearthed in social science—that present even a prima facie threat to our status as intentional agents.

One set of regularities will be causally continous with another if two conditions are realised. First, it introduces factors that are related by causal laws to the factors involved in the original regularities; it directs our attention to actual or potential influences on the original factors, influences that are not tracked in the original regularities themselves. Second, it does this without thereby putting the original regularities in jeopardy: it may indicate a limit on the conditions under which the original regularities operate reliably but it does not give the lie to those regularities; they remain fundamentally sound. In this case we may say that the extra set of regularities constitutes an extension of the original set; it is a set of regularities that bears on the same realm as the original set, as we might put it. Where the original set relates factors of kind X and Y, for example, the extra set introduces regularities that relate one or both sorts of factors to factors of kind Z.

It should be clear that many social regularities will be causally continuous with intentional regularities in this sense (Jackson and Pettit 1992c). Consider the regularity, if it is one, whereby the fact that those of high status in a group hold by a certain opinion means that that opinion will tend also to be held by most others in the group. Such a regularity counts as social, since it governs a social property: the forming of an opinion by certain members of a group. But it identifies a factor that influences the formation of opinion in the group in a way that need not undermine our status as intentional agents. It suggests that as people become aware of the opinions of those of high status, they often adapt their own views to fit. It points us to a sort of cause that we can recognise in operation at the intentional level. We may think of the influence involved in a flattering light, postulating that the testimony of high-rankers is taken particularly seriously by others. Or we may think of it in a more cynical way as an influence that warps the judgment of others and makes conditions unfavourable for the rational consideration of evidence; our intentional psychology allows that this sort of thing may happen, given the open-ended distinction between favourable and unfavourable conditions. In neither case should we see the regularity as one that jeopardises the rule of intentional regularities. The regularity complements or extends intentional regularities, enriching our intentional psychology; it does not undermine that psychology.

As social regularities may be causally continuous with intentional regularities, so they may be logically continuous with them too. Suppose that we have two sets of regularities, one bearing on entities of relatively small size, and the other on larger entities that they compose: one bearing on micro-entities, the other on macro. The macro-regularities will be logically continuous with the micro, I will say, if the fact that the micro-regularities obtain entails that the macro obtain, given that there is nothing to interfere with the macro-composition: there is no evil demon, or whatever, to disrupt macro-formations. Consider the micro-regularities that govern the movements of individual molecules in a gas, reflecting their exact position, mass, velocity, and size. And now consider the familiar, statistical gas laws that govern the movements of such molecules en masse, at the macro level. I assume that the macro-regularities in this case will be logically continuous with the micro. Given there is nothing to disturb macro-formations, the obtaining of the micro-laws will ensure that the macro also obtain.

The point can be put in terms of the notion of supervenience, which we introduced in the first chapter. Imagine two worlds that are identical in regard to the micro-regularities that obtain there. These worlds will also be identical in regard to statistical facts, since such facts are mathematical in nature and obtain at all worlds. And so, absent the evil demon we mentioned, if a given statistical law obtains at either world, it will obtain at both. The obtaining of the macro-laws is supervenient on the obtaining of the micro-laws. Notice that when I say that the same micro- or macro-laws obtain at both worlds, I leave open the question of whether they obtain emptily or non-emptily. A law obtains emptily at a world if the antecedent—the initial conditions of the law—is not fulfilled there but the world is such that if the antecedent were fulfilled, then the consequent would be fulfilled also. A law obtains non-emptily at a world if the case has a more positive cast and the antecedent is realised at that world.

What sort of social regularities are logically as distinct from causally continuous with intentional regularities? Here is a good example from Thomas Schelling (1969). Schelling identifies a regularity that relates the attitudes of parties in a mixed neighbourhood—the mix may be black and white, Catholic and Protestant, or whatever—to the emergence of total residential segregation. Each party has a maximum of eight immediate neighbours, as a square on a chess board has at most eight adjacent squares. Suppose now that each becomes very anxious not to be in a minority in the immediate group of neighbours. The regularity postulated by Schelling associates such a social configuration of attitudes to the emergence of more or less total residential segregation. That the regularity obtains can be shown by a little experiment on a chess board. Distribute black and white pieces at random over most of the squares on the board. Now move those pieces that will want to move, under the postulation about motives. The result will be that some pieces that were previously happy will become motivated to change position. So, once again, move the unhappy pieces. The result of this process will be that in a short time, the board will display a pattern of near-total segregation.

The regularity involved here is logically continuous with intentional regularities. Unlike the causally continuous law, it belongs to a different realm from the intentional regularities. But it obtains in virtue of such regularities. More specifically, it obtains in virtue of, first, the intentional regularity whereby someone who wants not to be in a minority among her immediate neighbours will tend to move if this condition is not satisfied; and second, the quasi-mathematical fact that each wave of removals will tend to put others in an undesired minority position, and occasion further removals, up to the point when there is nearly total segregation. We cannot imagine two worlds that are identical in regard to intentional regularities—and, as all worlds will be, statistical and mathematical facts—but that differ in regard to this regularity. Or at least we cannot do so, absent an evil demon who makes things go awry in the event of a mixed neighbourhood forming.

Social regularities that are causally or logically continuous with intentional regularities may be described as social-intentional. They have as good a claim to be regarded as intentional regularities as they have to be regarded as social, and so the name seems appropriate. Certainly these regularities represent no threat to the rule of intentional regularities and they offer no basis for raising the issue between individualism and collectivism: the issue of whether our autarchical status as intentional agents is compromised by the operation of social laws. Henceforth we shall have little more to say about them. In order to find social regularities that offer a prima facie challenge to intentional psychology, we need to go to a third category: to regularities that are causally and logically discontinuous with intentional regularities. Such regularities I will decribe as social-structural rather than social-intentional.[1]

As it is obvious that there are social-intentional laws, so I think it is obvious that there are social-structural laws. Consider the expectations that we naturally entertain with a group-agent: say, a political party or an economic firm. We expect that it will reliably act for certain ends or, equivalently, that the officials who act in its name will reliably pursue those ends. Whatever the motivation of the individuals, whatever their beliefs, we expect them to act appropriately if they are officials of the party or firm. The regularities that we associate in this manner with office-holding are social, because they require the display of certain intentional responses by a number of people. And equally the regularities are structural. They do not point us towards causal influences—say, influences of loyalty or whatever—that are supposed to impact intentionally on agents and produce the associated results; they are not designed to enrich our intentional psychology. And the regularities are also logically discontinuous with intentional laws. We can imagine two worlds in which the same intentional regularities obtain—people are the same kinds of beings—but which differ in regard to those regularities. Things are organised in one world so that the corporate regularities obtain and parties and firms thrive, whereas, for whatever reason—whatever reason of ideology, tradition, or institutionalisation, for example—such corporate entities fail to produce the same reliable behaviour in the other.[2]

Consider again the expectations that we will usually have formed if we postulate the existence of a certain sub-culture: say, a sub-culture of young teenage males. We expect that participants in such a culture will generally act according to certain regularities. They will be suitably dismissive of authority, aggressive towards outsiders, enthusiastic in support of certain activities, deferential towards acknowledged leaders, and so on. Such regularities are social, requiring intentional responses from members of the culture. And equally they seem to be structural. They may not point us towards any distinctive factors that are supposed to interact causally with the factors involved in intentional regularities—they may not point us to an influence that intentional psychologists might wish to countenance—so may not be causally continuous with intentional regularities. And they are not regularities whose obtaining is more or less entailed by the obtaining of intentional regularities, so they are not logically continuous either. We can imagine two worlds that obey the same intentional laws but in one of which sub-cultures display this conformity—for whatever reason—while in the other they do not: cultures amount there to something much more disjointed.

It is easy to make sense of the idea of social regularities that are causally or logically continuous with intentional regularities; they are regularities that causally or logically extend the set of intentional regularities. But how are we to make sense of the idea of intentionally discontinuous social regularities? No problem. Think of discontinuous regularities as regularities that fail, but only just fail, to be logically continuous with intentional: as regularities that are logically continuous, not with intentional regularities alone, but with intentional regularities as they operate under certain boundary conditions. Let people be the way we know them to be in intentional psychology and there is no guarantee that corporate agents or sub-cultures will display the patterns described. But such a guarantee comes into view if we say not only that people conform to the familiar intentional regularities, but that a certain boundary condition is in place. Suppose, for example, that corporate agents and sub-cultures are always subject to a gatekeeping-monitoring arrangement: an arrangement under which officials or members get elected only if they seem likely to behave appropriately, and an arrangement under which they are forced out of office or forced out of the group if they begin to behave inappropriately. Under such a boundary condition a world in which the intentional regularities obtain will be a world in which the social regularities under discussion will obtain also. But absent that boundary condition, the fact that a world is one in which the intentional regularities obtain is no guarantee that the social regularities will obtain. Thus those regularities count as logically discontinuous with intentional regularities.

The two examples I have given of social-structural regularities are both salient in our everyday perception of the social world: in our folk sociology. The regularities become perceptible to us, as we identify the sorts of enduring entities with which they are associated: the corporate agent in the one case, the sub-culture in the other. The salience of those entities is tied up with the fact, plausibly, that they conform to structural regularities. We said earlier that there is no social entity without a social property: a social entity is defined by its essentially realising a social property. We now add the observation that enduring social entities are also likely to be characterised by an association with certain social-structural regularities.

This observation suggests that there will be no problem in supplementing our two examples with others. As there are a variety of social entities we distinguish, in particular a variety of social continuants, so there are many potential areas in which we can look for examples of structural regularities. For instance, we can look to the following:

1. Groups, like the party and the firm, whose essence it is to have a mode of collective behaviour: these will also include clubs, unions, governments, and so on (J. S. Coleman 1974; Finnis 1980, chap. 6; French 1984).

2. Groups that may have only a non-behavioural collective identity like genders, races, and classes (Hindess 1987).
3. Instrumentalities whose essence is tied up with the behaviour of designated officers: for example, museums, libraries, and states (Stoljar 1973).
4. Instrumentalities, like the sub-culture of teenage males, whose essence is independently given: these will include cultures, territories, markets, and the like.

Let it be granted that there are social regularities to be found at work in human affairs as well as regularities of a social-intentional kind. I emphasised earlier that social-intentional regularities do not offer a prima facie challenge to our status as intentional agents and do not motivate the issue between individualism and collectivism. Do the social-structural laws that we have considered do any better in this regard? Do they suggest that the domain of intentional regularities is invaded from outside, with intentional responses being brought under alien control: under the control of a social-structural regime? Do they threaten our intentional autarchy?

It would be convenient for purposes of exposition if I could say that yes, such social-structural regularities offer a telling, prima facie challenge. We could then go on and consider the challenge, examining the collectivist case to be made in its support and the individualist case to be made against it. But the truth is that, as we have presented social-structural regularities, there is little or no temptation to be worried about their impact on intentional psychology. True, the laws are not causally and logically discontinuous with intentional regularities. But they are logically continuous with a package of factors that will not cause any anxieties for intentional psychology: the combination of the obtaining of the intentional regularities with the realisation of a suitable boundary condition, in particular a condition that does not itself threaten intentional psychology. The social-structural regularities obtain superveniently on that combination of factors, at least absent an evil demon that would disrupt social-structural formations.

In order to see why this package should not cause any anxieties, it may be useful to consider a parallel to the sort of dispensation described. The laws of chemistry bear on items that are subject also to the laws of physics and they are causally discontinuous with the laws of physics: they do not identify causal factors overlooked in physical laws. Are they logically continuous, then, with physical regularities? Again, almost certainly no. For the obtaining of chemical laws plausibly requires, not just that the laws of physics obtain, but that various physical constants assume certain values; it requires that the world in which the physical laws operate satisfies certain fixed parameters, certain fixed boundary conditions. Yet this logical discontinuity gives us no inclination to think that the regularities charted in chemistry, albeit they bear on physical items, invade and usurp the domain of physical laws. Absent an evil demon that is liable to disrupt chemical formations, the regularities of chemistry obtain superveniently on the obtaining of the physical regularities together with the fulfilment of the appropriate boundary conditions. There is no question, under such an arrangement, of a usurpation of the physical domain.

As we do not think that the rule of physical laws is disturbed by the presence of chemical regularities, so we need not think that the rule of intentional laws has to be disturbed by the sorts of social-structural regularities countenanced in common sense. No matter how deep our commitment to intentional psychology, we need not be troubled by the presence of such structural regularities. But the fact that we need not be troubled does not mean that people have not actually been troubled in the past. And here there is something worth remarking, though it is orthogonal to our interests (Brown 1984).

In the period from the seventeenth to the nineteenth century, folk sociology, particularly folk sociology in the West, was confronted with a whole new range of social entities. The state emerged as a relatively novel sort of institutional entity; cultures became visible in the light of anthropological exploration, in a way in which they had not been visible previously; and attention became focussed on the variety of corporate persons, as a new economic order called banks and companies, exchanges and markets, into being. These shocks to folk sociology led many people to see the relevant social-structural regularities— the regularities associated with the novel entities—as threatening; they gave life to a collectivist vision.

The emerging conceptualisation of the state in modern Europe certainly served as one source of collectivism, a source that was well and truly mined in the explorations of Hegelian and post-Hegelian philosophy. Under that conceptualisation, as Quentin Skinner (1989, p. 112) has put it: 'the state must be acknowledged to be an entity with a life of its own; an entity which is at once distinct from both rulers and ruled and is able in consequence to call upon the allegiances of both parties'. Under that conceptualisation, as we might say, the state is assumed to involve the operation of a variety of social-structural regularities. The novel recognition of such an entity must have encouraged various social and political thinkers to adopt a collectivist vision.

But there were other sources of collectivism too. One was the discovery of exotic cultures and the associated emergence of notions like those of culture and tradition, nation and nationalism (Hobsbawm 1990). This development made an impression on folk sociology of the same order as the recognition of the state. It led to the notion of the *Volksgeist*, the spirit of the people, a notion that was easily reified into a collective agency, which supposedly acted through individuals, dictating all that they felt and did. If such an agency is admitted, then it is easy to think of the associated social-structural regularities as laws that compromise people's intentional autarchy.

Yet another source of collectivism was the growing juristic emphasis on corporate persons (Webb 1958). Talk of corporate persons goes back to mediaeval and even earlier origins, but it was sharpened by the emergence of novel corporate agents, under the influence of institutionalised capitalism, in the nineteenth century: agents like the limited liability company and the trade union (Coleman 1974). The focus on corporate entities motivated expressions of the most collectivistic kind among the theorists of law. 'Itself can will, itself can act; it wills and acts by the men who are its organs as a man wills and acts by brain, mouth and hand' (Maitland 1900).

The influence of these developments would have been reinforced in the eighteenth and nineteenth centuries by an organicist way of viewing social life, under which a society gets to be described as an organism and the history of a society gets to be seen as a life-process: a process in which we naturally look for cycles or stages or even meaning. With this metaphor in command of the collectivist vision, it became fashionable to deny individual autarchy. It became possible even to think of individuals in nightmare fashion, as the pawns or playthings or puppets of social structure and historical process (Popper 1960).

These comments on developments in the formation and recognition of corporate entities are not intended to motivate the issue between individualism and collectivism for us. As our folk sociology has accommodated states, cultures, and legal novelties, those entities have ceased to be felt as even a prima facie danger for our intentional autarchy. The associated social-structural regularities now constitute no more of a threat to our autarchy than the regularities considered earlier; they are essentially of the same type. My comments on corporate entities are intended only to try and explain why regularities that do not trouble our individualist intuitions may have been found troubling at an earlier period.

But we must now return to the main task, which is to find social regularities that motivate the issue between individualism and collectivism, giving life to the collectivist threat. Social-intentional regularities will not do the job. And neither, apparently, will the sorts of social-structural regularities that we have been considering. So where does this leave us? There is one remaining possibility for motivating the issue. We illustrated social-intentional regularities with examples taken from social science—the regularity involving the influence of high-rankers and the regularity bearing on residential segregation—but we have drawn only on folk sociology for examples of social-structural regularities. The remaining possibility is that there are social-structural regularities identified in social science that can give life to the collectivist vision, offering at least a prima facie threat to our intentional autarchy.

In order to explore this possibility it will be useful to say a little on the rise of social science. The sort of social science that has particularly focussed on social-structural regularities belongs to the broad tradition associated with the nineteenth-century French sociologist, Emile Durkheim. That tradition had its origins in the rise of social statistics in the early part of the nineteenth century. Responsive to electoral and utilitarian pressures, the governments of the newly industrialising countries sponsored the statistical mapping of a great variety of phenomena. 'It was necessary to count men and women and to measure not so much their happiness as their unhappiness: their morality; their criminality; their prostitution; their divorces; their hygiene; their rate of conviction in the courts. With the advent of laws of statistics one would find the laws of love or, if not that, at least the regularities about deviancy' (Hacking 1981, p. 13; also Hacking 1980, 1991).

This statistical mapping of large-scale society, this avalanche of numbers, had an important intellectual impact, because it revealed that social aggregates display surprising constancies, constancies that are often invariant over changes in the individuals, and in the mentalities of the individuals, involved. The statisticians constructed indices of divorce, suicide, crime, sickness, prostitution, and a variety of other phenomena. It soon became clear from their work that there were constancies over time in many such indices, and there were constancies in the correlation of different indices with one another. At the beginning of the nineteenth century it would have been common to regard such statistical patterns as epiphenomena of intentional action, which we should expect to be as variable as we seem to find one another. But from the earliest days of statistical mappling, it was obvious that the patterns were not as variable as that; it began to seem that they had a life of their own. 'The avalanche of numbers after 1820 revealed an astonishing regularity in statistics of crime, suicide, workers' sickness, epidemics, biological facts' (Hacking 1981, p. 13).

This revelation of unexpected pattern encouraged a sort of statistical determinism, one associated in the beginning with French writers like A. M. Guerry and Adolphe Quetelet. Already by 1832, Guerry could write: 'We are forced to recognise that the facts of the moral order are subject, like those of the physical order, to invariable laws' (Hacking 1990, p. 73). Such determinism came to be described as an astronomical conception of society, a conception in which people were driven by forces akin to a cosmic force like gravity; the fact that Quetelet was an astronomer lent credence to this characterisation of the view (Hacking 1990, chap. 15). The astronomical conception soon spread abroad, receiving a powerful impetus in the English-speaking world, and indeed more broadly, with the publication in 1857 of T. H. Buckle's *History of Civilization in England*.

The astronomical conception of society received perhaps its most dramatic expression in Tolstoy's *War and Peace*, which was written between 1863 and 1869. One passage will serve to give the tenor of the view.

> Ever since the first person said and proved that the number of births or crimes is subject to mathematical laws, that certain geographical and politico-economical laws determine this or that form of government, that certain relations of the population to the soil lead to migrations of peoples—from that moment the foundations on which history was built were destroyed in their essence. By disproving those new laws, the old view of history might have been retained. But without disproving them, it would seem impossible to continue studying historical events, merely as the arbitrary product of the freewill of individual men. For if a certain type of government is established, or a certain movement of peoples takes place in consequence of certain geographical, ethnographical, or economic conditions, the freewill of those persons who are described to us as setting up that type of government or leading that movement cannot be regarded as the cause.
> (Tolstoy 1972, pp. 1313–14)

The avalanche of numbers and the statistical determinism associated with it led to the development of the new science of sociology. True, the person who invented the term 'sociology', Auguste Comte, was averse to seeing the discipline as a merely statistical enterprise. But the rise of statistical mapping had a crucial impact on Emile Durkheim, who was responsible for establishing sociology as a respectable science. The first empirically based study that he undertook was an investigation of suicide, published in 1897, and this reveals both how deeply he was impressed by the statistical constancies of suicide and how naturally he moved towards the conclusion that those constancies had a life of their own, a life independent of the whirl and flow of individual choice. 'Each society is predisposed', he remarked, 'to contribute a definite quota of voluntary deaths' (Durkheim 1951, p. 51).

The statistical determinism occasioned by the avalanche of numbers was not always well motivated or formulated: whatever Tolstoy may have thought, for example, it is quite consistent with people's being free that their individual choices should display certain aggregate patterns. But, such misconceptions apart, the constancies revealed in the statistical mapping of society still offer a challenge to our prevalent self-image as human beings: to our self-image as intentional, autarchical subjects. As the Durkheimian tradition of sociology made clear, they point to novel social-structural regularities. They point to regularities that, unlike the social-structural regularities of folk sociology, raise at least a prima facie question about whether we really enjoy the autarchy with which intentional psychology credits us.

Durkheim accepted that aggregate constancies have a life of their own, remaining surprisingly invariant over relevant changes at the individual, intentional level. He postulated aggregate, social regularities to explain those constancies. An aggregate regularity may explain a constancy by subsuming it, as when a constancy in the correlation between two indices—say, the divorce rate and the suicide rate—is taken to reveal a regularity. Or an aggregate regularity may explain a distinct constancy, revealing a relationship between it and some independent factor: say, a relationship between an aggregate index or correlation of indices on the one side and a background factor like increased population density, an established current of opinion, growing urbanisation, or even the fact that the factor in question—say, the correlation—is socially functional in some way. Durkheim gives expression to the sort of regularities that he saw at the origin of statistical constancies, when he writes in the study on suicide: 'the relations of suicide to certain states of social environment are as direct and constant as its relations to facts of a biological and physical character were seen to be uncertain and ambiguous. Here at last we are face to face with real laws' (Durkheim 1951, p. 299).

In order to understand the sorts of regularities Durkheim envisaged, here are some examples; we shall take them to be sound for purposes of the present discussion (see Jackson and Pettit 1992c). They are not chosen on the grounds that Durkheim actually endorsed them or even considered them—he didn't— but because they represent a fair spread of the sort of sociological doctrine with which he was associated.

1. Increased unemployment leads to a rise in crime.
2. Urbanisation leads to a decline in religious practice.

3. Policies for increasing employment cause inflation.
4. Economic stratification comes about in order to secure personnel for crucial positions.
5. Capitalism is stable when it is optimal for developing productive power.
6. States act internationally in their best economic interest.
7. States decline in political influence as they fall behind in economic capacity.
8. Companies maximise expected returns.
9. The Protestant ethic facilitates the rise of capitalism.
10. Capitalism generates a breakdown in community values.

Regularities of these kinds are social-structural in character. They are social regularities, because they each demand that a number of individuals display certain intentional responses: they require that individuals commit crime in a certain proportion, give up attending church, and so on. And they are structural, because they are causally and logically discontinuous with intentional regularities. It is not supposed that unemployment leads to crime, or urbanisation to religious decline, because it causally impacts in some intentionally intelligible way on the antecedents of intentional regularities; it is not supposed, for example, that it is because people become aware of unemployment or urbanisation that they produce the behaviour required for the consequent. Thus the regularities are causally discontinuous with intentional regularities. And, as a little reflection reveals, they are logically discontinuous too. We can imagine two worlds that are identical in regard to the intentional regularities that obtain in each but which differ in regard to such regularities. We can imagine a world that is exactly like ours in the intentional regularities that obtain there but which does not conform to the regularity linking unemployment with crime. Durkheim would not be impressed at the suggestion but the difference might simply be that in that world the weather is so good that the unemployed can spend their time on the beaches. The contingent availability of that resource, that extra opportunity, might reduce their exposure to the temptations that would otherwise occasion crime.

It will become clearer in the next section how the sorts of examples given can exhibit causal and logical discontinuity with intentional regularities. For the moment, the important point is that to someone like Durkheim it would have seemed obvious that there are unexpected social regularities that are causally and logically discontinuous with intentional regularities. He would not have expressed the discontinuity by the use of notions like causal and logical continuity. But the formulation is broadly faithful to the spirit of his perception.

Do regularities such as those on our list serve to motivate the issue between individualism and collectivism? I shall assume that they do: more specifically, that they offer at least a prima facie threat to our intentional autarchy. The social-structural regularities countenanced in folk sociology, the regularities associated with the different social entities that we recognise, may not give us any pause about assuming that we remain faithful to the rule of intentional regularities. But these more or less unexpected laws may well give us pause. They may raise the spectre that we are compromised from above, that we are moved or constrained by social factors in a manner undreamt of within our intentional psychology.

In this connection, it is striking that Durkheim himself, and most of those who followed him, certainly took the sorts of regularities illustrated to give the lie to intentional psychology. Durkheim was the first important social scientist to draw a collectivist lesson from the avalanche of numbers and the regularities invoked to explain them. He saw those regularities as pointing us towards constraints that undermine or limit human autarchy. He saw them as vindicating, in essence, something like the so-called astronomical conception of society. The constraining factors to which social-structural regularities pointed, in Durkheim's view, range from the morphological features of a society, like the density of its population, to the norms or rules institutionalised there, to the currents of opinion and the enthusiasms that take over from time to time (Durkheim 1938; Lukes 1973, pp. 8–15). He described all of these as social facts and argued that social facts constrain intentional agents from without. 'Currents of opinion, with an intensity varying according to time and place, impel certain groups either to more marriages, for example, or to more suicides, or to a higher or lower birth rate, etc. . . . A social fact is to be recognised by the power of external coercion which it exercises or is capable of exercising over individuals' (Durkheim 1938, pp. 8–10).

In giving social facts this constraining role, Durkheim was explicitly challenging the individualist view—a view he naturally presents as an extreme position—that individuals enjoy intentional autarchy. Thus he writes of social facts that: 'when we define them with this word "constraint", we risk shocking the zealous partisans of absolute individualism. For those who profess the complete autonomy of the individual, man's dignity is diminished whenever he is made to feel that he is not completely self-determinant' (Durkheim 1938, p. 4).

Two Relevant Questions

I hope that this discussion of social regularities allows us to get a feel for the issue between individualism and collectivism, motivating it as a problem. But the discussion so far may not help much in providing us with a satisfactory formulation. Thus Durkheim's rhetoric of the coercion of individuals leaves a lot wanting in the way of clarity. The intention is clearly to say that social-structural regularities displace folk psychology in some significant manner, but the nature of the displacement alleged remains somewhat obscure. The vertical issue in social ontology, the issue joined by Durkheim on the collectivist side, remains itself obscure. We must now try to remedy this defect by isolating the different things that collectivists might be taken to be saying.

I distinguish two propositions that we might take collectivists like Durkheim to be defending, when they argue that social-structural regularities compromise the autarchy of intentional agents. Each of them involves a radical questioning of our intentional psychology and of the autarchy which that psychology ascribes to intentional, thinking subjects. The first and more challenging thesis is that social-structural regularities—or at least some such regularities—override the intentional regularities that we naturally take to govern human behaviour; the second is that social-structural regularities outflank the intentional regularities, enjoying a certain sort of edge or priority.

The thesis that structural regularities override intentional regularities is, very briefly, this: that there are circumstances in which instances of the two sorts of regularity are both engaged, and where they would lead to different consequences, and that in those circumstances the structural regularities prevail; the intentional regularities are suspended, being overridden by the aggregate. Imagine that social science claims to identify a structural regularity bearing on crowd behaviour and that there are occasions where it would lead us to expect something quite different from what the relevant intentional regularities countenanced in our intentional psychology would lead us to predict. The overriding thesis would suggest that in such a case we may expect to find that the intentional regularities are simply suspended, as agents find themselves behaving, as if under a hypnotic spell, in the manner that the aggregate regularity requires. Moreover it would suggest that this sort of phenomenon—a phenomenon involving agents going on the blink, so to speak—is typical of what happens under the rule of social-structural regularities.

Although he is never absolutely unambiguous, Durkheim sometimes comes close to suggesting that structural regularities do indeed override intentional regularities in this way. Here is one example.

> The great movements of enthusiasm, indignation, and pity in a crowd do not originate in any one of the particular consciousnesses. They come to each one of us from without and can carry us away in spite of ourselves. . . . Also, once the crowd has dispersed, that is, once these social influences have ceased to act upon us and we are alone again, the emotions which have passed through the mind appear strange to us, and we no longer recognize them as ours. . . . And what we say of these transitory outbursts applies similarly to those more permanent currents of opinion on religious, political, literary, or artistic matters which are constantly being formed around us, whether in society as a whole or in more limited circles.
>
> (Durkheim 1938, pp. 4–5)

The message of the overriding thesis is that intentional agency is an illusion, at least in those areas where social-structural regularities rule. The intentional subject, as it is sometimes put, is dead. We find the most extreme statements of this point of view in the work of structuralist social scientists in the mid-century. Thus consider this passage from Louis Althusser (1970, p. 180), in which he asserts that the motors of social life, as revealed in his Marxist vision of it, are not intentional subjects; the true motors or subjects are relations of production: it is the aggregate regularities in which they figure that dictate what happens.

The structure of the relations of production determines the *places* and *functions* occupied and adopted by the agents of production. . . . The true 'subjects' (in the sense of constitutive subjects of the process) are therefore not these occupants or functionaries, are not, despite all the appearances, . . . 'concrete individuals', 'real men'—*but the definition and distribution of these places and functions. The true 'subjects' are these definers and distributors: the relations of production* (and political and ideological social relations).

Claude Lévi-Strauss (1971, pp. 614–15) offers a more elegant, though no less extreme, statement of this opposition to our intentional psychology when he says that the intentional subject is an 'intolerable spoiled child who for too long has held the philosophical scene and prevented any serious work, drawing exclusive attention to itself'. Althusser and Lévi-Strauss are both social scientists working in a French tradition that is influenced perhaps more by Durkheim than by anyone else, Marx included. But the compromise of the intentional subject to which they testify is also found, though perhaps in less virulent forms, outside that tradition, where Durkheim has not had such an exclusive influence.

The most outstanding American sociologist of the twentieth century has probably been Talcott Parsons, a figure who was influenced as much by the work of the German social scientist, Max Weber, as by that of Durkheim. It is significant that while the influence of Weber leads him to stress the role of individual agents, still those agents tend to assume just cipher status in his theory, or at least in many interpretations of his theory. A. W. Gouldner (1969, p. 206) writes of the theory as follows.

While Parsons' voluntarism places great importance on man's efforts to realize certain ends, it is paradoxically true that these ends are no longer seen as derived from him; though they reside *in* him, they derive *from* social systems. Man is a hollowed out, empty being filled with substance only by society. Man thus is seen as an entirely *social* being, and the possibility of conflict between man and society is thereby reduced.

Perhaps we should not be surprised that even a relatively moderate sociologist like Parsons should drift, or should be taken to drift, towards a denial of the sort of autarchy asserted in intentional psychology. Among the social sciences, sociology has been most intimately associated with a focus on statistical constancies and laws, and such a focus leads easily to a way of thinking under which the platitudes about agency drift from view and are eventually rejected. C. S. Lewis makes the point nicely when writing in 1945 about a fictional sociologist.

His education had the curious effect of making things that he read and wrote about more real to him than things he saw. Statistics about agricultural labourers were the substance; any real ditcher, ploughman, or farmer's boy, was the shadow. Though he had never noticed it himself, he had a great reluctance, in his work, ever to use such words as 'man' or 'woman'. He preferred to write about 'vocational groups', 'elements', 'classes' and 'populations': for, in his own way, he believed as firmly as any mystic in the superior reality of the things that are not seen.

(C. Lewis 1983, p. 87)

I do not say that the overriding thesis is plausible; in the next section I shall be arguing that it is false. I do not even say that the thesis has often been unambiguously adopted, though it has certainly been loosely associated with the tradition of sociological thinking. But I do maintain that the overriding thesis represents a radical, collectivistic questioning of our intentional psychology and of the autarchy ascribed to the individual under that psychology. It means that the things we believe in conceiving of ourselves and one another as intentional, thinking subjects are in some measure false. The overriding thesis does not suggest that social-structural regularities may lead us just to modify our intentional psychology, pointing us towards unsuspected sources of perturbation. It takes the regularities to point to circumstances where intentional laws fail, circumstances that are of such a kind or extent that we must modify our commitment to the intentional-psychological understanding of human beings. That way of understanding may be of heuristic and predictive value but it leaves no room for the structural influences that allegedly move us.

It is clear that if there were a great many circumstances in which intentional regularities were supposed to be suspended, then intentional psychology would be displaced in this way. But I also allowed that the circumstances might be of such a kind, as distinct from such an extent, that we would have to modify our commitment to intentional psychology. Suppose the sort of circumstance in question is one that allows the agent to say afterwards only something of the kind: 'I don't know what came over me; I just found myself wanting to do this or that, for no reason that I can now endorse'. Suppose in other words that there is a first-person discontinuity between cases where agents behave in the intentionally explicable way—including cases where conditions are abnormal—and the cases where the structural regularities rule. In that event, it seems plausible to think of the alleged suspension of intentional regularities as being of a kind that really does displace intentional psychology; it is not the sort of suspension that is triggered in the commonly recognised kinds of abnormal conditions.

The second sort of thesis that might be ascribed to collectivists like Durkheim, and that is truly distinctive of a collectivist position, holds that the relevant structural regularities outflank intentional regularities rather than override them. This thesis is not as extreme or dramatic as the overriding doctrine. But it still has a significant impact, suggesting that intentional psychology is seriously in error.

Suppose that the following scenario obtains. For all that our intentional psychology says, the social world might have been in any of a given number of configurations, as people believe and desire and do different things. In some of these possible configurations—if you like, in some of these possible societies—the structural regularities identified by the social scientist obtain. In others they do not: people have such beliefs and desires as lead them to act in a manner that belies the regularities; for example, to return to our earlier case, they do not display the predicted patterns of crowd behaviour. But though these configurations are equally possible from our intentional-psychological perspective, it turns out that they are not equally possible in fact; there is some filter at work that ensures that the only likely configurations are ones where the structural regularities obtain. The filter means that any population that is not disposed to display those configurations, and thereby to underpin the favoured structural regularities, will not survive. Under the picture envisaged, things are rigged, and rigged quite dramatically, so that the structural regularities prevail. The structural regularities outflank the intentional, enjoying a decisive priority relative to them.

How could the members of two societies be alike in obeying intentional regularities, while the societies differ in regard to the structural regularities sustained in each? We have already seen how this can happen. If the societies differ in regard to a relevant boundary condition then, while the same intentional regularities obtain in each, the difference may mean that different social-structural laws obtain. The boundary condition may involve an institutional feature, as in the gatekeeping and monitoring arrangement deployed by certain corporate agents. It may involve an environmental feature, as in the difference of climate, which might break the connection between unemployment and crime. Or it may even involve a biological difference between agents: say, a biological predisposition to favour certain beliefs and desires. The people in the one society may find it natural, for example, to desire the good of the whole or to trust the testimony of leaders; the people in the other may be of a more antagonistic and sceptical disposition. If a filter favours intentional configurations where certain social-structural laws obtain, it will operate by favouring societies with suitable environments, institutions, and the like or, perhaps more plausibly, by favouring societies in which individuals are genetically predisposed to favour certain beliefs and desires. It will ensure that populations in which such factors are not present do not survive and, as I put it, that the social-structural regularities outflank the intentional.

The best way to appreciate the force of the outflanking thesis may be to see that it would represent the relation between social-structural and intentional regularities as directly analogous to the relation, as we described it, between intentional regularities and the naturalistic regularities that govern agents. Natural selection has ensured, so we may assume, that we human beings are more or less rational in how we form our beliefs and desires and in how we act on the basis of those states; the selection has meant that in general only human beings of this intentional kind have propagated and survived. The natural history of our species has meant that such, and only such, naturalistic regularities are instantiated in our make-up as ensure that we measure up to the rational model; the history has meant that the naturalistic world is rigged in favour of our rationality. The outflanking thesis about social-structural regularities alleges a comparable rigging in favour of the structural order that those regularities describe. It says that as nature is predestined to allow us to obey intentional regularities, so we intentional subjects are predestined to allow our societies to obey certain structural patterns.

As Durkheim sometimes seems to endorse the overriding thesis, so he sometimes comes close to supporting the outflanking thesis also. There is no inconsistency here, as the theses may be defended in respect of different sub-sets of social-structural regularities. Durkheim seems to support the outflanking doctrine in endorsing this functionalist idea: if a certain regularity—say, for our purposes, an imagined regularity of crowd behaviour—serves a function in increasing social solidarity, or whatever, then it is likely to have been selected for; it is likely that there will only be such individuals about, only individuals with such intentional dispositions, as behave in accordance with that regularity (Durkheim 1938, p. 96; Turner and Maryanski 1979, p. 97). If a regularity of a structural kind serves a certain social function, so the approach suggests, there will be some filter in operation to ensure that it survives, weeding out those who are not intentionally so disposed that their responses will support the regularity. The structural regularity, as I express it, will outflank the intentional regularities.

Durkheim and many other social scientists appear to be committed to something like the outflanking thesis. They support a brand of functionalism that suggests that certain functional regularities could not be sustained, short of the sort of selection involved in the outflanking scenario (Turner and Mary-anski 1979, p. 97; Van Parijs 1981, p. 87). But does the outflanking thesis displace our intentional psychology and compromise the autarchy of inten-tional, thinking subjects? It may seem that the thesis need not challenge the intentional-psychological view of ourselves as minimally rational creatures and that it ought not to be taken therefore as a distinctively collectivist proposition.

The autarchy countenanced in intentional psychology means that, where the intentional laws allow, we might well have thought or done otherwise than we did. We might have approached the actual evidence with different pre-sumptions or different weightings, for example, or we might have seen the actually relevant values in a different light: this, to the extent that rational pressures leave those different possibilities open. But these intuitions about our intentional autarchy are given the lie by the outflanking thesis. If we had been such that under the actual circumstances we might have thought or acted otherwise, then under the relevant survival story, we would not have been around. Or at least we would not have been around, if the difference had meant that the favoured social-structural laws did not obtain. We would have been filtered out by the mechanism, whatever it is, that favours those aggregate regularities.

I have distinguished two ways in which social-structural regularities might compromise the intentional autarchy of human beings. They might compro-mise autarchy by overriding intentional regularities or by outflanking them. Under the first possibility, the structural regularities represent more powerful forces than the intentional. Under the second, they represent a deeper order in the way intentional responses are organised. An analogy may help to make it clear that these two modes of compromising autarchy have a natural salience.

Imagine a schoolgirl at a hockey match who cheers for her school's team. We will say that such a girl enjoys autarchy only if certain rational pressures are liable to cause certain responses and only if, in the absence of such pressures, she is liable to respond in any of the allowable ways. We will say that she is behaving as a proper intentional agent, for example, only if evidence that her side is cheating can lessen her desire sufficiently for her not to cheer any longer; or only if evidence that the goal-keeper on her side is unnerved by cheering can stop her doing so. Such factors, we think, must be capable of making a difference.

We will hesitate to say that the girl enjoys autarchy under at least two sorts of circumstances. One is where she is predetermined to cheer in a manner that leaves her insensitive to the sorts of considerations mentioned: say, if she is under a hypnotically or neurally induced compulsion to cheer. And the other is where she is predestined to cheer. This will be the case if her school mistress had the say on whether she was to be at the match or not and, assuming sufficient foresight, would have prevented her being there in the event that she might not cheer.

The predetermination possibility obviously corresponds with the possibility envisaged in the overriding thesis and it explains why I say that that thesis denies the autarchy recognised in intentional psychology. The predestination possibility corresponds with the possibility countenanced in the outflanking thesis and so I hold that this thesis too compromises the autarchy of intentional, thinking subjects. I see both doctrines as ways in which collectivists may challenge the individualist's commitment to intentional psychology and in particular to the autarchy recognised in intentional psychology.

The identification of the overriding and outflanking theses as the propositions that divide collectivists and individualists lets us appreciate the nature of the vertical issue in social ontology. The issue is whether the intentional subject, as individualists hold, enjoys the control over herself, the capacity to have thought or done otherwise, which intentional psychology imputes to normal human beings; or whether that autarchy is compromised, as collectivists allege, by the presence of structural regularities of an overriding or outflanking character. In order to further the discussion of that issue we must debate the merits of these two propositions. We turn to those tasks in the two remaining sections of this chapter.

Five Distinct Questions

But before leaving this section we must distinguish the questions identified here from some other questions that often arise in debates between individualists and collectivists. These other questions do not directly involve the issue between the two sides: the issue of whether social-structural regularities compromise intentional agency. But they are frequently treated as if they did involve that issue; they are frequently treated as questions that engage the standing of intentional psychology and as questions that divide collectivists and individualists. The questions are of great interest and often loom larger in contemporary social theory than the issues raised by the overriding and outflanking theses.

The first question which it is important to distinguish from the individualism-collectivism issue is, naturally, the one that divides atomists and holists: whether the individual agent non-causally depends on relations with other people for the appearance of distinctive human capacities. In the next chapter I argue at some length that the question is distinct. But the point is probably obvious anyhow. Holists maintain that individual people are dependent on one another for the possession of distinctive human capacities: say, for the capacity to think. But this belief need not lead them to hold that people cannot therefore have the autarchy of subjects who act more or less rationally in the light of more or less rationally held beliefs and desires. As it is charted in the first two chapters of this book, intentional psychology is silent on the question of whether a capacity like that of thought requires communal resources. The holist assertion that thought does require those resources cannot therefore undermine that psychology.

The second question that must be distinguished from the individualism-collectivism issue has come to the fore in recent thought. Let us agree that people intentionally or non-intentionally produce—and reproduce—a variety of social entities by the patterns manifested in their attitudes and actions. They constitute the groups and instrumentalities mentioned in our discussion of folk sociology and they constitute the social properties that attach to certain individuals: the status enjoyed by this person, the power enjoyed by that. But here now is a question raised by such entities. Do the entities exercise a reciprocal influence on individual agents? Do they serve to determine, and indeed rationally determine, the opportunities, the motives and the perceptions of those agents, and thereby to shape their further intentional responses? There are some who think, or who think certain individualists have to think, that the answer to this question is no (see Ruben 1985, chap. 4; Taylor 1988). Those who defend an affirmative answer, therefore, will appear in the guise of collectivists. And this means, in effect, that most contemporary social theorists will be given a collectivist cast (see for example Giddens 1976, 1984, and J. S. Coleman 1990).

The affirmative answer to this question is surely the correct one; social entities exercise a reciprocal influence on agents, at least so far as agents come to form beliefs and desires that involve those entities in their contents.[3] But admitting the reciprocal influence of social entities in this manner does not in any way threaten the standing of our intentional psychology and does not involve a commitment to collectivism. The question of reciprocal influence is quite distinct from the question at issue between individualists and collectivists. I may be influenced by the objects in my social world, as I am influenced by the objects in my physical milieu, coming to form beliefs and desires in respect of them, but this does nothing to compromise my capacity as an intentional subject.

Why would anyone think otherwise? Why would anyone take the assertion of reciprocal influence to undermine our intentional psychology? One reason may be the assumption that intentional psychology can only admit attitudes with non-social contents: the assumption that intentional psychology is superceded at the point where social contents are acknowledged. But that assumption is absurd. We predicate of one another not just beliefs and desires about concrete objects and people, but also beliefs and desires involving networks and associations, cultures and countries, and a host of such distinctively social entities. Indeed we even take concepts from the treatises of social science and predicate beliefs and desires involving classes and elites, economic forces and the allegedly iron laws of history. It would be absurd to think that intentional psychology is superceded at the point where we admit such contents, for in that case intentional psychology would never get off the ground.

A third proposition that collectivists might be thought to be defending is no more damaging to intentional psychology than the holism or reciprocity doctrines. We saw in the first part of this book that a certain notion of normal conditions—even if it is not explicitly formulated—plays a crucial part in the application of our intentional psychology to one another. In the formation of our basic sorts of beliefs about things we do not take ourselves to be infallible authorities at any time; in particular, we bring the beliefs formed into question in the event of a divergence with others or with our own past judgments. In seeking convergence across persons and times in this way, we come to identify certain factors as perturbing and to discount beliefs formed under their influence. In effect, we recognise circumstances in which such factors are absent as normal conditions and we learn not to expect people to be reliably rational in their belief-formation, or in related practices, when conditions are abnormal. The third thesis that might be ascribed to collectivists is that structural regularities identified in social science or even in folk sociology challenge intentional psychology in the sense of pointing us towards unsuspected disturbing factors. It is a revisability thesis about intentional psychology.

Like the earlier doctrines ascribed to collectivists, the revisability thesis is perfectly plausible. Consider the sort of regularity that much of Erving Goffman's work is given to supporting: that people, in his language, are extremely loath to break frame; that they readily establish a set of shared assumptions about what is say-able and do-able in any interaction and that they generally conform to those assumptions, even at the cost of considerable frustration of their other desires or ideals (Goffman 1975). Consider in particular the evidence that Stanley Milgram produced in support, as he himself thought, of that regularity: that in a laboratory setting roughly two-thirds of volunteers are prepared to administer what are presented as painful and life-threatening shocks, contrary to their own feelings and ideals, so long as the man in the white coat gives the instructions (Milgram 1974). Anyone who has read the work of Goffman and Milgram is likely to conclude, quite reasonably, that here we have been alerted to a factor that is often going to disturb people's intentional performance: for example, the beliefs they form as to what is tolerable behaviour. The work of such social scientists bears out the truth of the revisability thesis ascribed to collectivists, that the identification of social regularities is likely to point us to conditions that disturb the operation or revelation of people's intentional processes and in that sense that it is likely to affect our intentional psychology.

But though the revisability thesis is plausible, it does not serve to displace intentional psychology in a collectivist manner. The frame law used in illustration is causally continuous with intentional regularities, in the sense explained earlier, and it leaves intentional psychology fundamentally intact. Intentional psychology is an open-ended practice; it remains possible at any stage for its practitioners—who are of course also its subjects—to revise their conception of favourable conditions. If social science can interact with intentional psychology to force such revisions, in particular revisions that can be internalised by participants in their practice, then that does not involve any displacement of intentional psychology; on the contrary, it serves as a way of enriching the practice.

According to the revisability thesis, the results of social science interact with intentional psychology only in the benign manner of certain discoveries in other disciplines. Consider the effect of the discovery in optics that a rotating disc with a certain black-white pattern looks to be coloured in different ways at different speeds. This does not undermine the intentional-psychological story of colour-perception but simply points us to a perturbing factor, rotation, that generally goes unnoticed but that is of a kind with factors—say, to do with lighting conditions—that are commonly recognised. A similar gloss will apply to the deliverances of social science, under the revisability thesis. And so that proposition should not be seen as something that individualists have to reject and that goes to the heart of their dispute with collectivists.

We have distinguished three plausible propositions from collectivism: the doctrines of holism, reciprocity, and revisability. In doing so, we have marked off the questions to which those propositions are answers from the matter at issue in the individualism-collectivism debate. But there are still two other sorts of claims that are liable to be mistaken for the assertion of collectivism, and two other sorts of questions that are liable to be confused with the issue that concerns us. The first claim amounts to a thesis about the dispensability of individual contributions to social life, the second a claim about the inevitability of those contributions.

The dispensability thesis I have in mind would say that while any socially significant event is going to come about because of the actions of this or that individual or set of individuals, still no individual makes an indispensable contribution. Had that individual not existed or not acted appropriately, still the phenomenon in question would have occurred. It is either the case that the individual's contribution was not a necessary part of the total cause of the phenomenon or that in the absence of that individual some other agent would have taken her place.

For an analogy to the dispensability thesis consider a closed flask in which water reaches boiling point; we discussed this example in the first chapter. One effect of the water boiling will be an increase of pressure on the sides of the container. With respect to that effect no individual molecule is indispensable, for no individual molecule is a necessary part of the total cause. A second effect of the water boiling, on which we focussed in our earlier discussion, will be that the flask cracks. With respect to this effect no individual molecule is indispensable either, though now for a different reason: not because no molecule is a necessary part of the total cause but because if the molecule that triggered the cracking had not done so, then another molecule would have played that part; I assume that at a certain point, perhaps after a build-up of stress, one particular molecule broke a molecular bond in the surface and thereby precipitated the cracking.

It is clear that Durkheim intends to defend a dispensability thesis about individual contributions with regard to the social constancies—the suicide rate, the divorce rate, and the like—on which he concentrated. The fact that these constancies survive variation in the individuals and the mentalities of the individuals involved shows that no one individual contribution is a necessary part of the relevant total cause. But in defending this sort of dispensability claim, Durkheim is doing nothing to displace intentional psychology, so that the claim could equally well be defended by an individualist. The social constancies in which he is interested are characterised at such a high level that the absence of one individual contribution would not make a difference to the obtaining of the constancy. An artefact of aggregation and abstraction accounts for the dispensability of individual contributions, not any radical questioning of intentional psychology.

What of the other sort of dispensability thesis: the thesis that even with social phenomena that do actually depend on particular individuals—say, something like a political revolution—that phenomenon would still have occurred even in the absence of a relevant contributing party; it would still have occurred, because there would have been someone else available to take the vacated place? This type of claim is not associated with Durkheim particularly but is to be found in other, more historically oriented traditions of social science. The main source is the Marxist tradition and here Engels is quoted as making a telling remark about Napoleon Bonaparte. 'That such a man, and precisely this man, arises at a determined period and in a given country is naturally pure chance. But, lacking Napoleon, another man would have filled his place' (Sartre 1963, p. 56; see also Plekhanov 1976, pp. 283–315). Is this sort of dispensability thesis inimical to our intentional psychology? Again, I think not, though I doubt that it is generally true (see James 1984, chap. 6; Williams 1984–85). All that is required for it to be the case that an individual's contribution is dispensable in this sense is the following: that a number of individuals have the capacity and motivation required to make the sort of contribution involved but that at most one individual can have the opportunity. To think that this was true of all individually necessary contributions to socially significant phenomena might be dogmatic, but it would not involve rejecting intentional psychology.

The fifth and last sort of thesis that might be ascribed to collectivists would fit better with less empirical parts of Durkheim's work than his study of suicide: for example, with his analysis of the division of labour characteristic of our sort of society (Durkheim 1933). I describe it as an inevitability thesis, as distinct from a dispensability thesis. What it says is that various features of social life limit the opportunities available to individual agents—they are structural constraints on feasible options—in such a way that it is inevitable that agents will act so as to sustain certain social constancies. 'In an extreme version this would mean that the constraints jointly have the effect of cutting down the feasible set to a single point; in a weaker and more plausible version that the constraints define a set which is so small that the formal freedom of choice within the set does not amount to much' (Elster 1979, p. 113). This could be what collectivists have in mind, at least some of the time, when they denigrate the place of the intentional subject. Certainly it is a thesis that many social scientists defend.

Like Elster, I think that the inevitability thesis, even in the weaker version, is generally unsound. But that is beside the point in our present discussion. More important, I do not see that it does any more than the dispensability thesis to undermine intentional psychology. It just might be the case, as the inevitability thesis has it, that our opportunities are tailored to the production of certain social constancies: that we are so limited in our range of options that it is inevitable we will act so as to sustain such constancies. But even if it were true, that would cast no serious doubt on the soundness of intentional psychology or on our status as autarchic subjects. It would not mean that we individual subjects were compromised from above in a manner ignored within our view of ourselves as intentional subjects. Our intentional psychology represents us as more or less rational subjects but it says nothing on the range of opportunities that are actually available to us as agents.

It will come as a surprise to some that I do not think individualism is challenged by the inevitability thesis or the dispensability thesis. Many self-described individualists would certainly want to deny the general truth of such propositions. They would want to endorse a sentiment that is nicely expressed in the words of a seventeenth-century writer: 'the minds of men are the great wheels of things; thence come changes and alterations in the world; teeming freedom exerts and puts itself forth' (Hill 1972, p. 219). I agree that most self-described individualists would want to deny the sort of dispensability and inevitability in question. I also agree that many self-described opponents of individualism would want to assert it. But I do not think that the matters in question go to the heart of the vertical issue in social ontology: they do not have an intimate bearing on the status of our intentional psychology; they do not suggest that our autonomy as intentional agents is seriously compromised, as it were, from above.

The questions raised by the five doctrines that we have discussed—the questions of holism, reciprocity, revisability, dispensability, and inevitability— are of great interest in the philosophy of social science and in social theory more generally. I do not wish in any way to suggest otherwise. But I do not think that the questions are intimately tied up with the central issue between individualists and collectivists: the issue of whether social-structural regularities compromise the autarchy assigned to individuals in our intentional psychology. Others may wish to taxonomise things differently, and there is really no right or wrong of the matter, but as I conceive of the issue between individualists and collectivists, these other matters are comparatively marginal.

They do not have the radical impact on intentional psychology that the overriding thesis or the outflanking thesis would have. I turn in the next two sections to the discussion of these two propositions. Whether we are to take an individualist or a collectivist view depends on our assessment of their truth.

Do Structural Regularities Override Intentional?

The collectivist assumes that unexpected social-structural regularities are postulated in social science and he then argues that such regularities must override intentional regularities or must at least outflank them. Let us look first at what is involved in the overriding claim and at how the collectivist and individualist may join battle over it.

The defender of the overriding thesis argues that structural regularities have a greater power than intentional regularities; the opponent denies this. The defender may go to the extreme of denying that intentional regularities are reliable in any domain; he may say that the satisfaction of structural regularities shows that it is an illusion that people ever conform to intentional. Alternatively, the defender of the overriding thesis may admit that in general intentional regularities are sound, arguing only that they are trumped by structural regularities whenever the two kinds come into conflict. Call these two positions the extreme and limited versions of the overriding thesis.

The opponent of the overriding thesis has a corresponding pair of salient positions available on his side. He may say, in extreme mode, that there are no social-structural regularities, in particular no regularities that are not supervenient on intentional regularities. Or he may make the more limited claim that there are structural regularities but that when they conflict with intentional, they are overridden. The two sorts of regularity are in conflict, according to the limited thesis, but it is the intentional regularities that are the more powerful of the two.

We have distinguished two claims in defence of the overriding thesis and two claims in opposition to it, two collectivist claims and two claims of an individualist stamp. The positions mapped by these distinctions can be set out as follows.

	Intentional regularities	*Social-structural regularities*
Extreme collectivist	Non-existent	Unopposed
Extreme individualist	Unopposed	Non-existent
Limited collectivist	Threatening	Triumphant
Limited individualist	Triumphant	Threatening

I propose to discuss these positions in turn, starting with the extreme claims and then turning to the limited ones. I shall provide reasons for thinking that all four are mistaken. The upshot, surprisingly, is not a stalemate between defenders and opponents of collectivism but a victory for the opponents, a victory for individualism. If there are both intentional and structural regularities and if neither sort defeats the other in cases of conflict, then of course the individualist is right to deny the overriding thesis: to deny that structural regularities override intentional. He denies that structural regularities override intentional, without asserting either the extreme or the limited version of the claim that intentional regularities override structural.

Beyond the Extreme Positions

The defence of the extreme version of the overriding thesis has the virtue of being straightforward in intent. It claims that the existence of certain social-structural regularities overthrows our intentional psychology entirely. It claims that social science is inimical to intentional psychology in the way that some philosophers have thought that neurophysiology is inimical. We saw at the end of the first chapter that intentional eliminativists have argued on the basis of recent models in neurophysiology—models that require an architecture in which there is no shadow of intentional organisation—that intentional psychology is, quite simply, false. Here we have an attempt to argue a similar, eliminativist conclusion on the basis of what social science allegedly reveals. The eliminativist version of the overriding thesis is hardly to be found in Durkheim, though it may be what some of his successors have in mind when they speak of the death of the subject.

What are we to say in response to the extreme version of the overriding thesis: in response to the thesis, construed in this eliminativist way? There is little to say, short of returning to the themes of the first part of the book. Part of the point of those chapters was to make clear how compelling is the postulation of intentionality and indeed thought: how naturalistically acceptable it is, and how empirically persuasive. It is outlandish to imagine, as the eliminativist overriding thesis has it, that the deliverances of social science should undermine our faith in the image of ourselves as intentional, thinking systems.

This is outlandish in some part, because of the light-weight quality of most aggregate social science: the fact that there are few results available there that are robust across different schools of thought, let alone results that suggest an overriding of intentional regularities. But the eliminativist overriding thesis is mainly outlandish, because of the weight that we attach, and ought to attach, to intentional psychology. It should take a lot to persuade us to give up thinking of ourselves as intentional subjects, given the fact that we have managed thereby to make sense of one another for a very long time. Our intentional psychology must represent one of the best-tested programs of understanding that has been devised by human kind.

One further thought serves to bolster this consideration. Not only have we managed to make sense of one another with the help of intentional psychology; we have also managed to build cultures on the basis of that sense-making. You can't play chess with a computer unless you adopt the intentional stance: move to an electronic stance under which you predict its moves from its electronic states and you are no longer playing the game (Dennett 1979, essay 12). The lesson extends to human beings and to the many activities that parallel chess in the appropriate regards. It is only because we can represent one another successfully as intentional subjects that we can play games, make conversation, and pursue the various forms of interaction that give life its interest. The eliminativist version of the overriding thesis would raise serious questions about the appropriateness of such cultural forms, given its rejection of the folk psychology that supports them. This surely ought to tell heavily against it (Macdonald and Pettit 1981, chap. 2).

Like the extreme version of the overriding thesis, the extreme version of the opposition is straightforward in intent. It holds that we are mistaken to imagine that there are structural regularities at work in social life, in particular that we are mistaken to think that intentional responses are ever governed by laws that do not supervene on intentional laws: the presence of regularities that are causally discontinuous with intentional laws is hard to deny—examples manifestly support it—and so logical discontinuity is the natural target for the extreme position. The standard tradition of recent individualism adopts this position (O'Neill 1973; see Mellor 1982). Individualists have argued that the only regularities postulated in social science, and indeed folk sociology, are reducible to intentional regularities and so are not genuinely social-structural—in particular, not genuinely structural—in character (James 1984, chap. 1). They are reducible in the sense that the social-structural properties involved in those regularities can be defined in the terms of intentional psychology and that the psychologically redefined laws can then be derived from familiar psychological laws (Nagel 1961, chap. 11).

Strictly, this tradition is too hard on itself. In order to mount an extreme form of opposition to the overriding thesis, it is not necessary to be able to provide an effective reduction in this sense of the regularities countenanced in social science. By our account a social-structural regularity must govern intentional responses without being logically—or indeed causally—continuous with them. All that is needed for the extreme attack on the overriding thesis is an argument that the regularities that are apparently social-structural prove on examination to be supervenient on intentional regularities alone, at least absent a complication like the evil demon we imagined earlier. Such an argument does not require a definition and derivation of the social-structural regularities of the kind mooted in the literature. It will be enough if supervenience can be defended; an effective reduction is not strictly necessary.

The extreme opponent of the overriding thesis will have to be able to argue that those regularities that appear to be social-structural turn out, after all, to be regularities of a kind that is logically continuous with intentional laws: regularities such as that which relates the general preference not to be in an ethnic or religious minority to total residential segregation. With such regularities we may say that while they alert us to interesting patterns, the patterns in question are ones that are fixed in place just by the fact that people conform to familiar intentional laws. The information provided in the statement of the regularities is information that is available in principle, though not necessarily in practice, from the information that people conform to those intentional laws. If we know that the world belongs to the class of possible worlds where we human beings conform to those laws, then learning about the higher-level regularities does not involve learning that it belongs to a narrower class still: the higher-level regularities obtain—though perhaps only emptily, as we saw earlier—at every world where the intentional laws are satisfied.

But the extreme form of opposition to the overriding thesis, even understood just as an assertion that apparently structural laws are suitably supervenient on intentional, is not compelling. Look at the sorts of examples we gave earlier. If they are sound, as we are assuming, then the world is one in which unemployment leads to crime; the social utility of stratification, to its continuation; the fact that an international initiative is in the economic interest of a state, to its adoption; the presence of a Protestant ethic, to capitalism; and so on. But the fact that the world is of this kind is surely not fixed just by the fact that familiar intentional regularities obtain here. So long as we remain the intentional sorts of creatures we are, the segregation regularity that we discussed is more or less bound to obtain. But surely we could remain subject to the familiar intentional regularities without things being so organised that these other linkages are preserved. We can imagine things changing in such a way that the linkages are broken without imagining that we human beings fall under the sway of a different psychology. We can imagine the weather being so much better, for example, that unemployment does not generate crime. And so on in the other cases. The point has been made before and will be illustrated at greater length later.

Beyond the Limited Positions

In view of the considerations mustered against the extreme version of the overriding thesis, I am prepared to countenance the existence of intentional regularities. And in view of the considerations that tell against the extreme opposition to the overriding thesis, I am happy to recognise the reality of social-structural regularities. We return then to the familiar scenario. Intentional regularities and structural regularities both govern people's intentional responses. So what is the relationship between them? The limited positions on the defence and opposition sides assume that the regularities, or at least some of the regularities, are in conflict and they maintain exactly contrary positions. The limited defence says that when the regularities conflict, the structural laws override the intentional; the limited opposition holds that when the two regularities conflict, the structural laws are overridden by the intentional.

I am not impressed by either of these positions, and for the same reason in each case. Both intentional and structural regularities are causal, in the sense that they govern causal sequences or processes. If they conflict with one another, as in the picture shared by the limited positions, then they represent rival causal dispensations: they represent competing varieties of causally relevant properties, however relevance is understood; if you like, to use a terminology that is often invoked here, they represent competing varieties of causal power. Now there are competing sorts of causal power, and therefore of causal regularity, as we all learned in our introduction to science. There are the powers associated with gravitational and electro-magnetic fields, for example, and these may pull against one another, as when the magnet lifts the object that would otherwise stay on the ground. But there is reason to deny that there are competing sorts of causal power, and competing causal regularities, in the sense required here: in the sense that the same antecedent is linked to different consequences by rival laws. This is, as I shall argue, that all non-physical causal regularities supervene on the regularities and related conditions that actually obtain at the physical level.

I need to say something on what I mean by levels, and on what in particular I mean by the physical level. Intuitively, the higher and lower orders of states that we discussed in the first chapter represent different levels. Thus the elasticity of an eraser is at a different level from the molecular structure that underpins it, given that it relates to molecular structure as higher-order to lower-order state: to be elastic is to be of a molecular structure—maybe this, maybe that—which ensures fulfilment of a suitable condition: bending under a certain pressure. Intuitively again, the different realms of scientific inquiry represent different levels: there is the physical level, corresponding to physics; the chemical, corresponding to chemistry; the biological, corresponding to biology. Equally, we may say, there is the intentional level, corresponding to intentional psychology, and the structural level, corresponding to aggregative social knowledge.[4] The properties in such realms may not be defined by reference to one another in the fashion of properties at different orders: say, as elasticity is defined by reference to molecular structure. But equally with properties at different orders, we tend to describe them as belonging to different levels.

Taking such examples as my guide, I make three defining assumptions about levels. The first is simply the base assumption that each level is characterised by the properties that figure there and by the states and other entities constituted by the instantiation of those properties. The second and third assumptions go on respectively to dictate what makes levels different and what makes one level lower or higher than another.

The second assumption is that properties belong to the same level, or to different levels, depending on their causal congruence with one another. Suppose that two properties are both causally relevant—by whatever test of relevance—to the same result. They will belong to the same level if that means that the factors they constitute always relate as parts of the same causal whole or as stages (or parts of stages) in the same causal chain. Otherwise they will belong to different levels. We saw earlier that properties of different orders are of different levels by this criterion: the elasticity and the molecular structure of the eraser are both relevant to the bending but they do not combine to produce a joint effect and neither factor is in the causal ancestry of the other. Equally properties from different scientific realms will belong to different levels on this account. One and the same thing can be explained in terms of physics or chemistry but the physical and chemical factors will not relate as parts of the same causal complex or as earlier and later stages of the same causal process. Again, one and the same thing can be explained in intentional and structural terms without the factors involved relating as parts of the same causal complex or chain; this is only what we would expect, given the causal discontinuity of structural with intentional regularities.

The third assumption dictates, not what makes levels different, but what makes one level lower or higher than another. I assume that one level counts as lower than another just in case the properties that figure there are causally relevant in a more basic fashion than the properties that figure at the other level. What is the appropriate metric of the causally more basic? This matter is better left open so far as the definition of levels goes. But we should remark that on the assumptions made in the first chapter, properties of lower orders are properties of lower levels. A higher-order state like elasticity is causally relevant to something's bending only so far as a suitable molecular structure is relevant; the elasticity belongs to a causally less basic, and therefore a higher, level than the molecular structure.

So much for levels in general. But I need to comment on what I mean by the physical level in particular. The notion of the physical level—the level at which physics operates—is an idealisation. Physics is conceived of as the general or comprehensive science, the science that deals with everything in the familiar world, unlike the special sciences that concentrate on entities large enough to have a chemical character, organised enough to count as living, and so on. I assume that everything in the spatio-temporal world is composed in some way of the sorts of entities that physics—strictly, micro-physics—charts. And I assume that physics, or at least the complete physics, encompasses all the relevant properties, intrinsic and relational, of those parts: all the properties relevant to how the parts behave. Thus physics says something important about everything—it addresses everything *qua* composed out of physical parts—even if, intuitively, it does not say everything about everything (Crane and Mellor 1990, p. 191).[5]

My working assumption will be that at the micro-physical level the entities to be found are those of which contemporary physicists speak and that the relevant physical powers are the properties physicists describe as mass, charge, spin, and the like. But this is strictly not necessary. I need not deny that the actual particles and powers may turn out to be quite different from those currently acknowledged. I need not deny even that there may be an infinite progression downwards in the particles, and perhaps also in the powers, to be identified, though if there is it would be better to speak of the physical realm than the physical level.

I hold that all non-physical regularities are supervenient on the actual physical regularities and other actual physical conditions. The claim is that we cannot imagine keeping the world unchanged in regard to relevant physical features, while altering the patterns that operate at chemical, biological, psychological, or structural levels. Fix those physical regularities and conditions and the other regularities will be fixed too. Or at least this will be so, provided that no non-actual, non-naturalistic stuff is added to the world: no stuff of a Cartesian kind that might affect the performance of psychological systems; and similarly, no stuff of a kind that might introduce autonomous chemical, biological, or structural forces. As things actually are, so the supervenience claim goes, once we have put the physical regularities and related conditions in place, we have done all that is necessary to ensure that the other regularities are also in position. If we consider any two worlds that are identical in obeying the actual physical regularities and in satisfying the actual physical conditions then, if we assume that neither world contains non-actual, non-naturalistic stuff, those worlds will also be identical in regard to non-physical powers and regularities.

Why do we say that the supervenience base for other regularities includes, not just the actual physical regularities, but also other, related physical conditions? The relevant point was made earlier, when we said that the obtaining of chemical laws plausibly requires, not just that the laws of physics obtain, but that various physical constants assume certain values; it requires that the world in which the physical laws operate satisfies certain fixed parameters, certain fixed boundary conditions. The point generalises to the obtaining of all non-physical laws. It is likely that certain non-physical regularities obtain reliably in the actual world, not just so far as the world operates in accordance with the actual physical regularities, but also so far as suitable boundary conditions are fulfilled. It is often said, for example, that life would not have been possible unless certain physical constants—constants like the speed of light, that are not fixed by any laws—had been as they actually are. On one interpretation this would mean that the biological laws could not obtain, even obtain emptily—that is, without there being any actual biological systems—unless the constants happened to assume certain values.

The supervenience claim connects in an obvious manner with the issue raised by the overriding thesis: the thesis that structural regularities override intentional. If we endorse the supervenience of non-physical regularities and powers on the actual physical regularities and conditions, then there is no room for the conflict between non-physical regularities—no room for the conflict between intentional and structural regularities—that is envisaged in the limited positions on the overriding thesis. The physical regularities form a coherent set and if their fixation, in the context of associated conditions, means that the intentional and that the structural regularities are both wholly in place, then those two sets of regularities cannot conflict with one another. The limited doctrines assume that there is such conflict, differing on whether the intentional regularities beat the structural or the structural the intentional. The supervenience claim undermines both positions, for it gives the lie to the assumption they hold in common.

What is there to say in favour of the alleged supervenience and against the assumption shared by the limited positions? I think that there are three grounds for endorsing the supervenience. One is that it is intuitively compelling. How could the world remain unchanged in regard to the actual physical regularities and other conditions, but come to differ in chemical or biological or psychological or structural patterns? We cannot begin to imagine how non-physical causal pattern could vary independently of the physical in the manner that a denial of supervenience would require.[6]

The second ground for the supervenience claim takes us back to the assumption, already noted, that everything else is composed in some manner from physical parts and that physics encompasses all the properties and relations of those parts, which are relevant to how they behave. This means that everything is composed of parts whose dispositions to interact with one another, and whose dispositions to act in concert, are comprehensively subsumed by the regularities tracked in physics. But if everything is composed of parts whose dispositions are governed by physical regularities then, miraculous overdetermination aside, all the non-physical regularities that govern things must reflect the effects dictated by the physical regularities, though perhaps only effects dictated in the context of certain boundary conditions. And so all the non-physical regularities that govern things must supervene on the physical regularities together with any relevant background conditions. Put the physical underpinnings in place and the non-physical regularities will follow.

The third ground for the alleged supervenience takes us back even further in our discussions, to the program model introduced in the first chapter. The program model was presented as an account of how properties and states at different orders can be causally relevant at the same time to a given effect. It solved the problem of explaining how those factors collaborate in the production of the effect, given that they do not collaborate as parts of the same causal whole, or as links in the same causal chain. It explained how the elasticity of an eraser can be relevant to a bending that is produced by its molecular structure, how the boiling temperature of water in a closed flask can be relevant to a cracking of the flask that is produced by this or that vibrating molecule, and so on. The higher-order state is relevant in each case because it more or less ensures that there will be a causally more basic, lower-order state present— maybe this, maybe that—sufficient to produce the effect in question.

The probem raised by causal collaboration across orders arises generally with causal collaboration across levels. Levels are distinguished by precisely the sort of causal insulation that raises the problem for orders, the problem that the program model is designed to solve. How are we to think of properties and states at different levels as collaborating causally in the production of something? In particular, how are we to think of them as collaborating with the physical properties and states of the physical antecedents of such an effect? If the program model solved the original difficulty, we naturally return to it to solve this more general problem. We will naturally suppose that when chemical or biological or psychological properties or states are causally relevant to something, something to which physical antecedents are also relevant, then they are relevant so far as their presence more or less ensures the presence of a suitable, causally more basic physical antecedent: an antecedent sufficient to produce the type of effect in question.

But if we do endorse the program model of causal relevance in the general case, then we must endorse the supervenience of all non-physical regularities on regularities of a physical kind. It is only if such supervenience holds, that we can expect non-physical factors to program for effects that are produced by physical antecedents. Under the generalised program model, the actual, physical regularities provide the resources in virtue of which higher-level regularities obtain; or at least they do this in the context of appropriate background conditions. It is not possible that one and the same set of physical regularities and associated conditions should leave it undetermined whether one or another set of regularities obtains at some higher level. We must expect that if we fix the world in the relevant physical respects, and if we add no stuff of a non-physical kind to the world, then we will also have done everything necessary to fix the non-physical regularities that obtain.

The upshot of all these considerations is that it is surely plausible to regard all non-physical regularities as supervenient on the actual physical regularities, together with certain background or boundary conditions. Indeed, more strongly, the upshot is that the physical regularities and associated conditions provide the basis in virtue of which the non-physical regularities obtain. There can be supervenience without this sort of dependence: any two worlds in which I am a philosopher are worlds in which the mathematical truths hold—they hold at all possible worlds—so that strictly the holding of the mathematical truths supervenes on my being a philosopher. But the supervenience for which we have argued is not of this kind. It is a bottom-up supervenience, as we may call it, rather than a top-down variety: it is a sort of supervenience that is explained by the relation between the subvenient level and the supervenient, not one that is generated, as in the mathematical case, by anything like the necessity of the supervenient truths.

The thesis that all non-physical regularities are supervenient in this bottom-up way on physical regularities and conditions means that the physical level is causally basic; by our criterion it is the lowest of all levels. Indeed it means that fundamentally, as we might put it, all the pattern that is reflected in the different regularities we discover is provided at the physical level. This is a doctrine of causal fundamentalism. Higher-level powers are powerful in virtue of the operation, perhaps under suitable conditions, of the actual physical powers. They may be otherwise similar—they may count as powers in the fullest sense—but they generate only patterns that are independently supported by those physical powers.

Our working assumption that current physics is on the right track means that physical powers do not progress infinitely downwards, that they come to an end at a certain level: say, with properties like mass and charge and spin. Under this assumption, physical properties—or at least the basic physical properties—are causally relevant in a basic, though perhaps not unique fashion.[7] The architecture of causality means that all non-physical powers are higher-level, causally less basic properties. They are causally relevant to what happens but not at the fundamental level at which the basic physical properties are relevant (see Pettit 1992).

We may return, finally, to the evaluation of the limited positions on the overriding thesis. Because I endorse causal fundamentalism, and in particular because I accept the physical supervenience of non-physical regularities, I can take a quick line with the limited positions defended respectively by individualists and collectivists in regard to the overriding thesis. Those who hold these positions both assume that while intentional and structural properties really can be causally relevant, while they both represent higher-level powers, the powers involved can come into conflict. The limited individualist position is that in such a case the intentional regularities triumph, while the limited collectivist view is that the structural powers prevail. Both parties are mistaken, for it is incoherent to think that higher-level powers might conflict in the manner envisaged.

Both sorts of powers, both sorts of regularities, obtain in virtue of the same physical regularities and related conditions, and so it makes no sense to think that they might pull in different directions. If they were to pull in different directions, then the physical powers as a whole would be pulling against themselves. In view of endorsing causal fundamentalism, we must reject both of the limited positions in regard to the overriding thesis. We must reject any views that would countenance intentional and structural powers, as they do, only to postulate that the powers can enter into conflict with one another.

A New Perspective

Where does this leave us? We have rejected the extreme individualist claim that there are no social-structural regularities and the extreme collectivist counterpart that there are no regularities of an intentional kind. And now we have rejected the limited individualist and limited collectivist claims, that one or other sort of regularity prevails in the case of a clash between the two. So what do we say on the issue between individualists and collectivists, in particular the issue associated with the overriding thesis. Is that thesis true, as collectivists think? Or are individualists right to say that social-structural regularities do not override intentional and so do not compromise the autarchy that intentional psychology assigns to individual subjects?

The upshot of our discussion is that individualists are right on this question, though not for the reasons put forward by the extreme or limited versions of individualism: not because there are no social-structural regularities, and not because intentional regularities override social-structural. There are social-structural regularities, as there are intentional; that is what emerged from our critique of the extreme positions. But social-structural regularities, as individualists say, do not override intentional, any more than intentional override social-structural; that is what emerged from our critique of the limited positions. So individualism wins after all. The physical powers fix the pattern of powers and regularities that rule at all levels, as causal fundamentalism teaches, and that means that there must be a harmony between levels. It cannot be the case that structural powers ever cause the intentional to be suspended or ever deprive individuals of the autarchy associated with the rule of the intentional.

It should be no surprise that the failure of all the four positions we have described makes a win for individualism. The individualist only has to deny the collectivist thesis that social-structural regularities override intentional regularities, both in its extreme and limited version. That both versions fail means that individualism is correct, even though the contrary doctrines also fail: even though intentional regularities do not override social-structural, whether in the extreme or limited manner.

Can the collectivist respond to our line of argument? I see only one recourse. This is for him to deny the supervenience of the structural on the physical by postulating a non-physical, purely structural stuff that would give structural powers and regularities a primitive status of their own: a status akin to that which Descartes gives intentional powers through positing the existence of a purely mental sort of substance. The more extravagant utterances of some social scientists may seem to suggest that this is what they are doing in recognising social-structural regularities. But recourse to such a hypothesis is surely a last and desperate resort. There is nothing even remotely persuasive to be said for the idea of a structural stuff that might invade and subvert the intentional realm in the way in which the Cartesian stuff is held to usurp those regions of the physical world in which our wills have an effect.

Strictly, it is not necessary to say anything more on the overriding thesis. We have done all that is required, as I see it, to establish the soundness of the individualist insight. But it will be useful, before leaving the thesis, if we address one lingering puzzle. The social-structural regularities countenanced in social science are causally and logically discontinuous with intentional regularities, like the social-structural regularities salient to folk sociology. How then do they relate to such regularities? Are they logically continuous with the intentional regularities, as those regularities operate under suitable boundary conditions? In other words, do they fit the model exemplified by the social-structural regularities of folk sociology in relation to intentional regularities, and indeed by all non-physical regularities in relation to physical? Or do they relate to intentional regularities in some quite distinctive way? I shall argue with reference to some examples that the boundary conditions model applies here too.

Among the examples we gave of social-structural regularities, a first sort is illustrated by that which links an increase in unemployment to a rise in crime, or by that which associates urbanisation with a decline in religious practice. We can imagine a world in which intentional regularities are satisfied—say, non-emptily satisfied—but in which these structural laws do not obtain; that is why we regard them as structural. Things might be so organised in a society that when unemployment occurs it is not the case, as it is in most of our societies, that there is an increase in the motive or opportunity for crime. Perhaps the weather is so good that the unemployed flock, without shame, to the beaches. Again things might be organised in such a way that when there is urbanisation there is not the paucity of churches, the break with the seasonal basis of liturgy, or the greater immunity from peer pressure to attend church, which is associated with the urbanisation we know. Perhaps urbanisation takes place in a planned, socially engineered fashion.

We recognise intuitively that were things different in any of these ways, then although intentional regularities would be unchanged, the structural regularities in question would fail. But to recognise this is to admit something that parallels the folk sociology case: that the structural regularities require the satisfaction of the intentional regularities together with the satisfaction of various background or boundary conditions that normally go without saying. Absent evil demon complications, the structural regularities supervene on the intentional regularities together with such associated conditions.

A second sort of example that we gave of structural regularities is probably the least persuasive and would be rejected by many. It relates the fact that something is socially functional—say, economic stratification or capitalist organisation—to its existence or at least persistence. In the next section we shall be looking at some models of natural selection that would make such regularities supervenient on intentional regularities together with a certain history of competitive selection. These selectional models would mean that though structural regularities do not override intentional, they do outflank them. But short of adopting such selectional models—and we shall see reasons not to adopt them—there are other suggestions available as to how a functional regularity might relate to intentional regularities (see Van Parijs 1981). Here is a suggestion that fits rather well with various points made in this book, in particular with points to be made about rational choice explanation in Chapter 5.

Suppose that something like economic stratification really is beneficial for a society, so that if any society begins to move away from that sort of organisation then many of its members suffer in some measure. Just the fact that people satisfy intentional regularities does not ensure that someone will monitor the move and call an effective halt, pointing to the disadvantage experienced. Apart from anything else, there will be a collective action predicament facing anyone who tries to call a halt: each may find the onerous task of protesting about the move unattractive and may want to leave it to others. But there might be institutional arrangements in the area of individual cooperation that would make alarm-sounding attractive and effective. And if that were the case then the satisfaction of intentional regularities, together with the satisfaction of such conditions, would ensure that the structural regularity obtains. Again the sort of picture projected in our hypothesis would be vindicated.

A third sort of example that we gave of structural regularities involves the action of group agents and is close enough to folk-sociological examples to fit the sort of story already told in those cases: the story invoking a gatekeeper-monitor arrangement as the relevant boundary condition. But there are other sorts of stories that may apply in such cases too. One is a local selectional story of a kind that parallels the global selectional stories we will be considering in the next section. It represents a possibility that might also sustain certain functional regularities of the kind just discussed.

How do certain economic firms come to be good at decision-making; how do they come, as is sometimes alleged, to maximise expected returns, or at least to maximise returns subject to certain constraints? A number of economists have argued that the decision procedures of firms in a competitive economy are selected for conferring economic success on those who use them and that this is what sustains the success of firms (Alchian 1950; Nelson and Winter 1982). The most successful firms preserve the decision procedures that brought them success, the less successful ones do not: indeed, if they do not go out of business, the less successful firms may well copy the procedures of the successful or inherit them in the course of a takeover. That being so, we can see that the procedures that bring success, the procedures that facilitate the maximisation of expected returns, may be selected for in the economic competition between firms. Or at least they may be selected for, if the competitive environment remains sufficiently unchanged over the period needed for competition to do its pruning. The social-structural law, if it is a law, that economic firms maximise expected returns obtains superveniently on the obtaining of intentional regularities, given as boundary condition the sort of competition described.

We have considered three sorts of social-structural regularities in an attempt to show how they can supervene on intentional regularities together with associated, boundary conditions. A last kind of example used to illustrate structural regularities is that which relates Protestantism to the rise of capitalism, or capitalism to the breakdown of family values. We can imagine a world in which the intentional regularities rule but in which these structural connections are not borne out: a world, for example, in which the domain of religion is insulated intellectually from the world of commerce and the world of commerce from family life. But in our world we do not find the intellectual insulation that would break the connections. On the contrary it is part of how things are organised with us, and may even be part of our biological constitution, that we carry over ideas from domain to domain, so that a concern with individual religious salvation can generate a concern for individual commercial success, and a ruthless habit of commercial calculation can generate a similar habit in the area of family relations. Thus our preferred hypothesis looks plausible here too. It appears that the structural regularities in question do not supervene on intentional regularities alone, because they supervene on the satisfaction of the intentional regularities together with suitable associated conditions.

This completes our discussion of the overriding thesis. The extreme individualist and collectivist doctrines, that there are no structural or intentional regularities, are both false. So are the limited claims that though there are such regularities, they occasionally come into conflict and then one or the other type prevails. There is no possibility of such conflict, because all non-physical regularities supervene on physical regularities and related conditions, as causal fundamentalism implies. This means that the individualist claim that structural regularities do not override intentional is sound after all; it is sound, but not for the reason canvassed in the extreme or limited doctrines. How do structural regularities relate to intentional, then? Absent evil demon complications, they obtain superveniently on the obtaining of intentional laws under appropriate boundary conditions.

Do Structural Regularities Outflank Intentional?

But if social-structural regularities are not overriding, do they perhaps compromise intentionality in the other way that we mentioned, by outflanking intentional regularities? That is the question we must address in this final section. I shall argue that there is no good reason to think that social-structural regularities have this outflanking effect.

The outflanking thesis is easy enough to introduce as an abstract proposal. We imagine all the psychologically possible configurations of intentional attitudes and actions: that is, all configurations that are possible so far as intentional psychology goes. In some of these configurations—in some of these possible worlds—intentional responses will be such that certain social-structural regularities will obtain, in others not. The proposal is that the social-structural regularities outflank intentional in the sense of having been selected for in a survival-based way. Some filtering or pruning mechanism is at work that ensures that although it is psychologically possible to find configurations under which the regularities do not obtain, this is not possible in fact. The mechanism ensures that the sorts of agents who would generate such configurations do not survive: they are filtered out. It ensures that the only people around are agents who will have the attitudes and display the actions required to sustain the social-structural regularities.

If social-structural regularities outflanked intentional in this selectional way, then there would be a sense in which people lacked the autarchy assigned by intentional psychology. There would not be any predetermining pressures at work on them of the sort that might be thought to override intentional regularities. But there would be predestining factors in operation that would give the lie to the autarchy that we naturally ascribe to them within the intentional-psychological perspective. For example, we naturally assume of intentional agents that if rational pressures leave a number of rival belief-sets tenable, then the subjects will be liable to hold any candidate set. But the spectre of outflanking structural regularities suggests that the agents may be predestined by the operative selection process to hold only one of the allowable sets; they may be predestined to favour that set as the default option: as the option to hold in the absence of intentional reason to the contrary. It may be, to put the matter otherwise, that if they did not favour that set of beliefs, then they would not be around.

We saw in the last section that there are various structural regularities that may be supported by the presence of a sort of filtering mechanism. For example, the regularities characteristic of the behaviour of officials of a political party may be supported by a gatekeeping arrangement that guards against the induction of uncongenial members and a monitoring arrangement that weeds out those who fail to behave appropriately. Again, the regularities characteristic of successful economic firms may be supported by a competitive history in which successful decision procedures are selected for: less successful alternatives are filtered out. Such filtering mechanisms do not compromise intentional autarchy, and the structural regularities they support do not outflank intentional regularities. The mechanisms affect the behaviour only of certain people, not of all the members of the society. They do not mean that only intentional subjects who are disposed to support the structural regularities will be around; they mean that only intentional subjects who are disposed to support the regularities will be found in the appropriate positions: that is, holding office in the political party or the economic firm. If filtering mechanisms are to compromise intentional autarchy and support an outflanking thesis, then they must mean that the surviving population in the society as a whole are disposed intentionally in such a way that the relevant structural regularities obtain.[8]

In order to understand the outflanking thesis properly, we need to move from the level of abstract proposal to that of specific hypothesis. We need to consider the different sorts of selection process that might be expected to support an outflanking, social-structural regularity. It turns out that only a very specific sort, and a sort that it is implausible to imagine at work, would ensure the compromise of intentional autarchy that collectivists allege. Other sorts of selectional process, and the social-structural regularities they sustain, can be countenanced by individualists and collectivists alike, because they do not offer any threat to people's autarchy.

The most familiar sort of selection is natural selection, as described in Darwinian evolutionary theory (Dawkins 1976, 1986; J. Smith 1986). Under this theory, the items that are selected are genes; we may ignore the question of how exactly they are individuated (Sterelny and Kitcher 1988). But though it is genes that are selected, they are usually selected for the effects they have, not in direct gene-gene competition, but in competition at higher levels: usually in the competition between individual organisms (Sober 1984). It is usually the fact that a gene benefits the individual to which it belongs, that ensures that the gene is selected for. Why is it such high-level effects that tend to govern the selection of genes? Because those effects are the most important influences on the gene's rate of replication in the given environment. The genes with the best rate are the ones that are selected for; that is a matter of definition. And the genes with the best rate are usually those that benefit the individual organism: specifically, as it is said, those that increase the organism's inclusive fitness.

Generalising from this familiar picture, we can see that in any selection process certain replicators will be selected in virtue of having replication-maximising effects at a certain level of competition. We can therefore distinguish between selection processes on at least two different dimensions. We can distinguish between them on the basis of the sort of replicator—and correspondingly the sort of reproduction and variation mechanisms—in question. And we can distinguish between them on the basis of the relevant level of competition.

The gene is one possible replicator. The only other sort that is relevant for our purposes is described by Richard Dawkins (1976) as a meme: the word comes from the same root as mimesis or imitation. The meme is the unit of cultural transmission in the way in which the gene is the unit of biological transmission. 'Examples of memes are tunes, ideas, catch-phrases, clothes fashions, ways of making pots or of building arches' (Dawkins 1976, p. 206). Examples might also include decision procedures, modes of punishment, routines of hospitality, arbitration devices, and other institutional constructions. And, extending into associated systems, they might even include patterns in the environment that distinguish a culture: say, the pattern whereby loss of employment means more time for the beach, the coming of rain creates a motive for rejoicing, tremors underground signify volcanic activity, and so on. Whereas the genes are biological replicators that reproduce themselves chemically and that vary with externally induced mutations, memes are cultural replicators that reproduce themselves via imitation, tradition and the like, and that vary with spontaneous or planned innovations or, of course, with exogenous shocks.

So much for different replicators. What now of the different levels of competition at which genes and memes may be selected? There are three salient possibilities: that genes—or, with appropriate matters changed, memes—are selected for their effects at a sub-individual level, in the competition between genes; that they are selected for their effects at the individual level, in the competition between organisms; and that they are selected for their effects at a super-individual level, in some competition between groups. The group-level possibility is particularly relevant to our interests, as we shall see. We need to illustrate what it would involve both with gene-based and meme-based selection.

Here is a scenario under which genes would be selected for their effects in group competition. There are a number of distinct groups in competition for survival; members of the groups are bound to stand or fall, disappear or survive, together. Differing genes make for differences between the groups, in particular differences that affect the groups' survival and the genes' replication. Those groups and those genes survive that have a reproductive advantage; the others die (Wynne-Edwards 1986). For example, it might be that the groups that survive are those in which people are prepared to go along with a rigidly hierarchical mode of organisation, giving total obedience to a single ruler. All non-hierarchical groups might die, and only hierarchical ones survive, in the intergroup struggle for a place in the sun.

The population-survival story is the only plausible scenario under which we could have gene-based selection at group level. What of meme-based, group-level selection? One possibility would mimic the sort of population-survival story just described, with memes replacing genes as the relevant replicators. An example, albeit a controversial example, is available in real-life social science. Stuart Piddocke has argued that the potlatch institution of the southern Kwakiutl, involving a ceremony in which individuals from different groups compete to confer valuable goods on one another, is an institution—a meme, in our terms—that, however it emerged, accounts for the survival of some populations rather than others; it assures the flow of food resources in time of need. This is to say that the meme is selected for at a level of group competition. Just those groups where the potlatch has a hold survive and in surviving they replicate the meme across generations. 'The potlatch has adaptive value in allowing the Kwakiutl to preserve their entire population, even when some groups faced extinction through starvation. The potlatch was thus "selected" and retained because it facilitated, and now promotes, survival' (Turner and Maryanski 1979, p. 86; for another example, see Van Parijs 1981, p. 86).

There are other sorts of meme-based, group-level selection but only the kind involving the survival or non-survival of populations will be suitable to sustain an outflanking thesis; only this kind of selection will offer a collectivist challenge to intentional autarchy. For an example of another variety of meme-based, group-level selection we need only consider the story mentioned in the last section as to how successful decision-making procedures may get selected for in economic competition between different firms. But, as already mentioned, that sort of story would do nothing to compromise intentional autarchy. It would mean that intentional agents who resist following suitable decision procedures are not likely to be found in board rooms, but it would not mean that agents of that kind are not likely to be around.

We have done a quick inventory of relevant selectional processes, dividing them up on two distinct dimensions. The upshot is a taxonomy with six categories, as in the following matrix.

	Individual-level	Unit-level	Group-level
Gene-based	1	2	3
Meme-based	4	5	6

I assume that this taxonomy catches all the selectional processes that might possibly be relevant to our concerns. The question before us is which, if any, of these processes might serve to give social-structural regularities the sort of outflanking status alleged by collectivists, the sort of status that would compromise the autarchy of intentional subjects. I shall argue that the only process capable of doing the job is a group-level, gene-based sort of selection.

In order to have a social-structural regularity that is outflanking in the manner required, we must have a group-level process of selection. The regularity is a characteristic of a group, not an individual, and if it is to be selected for—if genes or memes are to be selected for the fact of generating that characteristic—then the selection has to occur at the level of competition between groups. It must be that the groups that are distinguished by the regularity are, for whatever reason, the only groups that survive and thrive.

The point is best defended by example. Imagine an individual-level process of selection that had the effect, consistently with their obeying intentional regularities, of channeling people into particular patterns of response: say, a process of selection that induced people generally to accept the instructions of those who are perceived to be in authority. Imagine, further, that the fact that this characteristic is selected for ensures that a certain aggregate pattern obtains: say, a pattern linking certain claims to authority with displays of obedience. We would not say in such a case that the aggregate pattern outflanks the intentional psychology of the individuals involved. What outflanks that psychology, if it is outflanked, is the regularity whereby deference is individually beneficial, or was beneficial in the period of selection. The aggregate pattern is an incidental by-product of that individual-level regularity.

So much for the claim that a selectional process capable of giving social-structural regularities an outflanking status would have to operate at the level of groups. Now for the twin claim that it would have to be a gene-based rather than a meme-based sort of process. What sort of group-level process is required to show that individuals are predestined to act according to certain intentional regularities: such regularities as will sustain an outflanking social-structural regularity? What sort of process will allow us to say of the individuals involved that if they were such as possibly not to act as required by the structural regularity, then they would not be around: they would not be in existence?

Two premises would seem to be required to derive that conclusion (Jackson 1977). First premise: if the individuals were such as possibly not to act as required, then their ancestors would have been such as possibly not to act as required. Second premise: if the ancestors had been such as possibly not to act as required, then they would not have propagated. These propositions will support the required conclusion that were the individuals such as possibly not to act in accordance with the regularities, then they would not be in existence: they would have had ancestors that did not propagate. So the question is, what sort of group-level selectional process would ensure the truth of both of the premises?

Both gene-based and meme-based processes could support the second premise: they could indeed mean that if the ancestors in question had not been such as to act—more or less reliably—in accordance with the intentional regularities associated with the relevant gene or meme, then they wouldn't have propagated; the Kwakiutl example may help to make the possibilty vivid. But it turns out that only a gene-based story would support the first premise, as a little reflection will make clear. The first premise says that if the individuals involved now did not have a certain trait, then their ancestors would not have had it. But this means that the trait must be such that if the ancestors had the trait, then the individuals who are their descendants must have it; otherwise we could not argue from non-possession by descendants to non-possession by ancestors. And while the gene could serve to generate such a trait, the meme could not. With the meme we can imagine that a later generation moves away from it and ceases to display the responses—the trait—involved; thus we can envisage non-possession of the trait by the later generation without being forced to imagine that the earlier generation must also have lacked it. Under normal assumptions, we cannot imagine that a later generation moves away from a gene and so in this case we can argue that if a later generation lacks the genetically induced trait, then an earlier generation must also have lacked it.

The upshot is that as any selectional process fit to support the outflanking thesis must be a group-level process, so it must be a gene-based process too.[9] If there are social-structural regularities that outflank intentional regularities, and which therefore compromise the autarchy of intentional subjects, then they must be supported by a group-level form of genetic selection. It must be that in certain past societies, people were selected for the genes required to ensure that those societies had a certain benefit in the competition with others: that, unlike those others, they were characterised by the social-structural regularities in question.

What sorts of regularities might have been selected in the manner described? They will all have to be implicitly or explicitly functional, though perhaps only functional relative to the society at the time of selection. Suppose that the selectional process ensured that the sub-populations of the society are all disposed to respond in a certain way, with a social effect, S, under certain circumstances, C. Suppose, in other words, that the regularity selected relates C to S and does not look particularly functional. Even in this case we can see that the regularity is implicitly functional. For if C is related in a lawlike way to S, that is because circumstances C are just those circumstances in which it was functional at the time of selection for S to appear. Under the selectional story, the C-S regularity derives from a more basic regularity, which says that S occurs when it is, or at least was, functional.

We need an example, however fanciful, of a social-structural regularity that might have been selected in the manner required for the outflanking thesis. Consider the putative regularity whereby the members of a particular society tend to adopt a hierarchical mode of organisation in the different associations they form. This will be a social regularity, because it governs the intentional responses of a number of people. It will be a social-structural regularity, if it is not causally or logically continuous with intentional regularities. It is not causally continuous, for we may assume that it is not intended to identify a causal factor overlooked in intentional psychology. And it is not logically continuous, for we can imagine peoples who obey the same intentional regularities and differ in regard to this tendency; we need something more than the obtaining of intentional regularities to account for the regularity: something like a suitable selectional history.

Here then is a selectional story that might underlie the regularity. Let us suppose that in some past competition between groups, those groups thrived in which individuals were genetically predisposed to adopt a hierarchical organisation; in other words the better adapted groups were those in which this regularity ruled. If that sort of competition occurred, if it ensured that the alleged regularity obtained in those groups where it was functional, and if it accounts for the continued satisfaction of the regularity—this, because the society in question is peopled by descendants of the members of those groups that survived—then the regularity will outflank intentional regularities and compromise people's autarchy. We may think that members of the society are such that they might well have had a different organisational tendency. But if this regularity obtains, then we would be wrong. Were members such that they did not have that tendency, then under the story told, they would not be around at all. They would have had ancestors who would have failed to propagate in the period of group selection.[10]

I hope that enough has been said to show that if there are social-structural regularities that outflank intentional regularities and which therefore compromise people's autarchy, then they must be the product of gene-based, group-level selection. I hope further that the fanciful example suggested will give an idea of how outflanking regularities might have emerged, and might diminish intentional autarchy. It remains now to ask whether there are likely to be many such social-structural regularities satisfied in our societies today. I shall argue that there are not.

There are not, for two simple reasons. One is that the conditions required for the sort of group-level selection specified are very special and are very unlikely to have been fulfilled. It is unlikely that there were distinct human groups that competed for survival over a suitable evolutionary period. Perhaps such conditions were fulfilled for the ancestors of some of today's human beings. But that may not help much, unless there also happen to be some societies today where all the inhabitants are descendants of just such ancestors; in mixed societies the social-structural regularities required will not reliably obtain, let alone have an outflanking status. These considerations combine to throw serious doubt on the empirical presuppositions of the outflanking thesis.

The second reason why we are not likely to find many social-structural regularities of the kind postulated in the outflanking thesis is this. Even if we assume that there once was the required competition between human groups, we must recognise that the regularities that would have been selected for in that process of competition are probably few in number, given the simple nature of the groups that would have been involved; and more than that, we must also recognise that there may be no occasion in our advanced societies for such regularities as would have been selected for to come into play. There could not have been selection for most of the significant regularities that appear with complex societies: say, regularities to do with stratification, urbanisation, minority status, political affiliation, and the like. And of the regularities selected many would have little relevance under life in complex societies: for example, there may be little occasion in complex societies for manifestation of the sort of hierarchical tendency mentioned in our example.

This brings us to our conclusion. The outflanking thesis looks to be no more promising a position for the collectivist to adopt than the overriding thesis we rejected in the last section. Once we see what it requires, once we see in particular that it requires an unusual gene-based, group-level selectional history, then we should recognise that it is not a sensible way to go. The upshot of our discussion then is that there is no reason why we should adopt the collectivist position on the vertical issue in social ontology. There is every reason, as the first part of the book will have made clear, why we should credit human beings with the sort of autarchy that intentional psychology ascribes.

But while there is every reason to be individualists in this way, the individualism we have adopted may not be altogether uncongenial to those who describe themselves as collectivists. We acknowledge the reality of the social-structural regularities that the collectivist cherishes; we break in this regard with perhaps the dominant individualist tradition. We acknowledge that individuals may depend on their relations with one another for the appearance of distinctive human capacities; that the social entities individuals constitute may exercise a reciprocal influence on those agents, structuring their perceptions, motives, and opportunities; and that the discoveries of social science may lead to revisions in our intentional psychology, shaping our sense of what count for example as normal conditions. And we acknowledge, finally, that there may be some truth in certain versions of the dispensability and inevitability theses espoused by many collectivists. Thus, while our individualism is a distinctive commitment, it is not as controversial a commitment as it may at first seem. It requires us to turn away from the overriding and outflanking theses that we have been discussing. But the loss of those doctrines may not look like a serious deprivation, even to those who have been loudest in proclaiming themselves opponents of individualism.

Notes

1. Jon Elster (1985, p. 18) distinguishes between sub-intentional and supra-intentional causes and there is a loose mapping between his categories and mine. It may be worthwhile offering a brief comment. Where causally continuous regularities obtain, they identify potential sub-intentional causes; specifically, they identify potential causes of social conditions—conditions involving intentional responses on the part of a number of people—and causes that do not undermine intentional regularities. Where causally discontinuous regularities obtain—be they logically continuous or not—then what Elster describes as supra-intentional causation may be involved: it may be that the regularities track the outcomes of individual actions, or of individual actions combined with certain background conditions. But that is only one possibility: alternatively, for example, the regularities may track the outcomes of intentional attitudes as distinct from actions.

2. I assume with the party, and with the sub-culture that I go on to discuss, that it is not definitional that the members will behave according to the regularities: that we can have the social entity without such regularities being fulfilled. I assume, in other words, that such entities are not functionally characterised relative to the intentional, as intentional states are functionally characterised relative to the physical. On a functional characterisation, a group would be a political party, for example, through its members being intentionally such as to satisfy the associated regularities.

3. I have made mistakes, or at least expressed myself badly, on this issue in previous writings. Ruben (1985, appendix) rightly takes me to task for some remarks in Pettit 1980. The position that I wished to defend there is better set out, and indeed set out in a way that I still find tolerable, in Pettit 1984.

4. There will also be levels within a realm like the intentional, if we allow that intentional states may relate as personal and sub-personal states are envisaged relating at the end of Chapter 1, section 2.

5. Under a plausible supervenience doctrine, the complete physics will chart a set of truths about the actual world that entails all other truths about the world. But even this doctrine does not mean that in an intuitive sense physics says everything about everything. To give the coordinates of its corners may be to entail that a straight-sided figure is a triangle of such and such an area, but it is not intuitively to say that it is a triangle of that area.

6. Someone may respond: most laws are idealised, abstracting from various, inevitable real-world effects—effects like friction and the like—and if the laws at different levels abstract from different sorts of effects, then the alleged supervenience may not hold. I agree that the scientific formulations of laws are often idealised in this way but I assume that the idealised laws formulated at different levels for a given world all point to objective patterns that obtain at that world; if we articulate these patterns in

idealised formulae, that is because they are so complex. My supervenience claim is meant to bear on the patterns, not on the formulae. I was helped by conversation with Peter Menzies about this point.

7. If the assumption is wrong there will still be causal relevance, contrary to what Block (1990a, p. 168) claims. There will still be programmatic causal relevance, with properties at each level being causally relevant in virtue of the relevance of properties at a lower. Consider an analogy. An aggregate-level object is positioned in space in virtue of the spatial position of its parts. But if it turns out that there is an infinite progression downwards in parts, that will not mean that there is no such thing as spatial position.

8. What of the case where a structural regularity does not require everyone to fall in line, only a certain percentage: say, two-thirds? This would not tell against the autarchy of those who do not fall in line but I presume that it would count against the autarchy of those who do. Thus I shall not discuss the case separately.

9. Here I take a different path from that taken with regard to what is described as the preemptive thesis in Jackson and Pettit 1992c: this paper allows the sort of meme-based selection illustrated by the Piddocke claim to represent a preemption of intentional psychology.

10. There is a selection for worlds where intentional regularities apply—initial conditions are fulfilled—which mean that the appropriate consequent follows: people continue to trust their leaders. We can say that the structural regularity is a selection principle for initial conditions. Putting the point in this way, we can see a parallel with proposals in physics. Consider, for example, the recent proposal that the second law of thermodynamics is a selection principle that ensures that only such worlds obtain as go to equilibrium and entropy: 'the second law becomes a selection principle of initial conditions. Only initial conditions that go to equilibrium in the *future* are retained' (Prigogine and Stengers 1984).

4

For Holism, against Atomism

At the beginning of the last chapter, we distinguished between two issues that are often confused in the discussion of how things stand in the social world: that is, in social ontology. One of those issues is the vertical issue, as we characterised it, of how far participants in social life are affected from above: in particular, how far aggregate social regularities compromise their intentional-psychological status as autarchical agents. The second of the issues we distinguished is the horizontal one of how far participants in social life are affected, not by higher-level forces, but rather by one another. We turn to that issue in this chapter.

I argued for the individualist rather than the collectivist position on the vertical question. In this chapter I shall be arguing for what will seem to many like a reversal of standpoints on the horizontal question: for a holist rather than an atomist position. The upshot of the two chapters will be a social ontology that may be described as holistic individualism. This social ontology derives from the philosophical psychology defended in Part I of the book and it supports the line on social and political theory that will be developed in Part III.

The discussion in this chapter falls into three sections. In the first section I discuss the horizontal issue itself, arguing for its distinctness from the vertical issue. In the second section I present my argument in support of social holism, drawing heavily on our discussion of thought in the second chapter. And in the third section I look at how far the sort of holism defended is compatible with a realist view of the world: in particular, how far it leads towards a relativist vision.

The Issue

Background

The collectivist view that individual agents are compromised by aggregate social regularities is a creation, as we have seen, of the past two hundred years. It was given perhaps the most important boost by the avalanche of numbers in the nineteenth century, as social statisticians charted every aggregate variable they could imagine and began to discover regularities in the correlation of such variables with one another. But collectivism was also encouraged by the discovery, as we saw, of social entities that were relatively new to folk sociology and that pointed towards novel kinds of associated regularity: entities like national states, the cultures of different peoples, and corporate persons that have rights and obligations in their own name.

It is of some interest that the holistic view that individuals depend on one another in distinctive ways is also a creation of the past two hundred years or so. Aristotle insisted of course that the human being was a *zoon politikon*, a social animal, and this theme recurred through the long period when his influence was paramount. It appeared for example in phrases that became almost proverbial in status: 'Morals makyth man', 'No man is an island', and so on. But though the received wisdom of two millennia had been that human beings were essentially social, still this theme only achieved the status of a salient philosophical doctrine from the seventeenth and especially the eighteenth century on.

Perhaps the reason that social holism first became a salient doctrine in this period is that it was only in that period that a view appeared against which it might be pitted. I am thinking of the view that a human being could have the full status of a human being—however this is understood—even if she lived outside of society. This view appeared probably for the first time in the seventeenth century, with the notion that political and social order, if it was legitimate, had to be the product of some tacit contract between presocial individuals. Social atomism was forced on many thinkers by the espousal of such a radical contractarian vision, at least in the contractarianism of a thinker like Thomas Hobbes; the vision only made sense if human beings did not depend on society for their status as human beings and therefore as potential contractors (Hobbes 1968).

Such social atomism was almost certainly encouraged by the discoveries of people who seemed to many Europeans to live more or less in the wild. Those discoveries nurtured the view that actual society must have evolved from a contract made by individuals in a state of nature. It may be no great accident that, in Charles Taylor's words, 'the great classical theorists of atomism also held to some strange views about the historicity of a state of nature in which men lived without society' (C. Taylor 1985, p. 190)

Perhaps holism could only become a prominent philosophical doctrine once there was an atomistic picture of human beings available to which it might offer an alternative. The fact is, in any case, that it first came to prominence among seventeenth and eighteenth forerunners of German romanticism like Vico and Rousseau and Herder (Berlin 1976). These were all thinkers who were familiar with the atomistic vision of individuals and society, and they self-consciously emphasised a thesis that challenged such atomism. They held that people were dependent on language for the capacity to think and that the language on which they were dependent was essentially a social creation (Wells 1987). They maintained that people depended on one another's presence in society to be able individually to realise what is perhaps the most distinctive human ability.

In stressing the dependence of thought on language and of language on society, these thinkers were taking issue directly with Hobbes. For someone like Rousseau it was self-evident that language, which he assumed to be social, was required for thinking. This is what created for him the famous chicken-and-egg problem: 'which was most necessary, the existence of society to the invention of language, or the invention of language to the establishment of society' (Rousseau 1973, p. 63; also see Wokler 1987, chap. 4). But where Rousseau thought that language—and therefore society—was essential in the development of thought, Hobbes saw it as something that was only of instrumental value, albeit of enormous instrumental value; it served a crucial mnemonic and communicative role (Hobbes 1968, pp. 101, 111; also see Hampton 1986, chap. 1). Thus Hobbes could write: 'besides Sense, and Thoughts, and the Trayne of thoughts, the mind of man has no other motion; though by the help of Speech, and Method, the same Facultyes may be improved to such a height, as to distinguish men from all other living Creatures' (Hobbes 1968, p. 99).

The romantic thesis that thought is dependent on language and that language is an essentially social creation came to fruition—and, arguably, overripened—in Hegel's notion of the *Volksgeist*: 'the spirit of a people, whose ideas are expressed in their common institutions, by which they define their identity' (C. Taylor 1975, p. 387). It came thereby to influence a variety of thinkers, from Marx to Durkheim to F. H. Bradley, who all stressed the social constitution of the individual. They claimed that the individual's relations with her fellows were not entirely contingent or external; some of those relations were internal or essential, being required for the individual to count as a full person. As Bradley (1876, p. 173) puts it: 'I am myself by sharing with others, by including in my essence relations to them, the relations of the social state'.

This broadly Hegelian tradition of social holism encompassed many thinkers of the past century, including Durkheim himself, who would have been loath to acknowledge any Hegelian sources. These thinkers looked to history and sociology as the likely sources of enlightenment about human beings, where their atomistic opponents tended to look to economics. The tradition found expression in a commonly endorsed line, that social atomism casts the human being in an artificially, abstract mould—typically, in the mould of *homo economicus*—neglecting all the concrete social features by which human beings are distinguished.

'To know what a man is', Bradley (1876, pp. 173–74) put it, 'you must not take him in isolation. He is one of a people, he was born in a family, he lives in a certain society, in a certain state. . . . The mere individual is a delusion of theory'. Bradley and others in the tradition derided this abstract individual, setting it in contrast to the concrete human being. The concrete human being is intrinsically social, reproducing within himself the mind of his community: 'the tongue that he makes his own is his country's language, it is (or it should be) the same that others speak, and it carries into his mind the ideas and sentiments of the race (over this I need not stay), and stamps them in indelibly' (Bradley 1876, p. 172).

It is probably unfortunate that many of the defenders of the holist thesis in this loose Hegelian tradition were also, in different ways, collectivists; or at any rate, like Hegel himself, they were taken to be collectivists. This coincidence must have made it very easy to take the holist doctrine as equivalent to the collectivist thesis. It must have nurtured a confusion between the two positions.

The misfortune would have been compounded by the particular form in which romantics and Hegelians chose to give expression to their holism. They insisted that there was a *Volksgeist* abroad in any society, a *conscience collective*—the phrase is Durkheim's—on which individuals drew for the resources of their thought. But there is a notorious ambiguity in any such notion, for it may refer to a people's thinking or mind as well as to a people's thought: to a collective state or medium of consciousness as well as to a collective content. It would have been easy, even natural, to think that holists who were anxious to stress the importance of the collective content of individual thought were collectivists who wanted to emphasise the role of a collective state or medium of consciousness. It would have been easy and natural indeed for holists themselves to become confused on this point. Thus there is some evidence that Durkheim was unclear about the distinction and that he slipped, in speaking about the *conscience collective*, between statements of a holist kind and statements of a collectivist intent (Lukes 1973, pp. 11-12).

This short account of the historical background to the issue between holists and atomists, like the historical account of the issue between collectivists and individualists, should serve to give us a feel for the question under debate. But a feel is probably all it can give for, again as in the other case, the terms of the traditional debate are not as well defined as we would wish. They nurture an engaging rhetoric, setting the abstract individual allegedly credited by social atomists in conflict with the socially concrete subject cherished by their holistic opponents. This rhetoric has a natural appeal and remains with us still. But the traditional terms do not define a dispute with the precision that invites an analytical resolution. This is already evident in the fact that those terms do not even serve to keep the issue between holists and atomists clearly distinct from the issue that divides collectivists and individualists.

There is one further point to make before leaving the historical background. Societies can be described as more or less atomistic or holistic in a sense that derives from that in which a social theory is atomistic or holistic. An atomistic society in this sense will be one in which institutions are devised, and norms are conceived, on assumptions that are more or less atomistic in character: at the least, they are assumptions it is easier to endorse if you are an atomist and think of individuals as self-enclosed monads (see C. Taylor 1989). Western society in the past couple of centuries is often said to be distinctively atomistic in this sense; it is contrasted in these regards with earlier Western society and with primitive and Oriental forms of community (Dumont 1986). If that claim is correct, and I do not doubt that it is, then this feature of the modern West must have had an important influence on the development of an atomistic vision of society; it must have helped to nurture the atomistic picture of individuals in community.

Explication

Charles Taylor (1985, p. 191) offers a contemporary account of what the holist thesis is.

> The claim is that living in society is a necessary condition of the development of rationality, in some sense of this property, or of becoming a moral agent in the full sense of the term, or of becoming a fully responsible, autonomous being. These variations and other similar ones represent the different forms in which a thesis about man as a social animal has been or could have been couched. What they have in common is the view that outside society, or in some variants outside certain kinds of society, our distinctively human capacities could not develop.

This explication, which remains broadly faithful to the Hegelian background, alerts us to two things that the holist must do in defending his position. First, he must isolate certain properties of the human being as distinctive and important. Second, he must argue that those properties require social relations in order to emerge and/or survive in an individual. Both of these moves require some comment.

What is it for certain properties to be distinctive and important: in Taylor's phrase, to constitute 'distinctively human capacities'? They must be properties that belong only or mainly to human beings, at least among the creatures with which we are familiar. And they must be properties such that we would count it a serious lack if an individual did not exemplify them.

Our discussion in the first part of the book nominates one obvious property that the holist might focus on: the capacity to think. Both intentionality and the capacity to think are important human properties. But intentionality may not be distinctively human, given our deflationary account of what it involves; it is a property that may belong to higher animals and even to artificial intelligences as well. The capacity for thought, however, the capacity intentionally to shape one's own intentional states, does appear to distinguish human beings among the intentional systems we know. Thus it would be natural for the holist, by Taylor's account, to try to argue that the capacity to think depends on social relations for its emergence and/or survival in an individual human being. This would be particularly natural, given that the romantics and their Hegelian successors all tended to return to thinking as the basic individual capacity that depends on society for its existence.

But now we come to the second move that the holist must make: he must argue for the dependence of a capacity like thinking on the enjoyment of social relations. One question here bears on the meaning of 'social' but I propose to take the same quick line on this matter that I took in the last chapter. Social relations are relations whose existence entails that the subjects related display intentional responses: they perform certain actions or exhibit certain attitudes. The main question raised by the holist's second move concerns the sort of dependence that he alleges between the capacity for thought, or whatever, and the enjoyment of social relations. Will it suffice for him to be able to argue for a causal dependence? Or must he establish a dependence of a non-causal sort?

Most traditional writing in defence of holism is vague on this crucial question. While many of the considerations adduced by someone like Bradley are more or less causal in nature, identifying the induction of the individual in custom, language, and the like as an important influence on the growing child, they lead him quickly to statements of a dependence that looks to be less accidental in character (Candlish 1978). So what are we to say? I propose a line that will keep the issue between holist and atomist as open as possible: a line that will not give the victory too easily to either side.

There can be little doubt that the human being is subject to important, even developmentally crucial, causal influences from parents and from others in the society where it grows up. If we allow that this sort of observation is sufficient to establish the truth of holism, then we make holism more or less platitudinous; equivalently, we make atomism more or less incredible. I propose that we should not regard it as sufficient for the holist to establish only a causal dependence of a capacity like thought on social relations. But what kind of dependence then will do?

Examples of apparently non-causal influence include the following sorts of cases (Kim 1974). Socrates's death made Xanthippe a widow. Tom's gain in weight makes him the heaviest person here. John's anti-bodies have made him immune to the disease. The queen's abdication makes her son, the prince, the rightful monarch. There is a question about how to analyse such cases but, intuitively, the connection is constitutive rather than causal (but see Sosa 1980). Given various background conditions, putting the antecedent factor in place means introducing all the elements required to constitute the consequent. Fix it that Socrates is dead and you will have done all that is necessary to make Xanthippe a widow. Fix it that John has the appropriate anti-bodies and you will have done all that is required to give him immunity.

The consequent in each of these cases is realised superveniently on the realisation of the antecedent, given the background conditions. The supervenience in question is bottom-up, to use a phrase from the preceding chapter: it is explained by the relation—a relation of constitution—between the subvenient and the supervenient levels, not by any peculiarities about the supervenient level. Thus we may say in each case that the realisation of the consequent is superveniently dependent, in the context of the background conditions, on the realisation of the antecedent. We naturally distinguish supervenient dependence from causal dependence. With causal dependence we think of the connection as involving, in Hume's phrase, distinct existences. With supervenient dependence, this is not so. We do not think of Xanthippe's widowing as displaying the distinctness from Socrates's dying that is required for a causal connection. And we do not think of John's immunity exhibiting this sort of distinctness from his development of appropriate anti-bodies.

I propose that what the holist must be required to show in respect of the thinking capacity of human beings is that it is superveniently dependent, at least in part, on social relations. This will not be something that it is easy to establish, unlike the causal dependence of the capacity to think on social inputs. But neither is it something that looks impossible to argue. Every individual who enjoys status enjoys it superveniently on the attitudes of others in the society; what it is to have status involves those attitudes intrinsically. And as it is with status, so it may be with thought. Thus we can cast supervenient dependence as being the issue between holists and atomists without doing a disservice to either side: without making either side look silly. The holist will say that the enjoyment of social relations is part of what is involved in the capacity to think; the capacity to think partly consists in the enjoyment of such relations. The atomist will deny this, arguing that the enjoyment of social relations is not an intrinsic component of the capacity to think.

Our examples of supervenient dependence make it clear that it may be partial rather than complete: the consequent in each of our examples is partly dependent on the antecedent, and partly on the associated background conditions; Xanthippe's becoming a widow is partly dependent on Socrates's death and partly dependent on her having been married to Socrates in the past. The subvenient determinant of something need not be sufficient for its realisation, as a causal determinant of something need not be sufficient either. Thus the supervenient dependence that the holist alleges, the dependence of the capacity to think on the enjoyment of social relations, need not involve a relation of sufficiency; there may be more to being able to think, as there surely is, than the enjoyment of social relations.

Some of the examples of supervenient dependence should also make it clear, at least on reflection, that as the subvenient determinant of something need not be sufficient for the realisation of that factor, so it need not be necessary either. Socrates's death may be the only way in which Xanthippe can become a widow and John's development of appropriate anti-bodies may be the only way for him to become immune. However, Tom may become the heaviest person here, not through a gain in weight on his part, but through a loss of weight on ours. And the prince may become the rightful monarch, not through his mother's abdication, but through her death or dismissal. Thus the subvenient determinant of something may be a non-necessary as well as a non-sufficient determinant.

The lesson bears, once again, on the content of holism. The supervenient dependence that the holist maintains need not involve necessity, any more than sufficiency. The holist may defend the view that thought is superveniently dependent on social life without thinking that social life figures among the conditions for the very possibility of thought. It may be possible, in his view, for thought to depend superveniently on something else in place of social life, though in actual fact there is nothing else that ever takes the required determining role. Thus, the holist who maintains the supervenient dependence of the capacity to think on the enjoyment of social relations may not hold that there is a valid transcendental deduction of the necessity of social relations for the possibility of thought.

The horizontal issue in social ontology, then, is whether the capacity to think, as it is realised among human beings, is superveniently dependent on the enjoyment of social relations by individual thinkers. The holist maintains such supervenient dependence, thereby defending the view that at least as things are—there may be other transcendentally deducible possibilities—human beings depend on their relations with one another for realising some of their distinctive capacities. The atomist denies the supervenient dependence of the capacity for thought on the enjoyment of social relations. He thinks that there is nothing incoherent—even as things are with human beings—in the notion of the solitary individual; such an individual is capable of realising all the capacities distinctive of human kind.

We spoke in the first chapter of a sort of intrasubjective holism that is entailed by the view of intentionality maintained there. Intrasubjective holism means that if it is fixed that I believe that p or whatever, then it will be simultaneously fixed that I have certain other intentional states, certain other beliefs and desires; no intentional state without an embedding set of intentional states. Intrasubjective atomism denies that sort of connectedness. The social or intersubjective doctrines that concern us now can be represented as parallel claims. Social holism means that if it is fixed that I have a certain thought, then it will be simultaneously fixed that other people have certain thoughts too; no act of thinking, at least as things stand with human beings, without an interactive context of thinking subjects. Social atomism rejects any such connectedness.

Independence

Before we leave discussion of the issue between holists and atomists, it will be useful if we go back to the question of how far it is independent of the issue between collectivists and individualists. I have asserted that the two issues are independent and I need now to say in greater detail why I hold by that claim.

The two issues cross in such a way that there are four possible positions that someone might maintain. Those positions are: the two pure positions, as it were, of atomistic individualism and holistic collectivism; and then the mixed positions of atomistic collectivism and holistic individualism. They are represented in the boxes of the following matrix.

	Individualism	Collectivism
Atomism	1	2
Holism	3	4

The pure positions are at the extremes of 1 and 4, the mixed positions are in between, at 2 and 3.

I assume that the extreme positions will be granted on all sides to be consistent. Those are the standard positions in the literature, given that people treat the two issues as if they were one: given that they think that going atomist and individualist on the one side, or going holist and collectivist on the other, comes to the same thing. The question of whether the two issues are independent turns, then, on whether there is any inconsistency involved in being an atomistic collectivist or a holistic individualist.

There is no obvious inconsistency involved in being an atomistic collectivist. If anyone counts as an atomist, then Hobbes must, for he certainly treats the idea of the solitary individual as a suitably coherent notion: the solitary individual never interacts socially with others, even if there are others around. But Hobbes would not have been plunged into any obvious inconsistency had he become persuaded that the individuals in society are not autonomous in the way that intentional psychology suggests: had he become persuaded, in particular, that there are certain aggregate regularities that override or outflank the regularities of intentional psychology. Such a collectivist turn of mind might have affected Hobbes's view of what are the distinctive capacities of the human being but he could still have maintained that the individual does not superveniently depend on her relations with others for the realisation of those capacities; he could have maintained that logically the solitary human being can display all distinctively human capacities.

The other mixed position apart from atomistic collectivism is the position maintained in this book: holistic individualism. There is no more reason to think that mixing holism with individualism generates inconsistency than there is to believe that mixing atomism and collectivism does so. Imagine that someone is a holist, believing that as things are, human beings superveniently depend on their relations with one another for the realisation of the capacity to think: with human beings the capacity to think, like the possession of power or status, involves the enjoyment of relations with other people. There is no reason why such a person cannot be an individualist, no reason why he cannot think that the aggregate regularities that characterise social life leave the individual uncompromised in her autarchical status. The endorsement of holism is entirely consistent with accepting the intentional-psychological picture of human beings; as the first two chapters of the book will have made clear, intentional psychology holds by nothing that the holist denies. Thus the endorsement of holism cannot entail the rejection of intentional psychology that is involved in collectivism. It must be consistent with an individualist outlook.

But if all the boxes in our matrix are open, if the horizontal and vertical issues in social ontology are independent, there remains a puzzle as to why this has not been made apparent in the tradition of the past two hundred years. How is it that atomism has been confused with individualism, holism with collectivism, and only the pure positions in our matrix countenanced as possibilities? I will conclude this discussion with some reflections on this question.

At the beginning of the last chapter I mentioned that there has been a tendency in social ontology to think of individuals and society on the model of parts and whole. One reason that our two issues have been run together may derive from this tendency. If we think in the part-whole way of individuals and society, then we will see the holist as saying that individuals are changed through becoming parts of the whole. If we see him as saying that, then we will easily take his thesis to be that the whole is greater than the parts. And if we take the thesis in that way, we will easily confuse it with the collectivist thesis, since this lends itself to expression in just those terms: the whole is greater than the parts in the sense that the parts are affected in some sense from above.

Bradley exemplifies the tendency, with all its dangers of confusion, to think of the issue between holists and atomists in terms of the part-whole model. Thus he casts the atomist claim, a claim he vehemently denounces, as follows: 'the community is the sum of its parts, is made by the addition of parts; and the parts are as real before the addition as after; the relations they stand in do not make them what they are, but are accidental, not essential, to their being; and, as to the whole, if it is not a name for the individuals that compose it, it is a name of nothing actual' (Bradley 1876, p. 164). In rejecting this atomism Bradley asserts that the whole is greater than the sum of its parts, society greater than the sum of the individuals who compose it. But when he asserts that, he is easily taken—and not least by himself—to be making a comment of a collectivist character: to be denigrating the status of individuals, as we know them in our intentional psychology.

Our discussion so far also suggests a second reason why the horizontal and vertical issues should have been confused with one another in social ontology. We mentioned that perhaps the original difference between holists and atomists goes back to the fact that holists take thought to be dependent on language, that they picture individual human beings as managing to think things only in virtue of having access, as it were, to the common fund of ideas represented by language. Holists say that there is a common fund of concepts, a shared body of thought, on which individual thinkers rely; atomists deny this. But what holists say is easily confused with the claim that there is a common centre of consciousness, a shared state of thinking, operative in social life. And that claim has a collectivistic character: it suggests that individual thinkers are not what they seem in intentional psychology to be. What atomists say is as easily taken then to be the denial of this claim, to be the assertion of the autonomy of individual thinkers. The upshot is a total confusion of atomism with individualism, holism with collectivism.

But apart from these two considerations, there is a further reason why atomism should have been confounded with individualism, holism with collectivism. This relates to the metaphors in which the doctrines have been traditionally expressed. The received individualist metaphor is mechanical, with the working of a society being compared to the working of a clock or whatever. The standard collectivist metaphor, as we mentioned in the preceding chapter, tends to be organic, with society being cast as an organism, often as an organism that displays characteristic patterns of growth and decline, perhaps even of recurrence. But the mechanical metaphor not only emphasises the desired individualist point, that the origin of social movement is in the micro-structure, among the individuals depicted in intentional psychology; it also suggests, in atomist vein, that the elements of that structure are as detachable without remainder as the bits of a clock. Thus the individualism it expresses gets to be confused with atomism. On the other side the organic metaphor not only highlights the collectivist view of society as involving regularities that compromise individual psychology from above; it also serves to express the holist view that individuals depend on one another for significant properties, in the distinctive way in which the parts of a living organism depend on one another. Thus the collectivism expressed by this metaphor equally gets to be confused with holism. The rhetoric employed militates against the distinction we have tried to put in place.

We have explicated the issue between holists and atomists and distinguished it from the issue, discussed in the preceding chapter, between individualists and collectivists. It is time now to consider how the issue should be resolved. In the next section I develop a line of argument that goes back to the discussion of thinking in the second chapter and use it to support holism against atomism. In the final section of the chapter I look at the significance of the holism defended for other questions, in particular for the question of whether we always see things just from the relativised viewpoint of our society; I argue that the holism defended does not entail such a relativism, though it does have some strikingly anthropocentric consequences.

The Argument for Holism

The argument that I shall offer for holism begins at the point where the discussion of thought left us at the end of Chapter 2. At that point we had claimed to offer a solution to the problem that appears to make the very possibility of thinking problematic, namely, the problem of rule-following. It will be useful to recapitulate on that problem and on the solution we offered, before getting into the argument for holism. The argument for holism involves a further development of the line of thought advanced in that solution. (On another argument, see Burge 1979, Pettit 1983.)

Thinking involves the intentional attempt to shape one's intentional attitudes, one's beliefs and desires, with a view to having them satisfy various conditions of rationality. We distinguished conditions of evaluative and deliberative, deductive and inductive, rationality, acknowledging that different traditions may offer different explications of these conditions. There are two independently important requirements that have to be met if thinking in this sense is possible. First, the subject must be able to have beliefs about the contents of intentional attitudes, about propositions. Second, the subject must be able to treat those propositions as rules, in particular as rules governing the formation of intentional attitudes: rules that determine when it is appropriate to form various beliefs and desires.

A subject treats something as a rule in the appropriate sense if and only if a number of conditions are fulfilled. On the objective side, the factor in question must be a normative constraint over an indefinite variety of cases, a factor that determines what the appropriate response is in each of those cases. On the subjective side, the agent must be able to identify the rule independently of its application in any instance and must be able to read off its requirements directly and fallibly.

A proposition is a normative constraint that determines what it is appropriate for a subject to believe in any of an indefinite number of possible situations. The proposition 'Paris is in France' makes it right in the actual world to believe that Paris is in France, for example; the proposition is such—in particular, its truth-conditions are such—that that belief is true and rational at the actual world. The belief is true, because things at the actual world are as the proposition requires; the belief is rational, because things at the actual world show evidence of being as the proposition requires.

In view of the normative constraints that propositions represent, a thinking subject will need to be able to treat a proposition as a rule for belief-formation. Propositions are non-basic rules in the sense that someone can be intentionally faithful in her thought to the constraint of a proposition only through being similarly faithful to more basic rules, such as the rule represented by an attribute or property. With a property the subject must be able to identify it independently of any instance and, being able to read off its requirements directly and fallibly, set herself intentionally to respect the constraint it involves in the intentional attitudes she forms. Take the property expressed by the predicate 'is a box', for example. The subject who forms box-beliefs in a thinking fashion must be able to set herself intentionally only to form the belief that it is a box of something that counts, by the relevant property-rule, as indeed a box. She must be able to fix on that rule as something by which to take her bearings in the course of forming beliefs.

But it turns out that the possibility of such rule-following is problematic. The problem is how something that is indefinitely normative—normative over an indefinite variety of cases—can be suitably accessible to a finite mind like ours: can be independently identifiable and directly and fallibly readable. This problem is at the centre of many of Wittgenstein's concerns and has been formulated in a particularly trenchant fashion by Saul Kripke (1982).

The solution that we offered to the rule-following problem goes like this. Certain instances of something like a property can exemplify the property to particular subjects, if those subjects are disposed on exposure to the instances to go on in a suitable way in the identification of other instances; certain boxes can exemplify the property of being a box to people who become inclined, on exposure to examples, to pick out suitable other things as boxes. The disposition or inclination in question need not come into the consciousness of the subjects. It serves a background enabling role, making it possible for certain instances to exemplify a rule for the subjects. In particular, it makes it possible for the subjects to pick out the rule independently and to read it off directly and fallibly.

The disposition will enable the subjects to pick out the rule independently, so far as no particular instance is vital for exemplifying the rule. It will enable them to pick out the rule directly, so far as it gives them each an immediate sense of whether some new object presented is a box or not; I assume that the box-rule is a basic one. But how will the disposition enable the subjects to pick out the rule fallibly? That is the crucial question. The disposition can do this too, I argued, provided that it is exercised under certain assumptions on the part of the subjects: assumptions that mean, as we theorists will put the matter, that something exemplifies the property and corresponds to the rule only if it attracts the dispositionally based response *under normal and perhaps ideal circumstances.* We may restrict our consideration here to normal circumstances; the extension to ideal is straightforward.

The assumptions ascribed to subjects in this story postulate a certain sort of intrapersonal and interpersonal constancy. The intrapersonal assumption is that something is amiss if I find myself reliably inclined to make different judgments at different times—in particular, judgments different by my own lights—without any justifying difference in collateral beliefs or whatever. The interpersonal assumption is that something is amiss if you and I find that we are inclined to make different judgments—again, judgments different by our lights—without any such justifying difference. If the subjects make assumptions of this kind, then they are going to discount some members of any set of responses that conflict intrapersonally or interpersonally. They had better discount in a systematic way, by reference to the same allegedly perturbing factors, if they are to achieve convergence across times or individuals. And if they discount systematically and successfully, that will enable theorists to say of the property they track that it is the property identified by the dispositionally based response under normal conditions: under conditions where perturbing factors are absent. That being the property participants track, it is a property that they each fallibly track in any instance. It is possible in any instance that there will be divergence across times or individuals and that in the course of negotiating about this the subject will come to see that her response was misled.

The essence of this view of thinking, and more generally of rule-following, is summed up in our description of it as ethocentric. The Greek word *ethos* means either habit or practice. And the crucial ingredients invoked in our story are, on the one hand, the spontaneous habit of extrapolation in virtue of which subjects identify and read off a rule; and, on the other, the practices of self-correction, in virtue of which they distance themselves individually from the rule, making it into something that any one of them can misread.

So much by way of summarising the argument of Chapter 2. The crucial upshot, from the viewpoint of our present concerns, may be described as the interactive thesis that the argument supports. I mean the thesis that rule-following, and in particular thinking, is an essentially interactive enterprise (Pettit 1990a). The rule-following subject must be in a position to interact with other bearers of the relevant inclination or disposition: herself at later times, or other persons. Without such interaction there are no normal or related conditions on the realisation of which the inclination can depend for successfully guiding the subject. Without such interaction the success of the rule-following cannot be contingent in the fashion required for fallibility. The subject will be incapable of being misled by her disposition, for the property which it leads her to recognise will be not the property that answers to the firing of the disposition under normal conditions, but just whatever property corresponds to its actual firing. If the subject is incapable of being misled by her disposition, of course, then the disposition cannot underpin an enterprise like thinking. It is of the essence of thinking that we try, without any guarantee of success, to get things right.

This upshot can be made a little more vivid as follows. Imagine the atom-for-atom replica—the *doppelgänger*—of any rule-follower, isolated socially from all contact with others at any past, present, or future time and isolated in time itself from relevant past or future counterparts. Such a creature of the imagination would lack the conceptual resources for thinking or, more generally, for rule-following. It might avail itself of certain inclinations to refer to *this* or *that* rule, as exemplified by certain examples, but even if there is a fact of the matter as to which rule is in question—even if the present disposition fixes this—the rule will not be one about which the inclination might ever be misleading. The creature will not have the distance from the rule that would make sense of the idea of its trying to be faithful. Rule-following, like keeping your balance, is an interactive project and the creature imagined lacks the social or temporal context of interaction that would enable it to pull off such a feat.

What sort of requirement is in question, when we say that the capacity to think requires interaction, whether interaction with oneself over time or inter-action with other people? Certainly not just a causal sort: the thesis is not that thinking cannot get going or stay going, as it were, without interactive jolts. The requirement relates rather to supervenient dependence. The capacity to think is superveniently dependent, at least in part, on the relations involved in intrapersonal or interpersonal interaction. Those relations are part of what goes to constitute the capacity to think. The subvening infrastructure of the capacity to think must include interaction with other people or interaction with oneself at other times. Without such interaction, no fallibility. Without fallibility, no thought.

It should not be surprising that the interaction invoked in our story about thought and rule-following, and for that matter the other materials invoked in that story, should be part of the subvening infrastructure of such capacities. In describing the task of accounting for rule-following, I said in Chapter 2 that the aim was constitutive rather than causal explanation. What we set out to do was to take certain plausible naturalistic capacities and connections as given, and then to show that they are sufficient to ensure that agents can practise rule-following. We sought an explanation that would show that certain naturalistic materials serve to constitute the capacity for rule-following and, more generally, thought.

We are now in a position to see that the interactive thesis connects closely with the holist claim. The holist claim is that the individual human being's capacity to think superveniently depends on her enjoying relations, in particular social relations, with other creatures. The interactive thesis is a weaker, disjunctive claim that the individual's capacity to think superveniently depends on her enjoying either interpersonal—and, inevitably, social—relations with other individuals or intertemporal relations with herself at other times. Why will the subject's interpersonal relations inevitably be social? Because they will require the people involved to display certain intentional responses, and that is all that 'social' connotes in our usage.

Given that the interactive thesis is a weaker, disjunctive doctrine, is there any way to move from it to the stronger, holist claim? There are two routes conceivable. If we could provide some reason for believing that interaction with oneself across time was necessarily unavailable, or was unsuitable on its own for supporting thought, then we would have the firmest possible ground for deriving the holist principle from the interactive thesis: for deriving the conclusion that the capacity to think is superveniently dependent on interaction with other people. This derivation would be transcendental in character, for it would give a basis for holding that thought would be impossible in the absence of social relations.

But this is not the only way of moving from the weaker to the stronger claim. We would have an adequate basis for deriving the holist principle from the interactive thesis, if we could provide reason for thinking that as things actually are with human beings, interaction with other people is always present in the subvenient infrastructure of thought. It may be that interaction with oneself across time is in principle enough to underpin the capacity to think but that interaction with other people is always part of the underpinning that is available in practice.

It would certainly be exciting if we could support the holist thesis transcendentally, arguing that social interaction with other people is necessarily required for thought. But I see no sure way to rule out the in-principle possibility of the socially isolated, intertemporally interactive thinker (Blackburn 1984b). I see no way of arguing that a Robinson Crusoe, even a Robinson Crusoe isolated from birth, could not think or follow rules (McGinn 1984, p. 195; Wright 1986, pp. 213–14). I am prepared to abandon the search for such big game, however, since I do see a way to secure the lesser prize just mentioned. I think there are solid grounds for believing that even if intrapersonal interaction might be sufficient in principle for thought, still interaction across persons is always involved in practice. If such interaction is always involved in practice, this will be enough to vindicate the holist thesis. We saw in our earlier discussion that as the holist may defend a non-sufficient variety of subvenient determination, so he may defend a non-necessary sort of determination. It will do if the capacity to think is superveniently dependent, as things are, on the enjoyment of social relations; it does not matter if there is an abstract possibility of solitary thought, a possibility unrelated to how human beings actually are.

The solid grounds for believing that interpersonal interaction underpins the human capacity to think derive from the fact that human thought satisfies a certain sort of publicity condition. I shall argue that any form of thinking, and more generally any form of rule-following, that satisfies this condition has to be subveniently determined by interpersonal interaction, not just by intertemporal.

Publicity conditions on rule-following have been much discussed in the Wittgensteinian tradition. 'Grasp of a rule must be manifest in what is interpersonally accessible—i.e. to others as well as to oneself—so that there can be no such thing as intrinsically *unknowable* (by another) rule-following' (McGinn 1984, p. 192; see also Wright 1986, pp. 209–10). Or again: 'if rule R is grasped by person x, then it must at least be logically possible for some other person y to grasp R' (McGinn 1984, p. 194). Publicity conditions may have attained prominence in the Wittgensteinian literature, because of their connection with the famous private language argument. This is the argument that there is an incoherence in thinking that rule-following can go on, in reference to private objects, without the logical possibility of access by others (Wittgenstein 1958).[1]

The publicity condition that I wish to introduce is not one of the various conditions that have been discussed in the literature that stems from Wittgenstein and it is important to stress this, in order to avoid confusion. Furthermore, the condition that I introduce differs in status as well as content, from the more commonly discussed candidates. I do not present it as a condition that rule-following or thought has to satisfy. I see it only as a condition that human thinking satisfies as a matter of fact. It is a condition that human beings meet in their performance as thinking and rule-following subjects but it is not a condition that all conceivable subjects are required to meet, on pain of some gross incoherence or implausibility.

So what is the condition I envisage? Briefly, that one human being can knowledgeably identify, as such, the rules followed by another and identify them as rules that she can follow herself: in particular, that one human being can knowledgeably identify the propositions and propositional elements that another targets in her thought. Not everyone may be able to know which rule is followed by the other, in this sense. But the point is that there is no block in place. There is nothing that prevents one individual, first, from coming to pick out, as such, the target of the other's attention and, second, from coming to pick it out as a target that she might follow in the same way herself. The mind of the other is scrutable.

How to describe the publicity that is involved here? The best way may be by recourse to metaphor. Commonable land is land that is or may be held in common by people, where being held in common is a condition that requires awareness of the sort of joint control exercised. By analogy, we might say that according to our publicity condition the rules that people individually follow, in particular the rules that they individually follow in their thought, are commonable. They are rules that another individual may knowledgeably identify as the rules followed by a given subject, and which she may identify as rules that she can follow herself in the same manner. They are rules over which no one individual has a monopoly; they are capable of being claimed as a common possession by any of the individual's fellows.

The interactive thesis combines with the thesis that the rules of human thinking are commonable to provide an argument in favour of the holist claim. The two propositions support the conclusion that the human capacity for thought, in particular the human capacity for commonable thought, is dependent on the enjoyment of social relations. Assume that there are or might be other individuals around. An individual human being will be able to think commonable thoughts, thoughts that others can or could claim as a common possession, only if she is actually involved in social relations with others. Or so at least I claim. No commonable thoughts, as we might put it, without actual community.

At first glance this holist claim may look more or less trivial. It may seem to have the form: if a person's thoughts are to enjoy a certain relationship to others—that of being commonable by them—then the person must enjoy a certain relationship to others: she must be socially related to them. But this appearance of triviality is misleading. Commonability does not involve a person in a relationship to others, as a matter of logic. Commonability is a property of rules that she follows: it requires them to be such that if there are others, and if they try to identify the rules she follows as rules they can follow themselves, then it is possible they will succeed: it is possible that they will knowledgeably identify the rules in that way. The property is identified by reference to other people, as the property of being soluble in water is identifed by reference to water. But there is no formal incoherence in the idea that it should be realised in the absence of other people, as there is no formal incoherence in the idea that something might be water-soluble in a water-less world. Thus it is a substantive and non-trivial thesis that a person can think commonable thoughts—thoughts that are suitably accessible or scrutable—only in a world where there are others and only in a world where she enjoys social relations with others.

The argument I offer in defence of this holist claim goes, schematically, as follows.

1. The interactive thesis. A human being can follow a rule only on the basis of interpersonal or intertemporal interaction.
2. The commonability thesis. The rules followed by a human thir:ker are commonable: they are rules that another can claim as a common possession.
3. A negative claim. If a human being follows a rule on the basis only of intertemporal interaction with herself, then that rule is not commonable.
   Conclusion. The rules followed by a human thinker are not followed on the basis of such intrapersonal interaction alone; they must be followed on a basis involving interaction with others.

The conclusion of this argument amounts to a statement of holism. The interpersonal interaction required is bound to involve social relations, as we have seen. And so the conclusion enunciates the holist message that there is no human thought without social relations. Given that it is commonable, human thinking must depend superveniently on the enjoyment, among other things, of such relations. The solitary but commonable thinker is an incoherent conceit.

The interactive thesis on which this conclusion rests is secure enough, in view of the argument of Chapter 2. But the other premises need considerable defence. So what is there to say about the commonability proposition that I have added to the interactive thesis, and about the negative claim that connects the two theses to the holist conclusion?

The first thing to say about the commonability thesis is that it is stronger than is strictly required for the argument to go through. It would be sufficient for the purposes of our argument to claim not that all the rules followed in human thought are commonable, but that at least some are. Thus it will not do any damage to the case that I am making for holism if someone can show that there are or may be some rules of thought that are not commonable. Why have I stated the commonability thesis in the strong form, if this weak form would do? Mainly, because I think it would be misleading to present the thesis in the weaker way. It would suggest that I have some examples of non-commonable rules in mind: presumably, examples that I cannot make properly available to readers, since to do so would be to show that they are commonable after all. And that suggestion would be very far from the truth.

But if the commonability thesis is unnecessarily strong in this respect, there are two other respects in which its strength is of the greatest importance. The thesis says that the rules followed in an individual's thought are rules that another can knowledgeably identify—and identify in a way that facilitates a similar rule-following on her part—as the rules followed by that individual. This thesis is strong in requiring, first, that the identification be knowledgeable and, second, that it be a mode of identification that facilitates a similar rule-following on the part of the identifier. Both requirements are important.

The first requirement means that it is not enough for the commonability of a rule that someone correctly believes, for whatever reason, that she has identified the rule followed by another in certain thoughts. However the notion of knowledge is to be analysed, the person must be in a position to know that she has identified the rule in question. Someone might believe that she has identified the rule followed by another, if she observes the other reacting as she does to certain examples; moreover, she might believe this correctly, and with some reason. But such belief would not amount to knowledge, on any current account of what knowing involves (Dancy 1985). When I say that a rule is commonable, I mean that it can be identifed as the rule that another follows in a much less conjectural way than this. I mean that it can be identified in a sufficiently privileged or reliable manner for the identifier to count as knowing which rule the other follows.

The second requirement on the identification involved in commonability is that it should facilitate rule-following of the kind practised by the other. Where the knowledge requirement is meant to rule out conjectural fortune, this requirement is designed to ensure that it is not enough for commonability that the identifier gets an oblique, parasitic fix on the rule that another follows. Suppose I see another person going through a procedure that looks like counting, up to the point where she reaches 1,000, but that goes beyond my ken after that: the person carries on from there with 1,002, 1,004, 1,006, and for all I can know, may be disposed to make equally surprising moves later. I am in a position to discriminate the rule followed by the other—if there is a rule involved—by reference to the fact that it is the rule that has her count up to 1,000 and go on in a strange way after that. But to discriminate the rule in this way would not be to identify it in a manner sufficient to facilitate a similar rule-following on my own part; it would not involve my cottoning on to what guides the other's operations, and to what might guide mine. That a rule is discriminable in this way is not enough to make it commonable in my sense.[2]

We have commented on three ways in which the commonability thesis is quite a strong claim. But one thing worth adding is that, strong as the thesis is, it does not rule out the possibility of idiosyncracy among human thinkers. Even if I know which rule you follow in certain thoughts, and even if I identify it as something that I can follow myself, I may not gain access to all aspects of your representation. Suppose I know which rule you follow in trying to believe of all and only appropriate objects that they are boxes, or triangles, or examples of adding, and that I see that rule as one I can follow myself. This still leaves it possible for me to recognise that there may be all sorts of images and associations that divide us when we think about such things. The criteria for determining whether I know which rule you follow will bear on my competence in regard to the sorts of factors invoked in the ethocentric account of rule-following: factors involving extrapolative inclination and corrective practice. And such criteria are bound to leave room for ignorance on my part about many facets of your mental set.

So much by way of elucidating the commonability thesis. It is time now to say something in its favour. I have two sorts of thing to say in support of the thesis. First, all the evidence suggests that as a matter of fact the rules that people follow in thinking are commonable. With any content that another person claims to grasp, with any proposition that she claims to believe, we take it to be possible for one of us to know which content or proposition she has in mind, and to know this non-parasitically. We rule out the possibility that the other person might inhabit a world of purely private, not knowledgeably accessible, rules and might form intentional attitudes with respect to contents defined by means of those rules. We do allow of course that we may not have got the mind of another right on some questions. But we always assume that the questions on which we seek to know the other's mind are questions in the public domain, questions that are in principle within the reach of people other than the individual herself.

Take the way a person goes on when she is trying to decide, as we will say, whether something counts as a box or triangle or as a sum of certain numbers. We do not countenance the possibility that such a person might actually be following a rule we cannot knowledgeably identify and identify in non-parasitic mode. We take it as axiomatic that if the person is a thinker at all, then she is a scrutable thinker: that if she is a thinker at all, then we can know which rules she is following and recognise them as rules that we might follow ourselves. We admit, of course, that some of us may suffer an incapacity that makes such identification impossible. Some of us are colour-blind and may not be equipped to share in a non-parasitic targetting of colour-properties; and some of us are poor at mathematics and may not be capable of sharing non-parasitically in the targetting of certain abstract operations. But we believe that there is no general block that would prevent someone from knowing in a non-parasitic way which rule another is following in her thought.

Might we be wrong in this general belief? Might we be wrong in the assumption that it is possible for one person non-parasitically to know which rule another thinker is following? Sceptics would say that we may well be wrong in this assumption, as we may be wrong in the assumption that it is possible to know where we are located at any moment, what our name is, or whether we have limbs. But sceptics deprive us of the ordinary notion of knowledge itself, refusing to countenance the regular paradigms of what it is to know something. And the important point is that, short of scepticism, it is very difficult to reject the assumption that one person can non-parasitically know which rule another is following. If I cannot know that what you mean by 'red' is, in my language, red, or that what you mean by 'adding' is, in my language, adding, then it is not clear that there is anything that I can properly be said to know.

I promised that there were two things I had to say in favour of the common-ability thesis. The first, just documented, is that there is considerable evidence, which is reflected in our everyday assumptions about one another, that the thesis is sound. The second thing I want to say is that people have reasons to make their thought commonable, and have the resources to make it common-able, so even short of such evidence, we should expect the commonability thesis to be sound. But this second claim is not one that I am yet in a position to defend. I will return to it later, after discussing other aspects of the argument for the holist thesis.

We have commented on the first two premises of the argument: the interac-tive and commonability theses. The third and last premise is the claim that if an individual follows a rule on the basis solely of interaction with herself, then that rule is not commonable. We are invited to imagine someone who exploits purely intrapersonal resources with a view to rule-following: with a view, in particular, to the sort of rule-following involved in thought. The person identi-fies a rule on the basis of certain exemplars, being spontaneously inclined to go on in a certain way in dealing with new cases. But she identifies the rule as an object of fallible rule-following, because she does not tolerate any discrepancy, as she sees it, in her own responses across time; she is disposed to negotiate about such discrepancies and to discount all but one of the discrepant re-sponses, faulting the conditions under which they appear. The rule she follows, then, is that rule which corresponds to her extrapolative inclination under the conditions that survive such intertemporal negotiation. The third premise in our argument says that a rule that is fixed in this way, on the basis of just intrapersonal interaction, is not commonable. Others are not in a position to know, in non-parasitic mode, which rule the person is following.

The claim is intuitively plausible. Suppose that you find a rule salient on the basis of certain examples, given that the examples induce in you an inclination to go on in a certain way. You let the inclination lead you in judging according to the rule except when you find that, by your own lights, it leads you in different directions at different times: such dissonance you resolve by identify-ing factors such that their presence allows you to discount certain conditions of response. But you are not disturbed by any dissonance between how the inclination leads you and how any allegedly corresponding inclination leads other people like me. The rule that you try to conform to is a rule fixed by reference to your own responses, and your own responses only. In that case, it seems impossible for me to know, in the non-parasitic mode, which rule you are following.

I may have a guess at which rule you are targetting, conjecturing that it is the rule that I find exemplified in such and such exemplars. But how can I know that it is really that rule you are fixed upon: that rule, rather than any of the alternative rules that are instantiated by the cases that you take as examples of the rule? I may hypothesise that you are like me and that you are almost certainly influenced by a similar extrapolative inclination, when you take certain cases to exemplify a rule. But that is all I can do: hypothesise. For all I know, your inclination may prove to be quite different from mine in various cases. And even if I knew that your inclination was similar to mine in relevant respects, I still would not know which rule you were following, at least not in the non-parasitic mode. For it may be that you are disposed to resolve discrepancies in your responses over time quite differently from how I would do so. It may be that you differ from me, not in the habit of extrapolation, but in the practice of negotiation whereby certain conditions get designated as favourable.

Objectively, the rule that you are following amounts to a rule-in-extension, a set of instances and responses, of indefinite extent; we may ignore the likelihood that any rule followed by humans is vague at certain margins. Which rule-in-extension you target is determined by which rule-in-extension answers to your extrapolative inclination, under conditions that count as favourable by reference to your practice of negotiation about discrepancies. But there is no way for me knowledgeably to pick out the rule-in-extension that you address. I do not know that I share your inclination, as I see a rule exemplified in the examples that move you. And I do not know that I share the practice of negotiation that engages with that inclination to determine which rule is in your sights. I may find myself confronted with a particular rule, as I put myself in your position, and ask which rule you are following. But I cannot know that the rule that stares me in the face is the rule that catches your eye. The rule you are following is not one I can claim as a common possession.

We have been supposing that you are aware of me and my responses and are indifferent to any discrepancies between how we each go. But suppose that you are unaware of me and that were you to become aware, then you would negotiate over any differences between us, acting on the assumption that there must be a common rule on which we are each focussed. If I knew this to be the case, would it enable me to know which rule you are targetting? No. It might enable me to say that were you to become aware of me, and were we to negotiate successfully with one another in the manner envisaged, then I would know which rule you were following. But it does not enable me to say that I know which rule you are following now. The gaps that deprive me of knowledge in the earlier scenario will still cause me trouble under this. All I will be in a position to say is that you assume that the rule you are following is a rule that anybody in your position would follow and that you are disposed to negotiate about discrepant responses with anyone like me who comes along. But, for all I can currently tell, we might find that negotiation about discrepancies was difficult and fruitless. We might find this, even if we are now exactly alike at the physical level, for we will each be subject to different inputs, come the time for negotiation; we will occupy different physical perspectives. We might each be forced to the conclusion that the rule you are following diverges radically from anything to which I can cotton on.

Under the scenarios envisaged it may well be the case that the rule you are following, given your habit of extrapolation and practice of negotiation, is exactly the rule that I am following, given my habit and practice. Equally, under the scenarios described, it may well be that the rule you are following is exactly the rule that we would each be following were we to attend to interpersonal discrepancies and negotiate successfully about those: negotiation might not take a different direction through going interpersonal as well as intertemporal. The important point, however, is that under none of these scenarios can I know which rule you are following, in non-parasitic mode. In all of them I am left on the outside, observing your responses, without being able to engage with you and negotiate about divergence. The data available to me on the outside are not going to be enough to determine which rule you are targetting.

The point connects with the familiar observation that any finite set of cases instantiates an indefinite number of rules. When I observe you following a rule that is fixed solely by reference to your own habits and practices, then I am in the impossible position of someone who is trying to identify an indefinitely relevant rule by the consideration of a finite number of cases. I am stymied by the oldest problem in the induction game. The problem will not stop me conjecturing about the rule that you are following. And it will not prevent me from later falsifying a given conjecture, as I find that you do not respond as the rule that I postulated earlier would require. But it will make it impossible for me to know which rule you are following, in the sense in which I am supposed to know this for commonable rules (see McDowell 1981, 1987).

We have now considered the three premises of the argument for holism: the interactive thesis, the commonability thesis, and the relevant, negative claim. Those premises are true and, patently, the argument in which they figure is valid. I conclude therefore that we should espouse holism. We should admit that human thought is superveniently dependent, in part, on social interaction between different individuals. There may be an abstract possibility of thought occurring on the basis of just intrapersonal interaction; there may be no transcendental necessity attaching to the dependence on interaction of the interpersonal sort. But as things actually are, and as things must be with any creatures whose rule-following satisfies the commonability condition, thinking superveniently depends on interpersonal interaction.

But our argument for holism still leaves something to be desired. It is one thing to argue, by exhaustion of cases, that an interpersonal foundation must support the commonability of rules. It is one thing to argue that commonable thinking cannot be supported by intrapersonal interaction, so that it must be supported by the only available alternative: interaction of an interpersonal kind. It is another to argue that there is positive reason to think that the interpersonal foundation can provide the support required. Happily, however, there is positive reason to think this.

Suppose that we each conduct our thinking on the basis, in part, of interpersonal interaction. You find a rule salient on the basis of certain examples, given that the examples induce an inclination in you to go on in certain ways. But the rule that you follow is not distinguished just by how the inclination happens to lead you, for you take pause and seek out discounting factors whenever you find that there is, by your lights, an intrapersonal discrepancy of response or a discrepancy across your responses and mine. Does this state of affairs ensure that the rule that you follow is commonable: that it is something I can claim as a common possession, knowing in non-parasitic mode which rule it is? I shall argue that it does.

I shall argue, more exactly, that if there truly is a rule you are trying to follow, then I can know which rule it is. But the qualification need not be taken to make a great difference. Knowledge, as already noted in Chapter 2, is fallible. While it is inconsistent with my knowing that p, that it should not be the case that p, still I am not infallible and, consistently with all the evidence available to me, it may not be the case that p. The fallibility of knowledge means that I can know which of my children is out in the garden, even though it is strictly consistent with all that I can see clearly that there is no child there, only a play of light in the trees. By analogy with this, the fallibility of knowledge means that I can know which rule you are following, even though it is strictly consistent with all the evidence available to me, or indeed to you, that there is no rule you are following. Thus there is no great difference between the claim that I can know which rule you are following and the claim that if there truly is a rule that you are following, I can know which rule it is.

I must now try to defend the claim that given you follow a rule on the basis of interpersonal interaction with me, the rule is commonable and I can know which rule it is. I shall defend the claim by arguing two points. First, that under this scenario there truly is a rule that you are trying to follow only if there is a negotiable convergence in our responses. And second, that if there is a negotiable convergence in our responses, then I am able to know which rule you are following, and in non-parasitic mode.

The first proposition imposes itself on very little reflection. You intend to follow a rule exemplified by certain instances. Which rule do you mean to follow? The theorist who examines your behaviour will offer the most perspicuous account: that rule, the one that your extrapolative inclination identifies under conditions that survive negotiation about intertemporal discrepancies and, in particular, negotiation about any discrepancies between you and me. While you pick out the rule to be followed by reference just to the exemplars, your intention embraces yourself at later times and your fellow-respondents: in particular, me. You intend to pick out that rule, the one identified in new cases by how you and I respond under conditions that count as favourable.

I say that there is a rule you are trying to follow only if there is a negotiable convergence in our responses. I take it that were our responses to diverge wildly, beyond the reach of negotiation, then you would see your project of rule-following as coming unstuck. You do not view my responses just as an external check on whether your own response goes wrong: on whether it goes wrong in a way that is likely to induce intertemporal discrepancy, where that is what really bothers you. You view a discrepancy in our responses with just the same degree of seriousness as a discrepancy in your own responses over time. You give my responses the same presumptive relevance and authority as you give your own responses, so that I am as intimately involved in your rule-following intentions and project as your other selves. Under this scenario, it should be clear that there is a rule you are trying to follow, there is a target there from which you are taking your guide, only if there is a negotiable convergence in our responses. If there is no such convergence, then your project is empty in the way in which it would be empty to hunt the snark or try to search out the present king of France; it fails to come off, because it lacks a target.[3]

So much for the proposition that there is a rule you are trying to follow only if there is a negotiable convergence in our responses. I now add the further proposition that if there is such a convergence, then I am in a position to know which rule you are following, and to know this in non-parasitic mode. This proposition is compelling, because in the scenario described our positions are perfectly symmetrical and if you can identify the rule appropriately, then so can I. You focus on exemplars and read off a rule, on the overt and successful assumption that you will achieve convergence with me. But I can do the very same thing. And so, if the rule is something that you can target, equally it is something I can share as a guide. The rule is commonable, it is something that I can claim as a common possession.

The two propositions just defended offer impressive support for the positive claim, that interpersonal interaction facilitates commonability. Our argument for holism showed, by exhaustion of possibilities, that commonable rules have to be fixed on a basis involving interpersonal interaction; they cannot be fixed on the only other available basis: that of intrapersonal, intertemporal interaction alone. The positive claim we have just defended serves to buttress that argument, for it shows that the possibility to which the argument directs us is a persuasive and congenial one; it is not a last and desperate resort.

The positive claim should not be surprising. We argued in defence of the negative claim that figures in the argument for holism, that I am not in a position to know which rule you are following, if the rule is fixed just by reference to your responses. We said that in such a case I would be observing you from the outside, and would face the impossible task of identifying a rule of indefinite range on the basis of a finite number of cases: the cases where I see how you go in following the rule. By contrast with that scenario, we can say that where the rule you follow is fixed by reference to our shared responses, then I am on the inside, not on the outside (see McDowell 1981, 1987). My own habit of response, and my own practice of negotiation, are involved in fixing the identity of the rule, if there is one, on which you are targetted.

Otherwise put, the success conditions for your project of rule-following involve me intrinsically. It should be no surprise that under this scenario I am able to identify that rule for myself. I am able to know which rule it is, and know this in non-parasitic mode.[4]

The lesson of the positive claim makes a connection with the hermeneutic tradition. How do we get to follow commonable rules, how do we get to understand one another's minds? It appears that we must form community with one another, each recognising the other as equally criterial for whether there are rules that either of us follows; we must invest one another with authority on what the rules are. If it yields knowledge, then interpretation has the sort of practical aspect that the hermeneutic tradition emphasises. No reliable interpretation, as things stand, without social interaction (Bubner 1981, chap. 1). No interpretation without incorporation.

Given the defence offered for the positive claim, we should be better disposed to go along with the argument for holism. But there is one thing more that we can do to buttress that argument. As promised earlier, we can provide a further line of consideration in favour of the commonability thesis.

Even if someone has doubts about whether we are always in principle able to know which rule another person is following in her thoughts—even if someone has doubts about the commonability thesis—they are likely to admit the following: that in general we each wish to follow rules that are suitably accessible to others; that, whatever ambitions of dissimulation we may harbour, we each wish to follow rules that can be claimed as a common possession by others. This is a reasonable admission, since it would be a sure sign of psychological derangement—or perhaps a warped philosophical motive—for any one of us to seek out purely private rules to follow. It is part of showing oneself to be sane that one does not seek the refuge of a thought-world that is epistemically sealed against the comprehension of others.

If it is granted that we each seek to follow such rules as are suitably accessible to others then, provided we have the ability to achieve what we seek, it will follow that our rules are indeed commonable. If we are each capable of identifying rules that are available to others, then it will follow that in principle we are each accessible to others. So the question is whether we do have the ability in question.

Our discussion of the basis for commonability provided by interpersonal interaction reveals that clearly we have that capacity. Just by acting on the assumption of interpersonal constancy, just by looking for convergence across people in the responses that interpret a rule, we can each ensure that if we manage to follow a rule at all, we will follow a commonable rule. By acting on the assumption of interpersonal convergence, by seeking out negotiation with others over any discrepancies of response, we invest others with such authority that they are capable, and capable in the same degree as ourselves, of identifying the rule we target. Or at least they are capable of this identification, if there is indeed a rule there to be targetted.

We have a second reason then for endorsing the commonability thesis that plays such a part in our argument for holism. Not only is it intuitively plausible, as we urged earlier. It is also a thesis that two independent observations make it reasonable to accept: that we human beings generally desire to follow commonable rules and that we are capable of doing so, just by investing one another's responses with a certain authority, just by incorporating one another in a shared community. We have reasons to make the commonability thesis true. We have the resources required to make it true. And so it ought to be no surprise that it is true.

It is time to take stock. The upshot of our argument is a holist thesis of just the kind that the eighteenth-century romantics espoused. We may merit the ascription of beliefs and desires, as many animals and robots do, in the absence of any distinctive interaction with our fellows. But if we are to be more than mere intentional systems, if we are to be systems that can intentionally work at the shaping of their own intentional states, and do so in a mutually accessible way, then we must rely on our community with others. In order to pursue the shaping of our intentional states, we have to be able to identify propositions and propositional elements as norms by which to guide ourselves. These norms give us our bearings as we seek out the rational things to believe, the rational things to desire, and the rational things to do. But where to find such norms? Where in particular to find such norms, if the norms are to be a common possession? Looking within ourselves, looking to our individual, extrapolative inclinations to give voice to suitable norms, will jeopardise our mutual accessibility. We have to look to what emerges in interaction with others, not to what remains locked in the intimate enclosures of the self. In divining the demands of the relevant norms, the relevant rules of thought, we have to give ear not just to our individual inclinations, but also to the inclinations of our fellows. We have to let ourselves be guided by the voice that emerges from the testing of those inclinations against one another.

Is this a dagger, or a box, or a game, that I see before me? I have seen examples of the property previously and I will certainly have a primitive inclination to say that it is a dagger, a box, or a game, or that it is not. Suppose that I follow that inclination and, in the tones of the expert, judge the object appropriately. Following the inclination will be a venture, and following it will bring a possible gain—it may lead me to get things right—only if there is some possibility of my erring. And that possibility will be secure and scrutable, only in the event that other people's inclinations are deemed to be as relevant as my own in determining whether this object is or is not a dagger, a box, or a game. My own inclination must count as an intimation of the communal voice that firms up in the convergence of different extrapolative dispositions. It must count as an intimation that is validated only in the achievement, perhaps after negotiation, of a concerted response.

In the preceding chapter we offered an account of what it is for a property to be social: the property must require intentional responses on the part of a number of subjects. The lesson of our social holism is that the property of being able to think, or at least to think commonable thoughts, must itself count as social. That capacity is realised superveniently, not just on the intentional responses of the thinker, but also on the intentional responses of others in her society. Contrary therefore to everything we might have been disposed to say, it is a social ability; it requires intentional responses on the part of a number of subjects. Thinking is not the purely private enterprise it seems at least sometimes to be: it never involves a total renunciation of the public forum, a complete seclusion in the cloisters of the inner self. The thinker may withdraw from social life but she will still carry the voice of society within her into her place of retreat. If that voice were absent, if there were no others to whom the individual thinker was answerable, then scrutable human thought would be impossible.

It is difficult to give expression to the holist thesis without going metaphorical, as in talk of a communal voice and a concerted response. This may explain why the romantic tradition became notorious for its heady language. But we can use the notion of externalism that we introduced in the first chapter to give a somewhat more sober account of the doctrine defended. The externalist about intentional attitudes holds that narrowly identical intentional subjects, subjects that are identical from the surface in, may differ in the things they believe or desire. If the subjects are to be in a position to believe or desire a particular content, things may have to be a certain way in their environment, and they may have to have interacted with that environment in a certain fashion. If I am to believe that that parrot is squawking, for example, then there really must be a parrot there and I must have had the contact required to be able to have it in mind; the narrowly identical counterfactual self which has not had contact with parrots, or at least not with that parrot, cannot share my belief.

Social holism involves an attitude-based, as distinct from a content-based, externalism. It says that if an intentional subject is to think commonable thoughts—if she is to have attitudes of a commonable kind, as distinct from attitudes with any particular contents—then she must occupy a certain sort of environment and she must have interacted with that environment in a certain fashion. She must occupy a world where there are other subjects and she must have interacted with other subjects in the manner required by our ethocentric story about rule-following. If her thoughts are commonable, then she must not only be a certain way internally, she must actively incorporate others in community with her. She must intend with respect to certain others, individually or collectively, that they shall matter—their responses shall count—for the determination of what she means to think.

In our earlier discussion of externalism, we distinguished between a weak and a strong version of the doctrine. The weak externalist holds that a narrowly identical subject in another world might not have the same intentional states as a given subject in this world. The strong externalist holds that the narrowly identical subject might not have the same intentional states even if she lived in this world: having those states may require, not just a suitably furnished world, but a suitable history of interaction with the furniture of the world. As the content-based externalism defended in the first chapter is of the strong variety, so is the attitude-based externalism defended in this: so is the attitude-based externalism involved in our social holism.

Imagine that you follow certain rules of thought on the basis of interaction with me, or with a collectivity to which I belong, and that the rules you follow are therefore commonable by me. Suppose now that a narrowly identical counterfactual self takes your place: the self that you would have been, had narrow inputs been exactly the same but had you not had any interaction with me or, more generally, with us. Are the rules followed by the counterfactual self, if one assumes she does follow rules, commonable by me? I say that they are not.

We are to imagine the self prior to any interaction with the rest of us. And that means that we are to imagine someone who has not yet invested us with any authority on the matter of what rules she follows. I may be confident that on making our acquaintance, this self will invest me with such authority, whether individually or as part of the collectivity, and that there will be a readily negotiable convergence of response between us. But as things stand prior to that acquaintance and incorporation, I cannot claim the rules she follows as a common possession. I may find her responses as unsurprising as yours, and I may be compelled to think of her as following the rules that we know you follow, but I cannot know which rules she follows. For all I can know, she may be disposed to go a different way in her later responses or negotiations, blandly ignoring the inclinations of the rest of us. I could claim to know which rules she follows, at least in non-parasitic mode, only if I was individually involved, or involved as part of the collectivity, in her rule-following projects: only if she had already interacted with us and formed the project of following rules that are borne out, after negotiation, in the responses we share.

We have done all that can reasonably be required in order to defend and articulate the doctrine of social holism. We have found that although the thesis is associated with the romantic tradition of thought, it can be articulated with the help of more or less familiar, analytical notions and it can be supported with more or less precise, analytical arguments. In concluding our discussion, there is just one thing more that I would like to add. This is that the sort of holism we have embraced, and in particular the idea that human thought may intrinsically involve social interaction, can be supported by considerations drawn from evolutionary theory. There are naturalistic reasons why we should be well disposed to the idea that people depend on interpersonal interaction, not just interaction of an intrapersonal kind, for the conduct of thought.

It is extremely unlikely that the capacity for thought should have evolved on the basis of intertemporal interaction of a person with herself. The sort of change involved in the shift from being a non-thinking intentional subject to being a thinking one would have required more than a single mutation. But it is hard to imagine what the series of changes could have been, in particular what the series of individually beneficial changes could have been, that would have led to the intertemporal but intrapersonal interaction required. The position is quite different with interpersonal interaction.

Vervet monkeys make one sound when a snake is seen in the grass, another when an eagle appears in the sky, and the noises affect other monkeys in the ways that the perception of the snake or eagle would have done (Dennett 1987, essay 7). We told a story in Chapter 2 that is a fanciful but naturalistically plausible elaboration of this sort of theme. The story was meant to explain how certain creatures might have attitudes towards propositions as well as attitudes with them; propositions become objects of thought for these creatures through the availability of noises that go proxy for real situations. The story serves in the present context to bolster the holist thesis: it gives us a naturalistic sense of how the rule-following required for thought might have evolved on the basis of interpersonal interaction.

Suppose that communicative speech developed from the sort of sign-making performed by vervet monkeys (McDowell 1980). If our ancestors came to use meaningful speech in such a manner, then they would have deferred to one another in the matter of whether it is appropriate on a particular occasion to sustain a given assertion. They would have developed the habit of sustaining an assertion only in the event of achieving a certain consensus: certainly, they could not have been individually indifferent to the discovery that others, even others as well placed as they are, do not share their inclination to make it. If they did not defer to one another in this way—if they did not seek convergence, as we will put it, on the content of the remark—then the assertoric practice would not have had any obvious utility. It would not have been possible for each to take the noise, as made by another, to signify what she would have signified by its use; it would not have been possible for each to treat the noise as a proxy for a familiar state of affairs. If you make the eagle-noise and persevere in making it even in conditions where the rest of us find it inappropriate, then there ceases to be a reason why we should treat the noise as a stand-in for that situation.

If our imagined ancestors came to use meaningful speech in the way envisaged in our story, and if they deferred to one another in regard to the content of that speech, then we can see how they might have come to speak—and think—in a public language. We can see how they might have come to think according to rules that are established in the public forum, under the assumption that there is something wrong if different individuals differ in their inclination to support a given remark or thought. The holist story that I have been telling is one that comes with a plausible—or at least a not wholly implausible—evolutionary gloss. There is a conjectural natural history available as to how it might have come to apply to our species, and the history does not offend against any of the maxims of evolutionary theory.

Holism and Relativism

The Relativist Connection

The holism we have defended is continuous in spirit with the holism that emerged among romantic thinkers in the tradition associated with Herder and the many thinkers he influenced. That tradition may have tended in its later development to run holism together with collectivism. Equally it may have involved some crucial obscurities in the formulation of holism. But at least in its original conception, the intended idea was clear and, to us, congenial: however dependence is to be understood, thought is dependent on language, and language is dependent on society, so that thought requires a social context for its realisation. In reaffirming the holist message we should have no hesitation about making a connection with this outlook.

But if we make a connection with that outlook, we should be prepared to consider how far our holism leads us towards a relativist vision of the kind associated with Herder and his heirs. Herder relished the variety of human cultures, as it is surely commendable to do, but he also tended to embrace a relativity that he may have seen as a corollary of that variety. He tended to embrace the view that the claims and standards of a culture are not assessable, are perhaps not even interpretable, outside it. This sort of relativism is a dubious doctrine and we need to see how far our holism commits us to such a position.

Isaiah Berlin emphasises both the variety of cultures that Herder cherished and the relativity that he seemed to take as a corollary.

> It was Herder who set in motion the idea that since each of these civilizations has its own outlook and way of thinking and feeling and acting, creates its own collective ideals in virtue of which it is a civilization, it can be truly understood and judged only in terms of its own scale of values, its own rules of thought and action, and not of those of some other culture: least of all in terms of some universal, impersonal, absolute scale . . . Herder may not be its only begetter, but the idea that variety is preferable to uniformity, and not simply a form of human failure to arrive at the one true answer . . . this radical departure is altogether modern. The ancient world and the Middle Ages knew nothing of it.
>
> (Berlin 1976, pp. xxii–xxiv)

How far are we pushed by the holism we have just endorsed towards the sort of relativism ascribed here to Herder? The question is worth exploring, because a commitment to relativism would make holism look suspect; it would suggest that holism conflicts with the realist attitude that most of us find spontaneously appealing. But the question about relativism is also worth exploring for other reasons. Looking into the relation between holism and relativism will help us to develop our understanding of where our holism leads. It is only as we see the relativistic and non-relativistic ramifications of the doctrine that we can fully understand what it involves.

The holism endorsed here rests on a theory about how our thoughts get to latch onto things in the world, a theory that does admittedly have a relativistic flavour. The theory is that, if we are to be able to follow rules of thought intentionally, and if we are to remain mutually scrutable in this enterprise, then our thoughts must be directed in the first place to those basic rules—those objects, properties, and other propositional elements—that are picked out by the extrapolative inclinations that we share with our fellows: in particular, by our extrapolative inclinations as they operate under conditions that are normal and/or ideal, surviving negotiation across times and persons. What do we think of when we form judgments involving a property like being red, or being a box, or being a game? We think of that property, whatever it is, whose extension is fixed by the way in which we actually extrapolate, under suitable conditions, from certain exemplars. To be red is to be such as to look red in favourable conditions. To be a box or a game is to be such as to look similar in appropriate respects, again under favourable conditions, to these or those examples.

It will help us in formulating and exploring this result if we introduce the notion of a concept. I have generally avoided talk of concepts up to this, because I think that in many contexts talk of that kind can be misleading. It can lead us to conceive of thinking as a matter of applying concepts to things, in the way in which speech involves applying words to things. And that way of depicting the enterprise of thought puts the emphasis in the wrong place: it suggests that the rules to be followed in thinking are not the objective constraints represented by properties, propositions, and the like, but conventional norms of correct inner speech. But now the danger of such misrepresentation is probably past and it will help the discussion if we put the notion of a concept into circulation.

A person has a concept of something, let us say, if and only if she is able to think about it. The ability just to form beliefs and desires that involve the item is not sufficient, contrary to many philosophical accounts. The person must be able to fix on the object or property or operation in question with a view to forming rational and true beliefs about propositions that involve it; she must be able to try and respect the requirements of that entity for the truth of those propositions. She will need the capacity to track the object through time, as she tries to determine if it is still thus and so. She will need to have the capacity to identify the property across different bearers, as she tries to decide whether something hitherto unencountered also possesses it. And so on in other cases.[5]

There are a number of comments that need to be added to this bare account of concepts, in order to guard against misunderstanding. First, it is possible to possess a concept parasitically on other individuals, as with the manner in which most of us possess the concepts of quarks, valencies, and genes, but in what follows I will always have non-parasitic concept-possession in mind (Putnam 1975, essay 12). Second, to possess a concept is inevitably to have certain beliefs about the item in question but there need not be a sharp boundary between those intimately associated beliefs and other beliefs about the item. Third, the words used by someone to express what a subject believes will presumably be fully appropriate only if they reflect the way in which the subject fixes on the items involved in the content. This last observation impacts on the relation between words and concepts. It means that two words or phrases may refer to the same object or property or whatever but reflect different ways of fixing on that item and so not express the same concept. If the concepts expressed are different—as presumably with the concepts expressed by 'Cicero' and 'Tully', 'human' and 'featherless biped'—then there will be an obstacle to the intersubstitution of the phrases, *salva veritate*, in ascriptions of belief; if the concepts are the same, then this obstacle will disappear. (For more, see Peacocke 1992.)

With the notion of concepts in hand, we can restate the relativistic sort of doctrine that is involved in our holism. The referents of our basic concepts—the concepts involved in our following basic rules of thought—are fixed in a way that involves the convergent inclinations of our community. It is a priori knowable, it is knowable just in virtue of understanding how the referent of the concept is fixed, that something is red if and only if it is such as to look red to normal observers in normal circumstances. Or at least something of that kind is a priori knowable; I do not mean to commit myself necessarily to a particular view of colours. More generally, where C is a concept that answers to a basic rule of thought, it is a priori knowable that something x is C if and only if x is such as to elicit a sense of salient similarity with x-exemplars under normal and perhaps ideal conditions. The point is that the referent of such a concept—the object or property or function onto which it latches—is fixed in a way that involves our extrapolative responses. The referent is that object or property or function or whatever that serves in the actual world to elicit relevant responses, under normal and ideal conditions.

We can sum up the position by saying that under our holistic view basic concepts are response-privileging or response-authorising (Pettit 1991b). A basic concept is response-authorising in the sense that its referent is picked out in such a fashion that the relevant response—the inclination to extrapolate—cannot lead us astray under favourable conditions. Under favourable conditions, under conditions that survive negotiation across times and persons, the response represents an authorised mode of access to the referent in question. It cannot fail to hit the target-referent; it is definitive of which entity is the referent of the concept.

I say that basic as distinct from non-basic concepts are response-authorising. But I should emphasise that what I mean by basic concepts are simply the concepts of basic rules: the concepts we form in becoming responsive to rules of thought that we do not intentionally follow by intentionally following other rules of thought. Basic concepts in this sense need not be ontologically or epistemologically basic and in making the distinction between basic and non-basic concepts I do not mean to commit myself to any corresponding form of foundationalism. Thus the concepts that are basic for a subject or culture at one time may cease to be basic at another, and the concepts that are basic for the subjects in one culture need not be basic for the subjects in another. The concept of a regular figure may be basic for a subject or culture at one time and come to be understood later on the basis of a mastery of other geometrical notions. People who rely on interpersonal interaction to fix the identity of the rules that concern them will presumably converge on the matter of which of their concepts are basic, which not. But otherwise there may be little stability in where the distinction falls.

People may want to reject the distinction between basic and non-basic concepts, not because of an association with foundationalism, but because it suggests that basic concepts can be mastered one by one; this suggestion would conflict with the intrasubjective holism defended in the first chapter. But again I should emphasise that the distinction to which I am committed does not conflict with such holism. It may be that although I follow a certain rule of thought other than by following certain distinct rules, still I could not follow it unless I was also able to follow other rules. It may be that although a certain concept I form is basic, still I could not be in possession of that concept without being simultaneously in possession of other concepts.

Why are only basic concepts response-authorising, on our view? Non-basic concepts will be mastered in a way that derives from mastery of a number of basic concepts. They may be mastered in such a way, for example, that the basic concepts can be used to provide an explicit definition: a bachelor is an unmarried man, and so on. This dependence on the mastery of basic concepts means that with non-basic concepts there is no single, distinctive response that is definitive, under favourable conditions, of the range of things to which the concept applies. And so the users of the concept need not ever enjoy an invulnerability to error or ignorance of the kind that is available, under favourable conditions, to the users of a basic concept.

The response-authorising view of concepts has a relativistic flavour, as already mentioned. But the matter requires further explication. There are at least three different ways in which it might be thought to connect with relativistic commitments and these need to be spelt out.

One radical way in which the response-authorising quality of basic concepts might seem to involve relativism is this. It might seem to carry the implication that in any discourse the subject-matter is constituted, in the last analysis, by the subjective responses that direct people in applying the relevant concepts. Talk about colour reduces to talk about colour-sensations. And, more generally, talk about any entities picked out by the concept C reduces to talk about the responses that serve, under favourable conditions, to identify the things to which C applies. This position may be described as reductivism. It is relativistic so far as it makes human responses the very subject-matter of any discourse to which they are relevant.

A second way in which the response-authorisation doctrine might seem to involve relativism is in suggesting that since our responses are definitive of what is red, or what is a box, or what is a game, they must be creative in the old sense intended by idealists. This suggestion would put us in the company, not of subjective idealists like Berkeley, but of those objective, more or less Hegelian idealists who make the community of thinkers into the creative subject: the subject such that if non-thinking things exist, they exist by virtue of its recognition. This sort of idealism is relativistic so far as it has human responses call into existence the subject-matter of any discourse to which they are relevant. Human responses are no longer the subject-matter of the discourse; if anything, they are the subject-maker.

A third and last way in which the response-authorisation doctrine might appear to support relativism is in making human beings, if not the makers of the things they recognise in thought, at least the measure of what is true and false. The fact that it is a priori that something is red if and only if it looks red in normal conditions means that there are limits on how far people can fall into error or ignorance in those conditions. Error is ruled out, because if something looks red then it is red. And ignorance is ruled out, because something looks red if it is red, given some plausible assumptions (see Holton 1992). The message is general. The fact that for any basic concept C it is a priori that something is C if and only if it looks saliently similar to C-exemplars in favourable conditions means that in such conditions there are limits on human ignorance and error about certain C-matters. The doctrine in play here is a sort of anthropocentrism, because it makes the world epistemically friendly to human beings (Goodman 1978, Putnam 1982). That it is relativistic appears in the fact that it makes human responses the measure of what is true in any discourse to which they are relevant. Man is the measure, as it used to be said: the measure, as distinct from the matter or the maker, which he would become under the reductivist and idealist doctrines.

Anthropocentrism Entailed

We must now explore the connections between the response-authorisation doctrine and these relativistic doctrines. Take the concept of redness and assume that it is a basic concept governed by an a priori biconditional of the form: something is red if and only if it looks red in normal conditions. Does the response-authorising quality of the concept mean that we have to be relativists about redness, and more generally about colour? Does it mean that we have to renounce realist intuitions that colour is a property of objects that we recognise in them, not anything of our own imagining or creation? The question is usefully raised with colour concepts, because they easily lend themselves, unlike many other basic concepts, to a non-realist construal. If we can argue that response-authorisation has a limited impact even on realist intuitions about colour concepts, then that should be particularly telling.

But before we consider the question raised about the concept of redness, we must rehearse the sort of story to which our ethocentric account would commit us with this concept. According to that account, people develop a competence with the concept of redness in virtue of having certain sensations in the presence of red objects—having certain habits of response—and in virtue of being involved in practices of negotiation about the circumstances under which those sensations reveal that something is red. The word 'ethos' that generates 'ethocentric' picks up both aspects of the account, since it may mean habit or practice.

The story about the concept of redness to which the ethocentric account would commit us carries no suggestion that people are immediately conscious in themselves of red-sensations and that they define the property of redness as the property possessed by something that produces red sensations in them under normal conditions: conditions such as those that prevail in good sunlight for people who are not colour-blind, and so on. Such a story would be extremely implausible, for a number of reasons. It requires people already to have the introspective, and relatively sophisticated, concept of red-sensations. It requires them already to have the complex concept of normal conditions. And it makes the concept of red things—the concept of redness proper—a non-primitive concept.

The sort of story that goes naturally with an ethocentric account of conceptual competence is much less reflective and intellectual. We already gave a sketch of it in Chapter 2 (see Pettit 1990a). People have red-sensations—things look red to them—as a matter of primitive experience of the world. Those sensations may not be the objects of introspective awareness but they will have an impact on what people find similar and on their extrapolative inclinations. They will make English postboxes, ripe tomatoes, and heated metals similar in a salient respect. This enables people to use such examples then to indicate a certain property, namely the common colour. What colour? *That* colour, they say, pointing at relevant examples. The examples make the property in which they are interested salient and the concept is ostensively identified by reference to those examples.

Or at least it is provisionally identified in this ostensive way. For it turns out that sometimes a ripe tomato looks different by their lights—and, no doubt, ours—from how it does at other times, and indeed that it looks different as between different people. This offends against a supposition that its colour is stable. The way people make sense of the variation, given the supposition of colour-stability, is to find a feature of the occasions when it looks different, or of the individuals to whom it looks different, that marks them off as not counting. Thus the colour they identify by reference to certain examples as *that* colour is not whatever colour-property the objects present, but whatever property they present under conditions that can be allowed to count.

Given this story about the concept of redness we can see how it can come to be a priori knowable that something is red if and only if it is such as to look red to normal observers in normal circumstances. There is no suggestion that those who master the concept do so by learning and applying that biconditional, as in the implausible story that we rejected. The biconditional belongs to theorists, not to participants in the relevant practice. Theorists register how the participants fix on the property that they refer to as redness and, introducing the concept of normal conditions to identify the conditions that survive the practice of negotiation, theorists use the biconditional to capture an important implication of how participants carry on. Although participants may have no notion of normal conditions in their repertoire, and although they may not even have reflected on the sensation of having something look red, their practice ensures that it is indeed a priori that something is red just in case it is such as to look red in normal conditions.

This ethocentric story of how people master the concept of redness directs us to conditions such that it is in virtue of satisfying them that users have a competence in the concept. The conditions, as we would expect, involve the two elements signalled by the word 'ethos': on the one hand, habits of response, on the other, practices of negotiation. The idea is that the participant mode of access to the concept is via the satisfaction of those conditions: via their satisfaction, rather than via an understanding of them. The story does not claim to give the theorists who understand it a competence in the concept of redness; it claims only to identify the conditions that make sense of how competence is secured. The story does not analyse the content of the concept and thereby make that content accessible; it explains how the concept gets to be used and its content to be accessed.

The ethocentric story just told for the concept of redness contrasts with the implausible story according to which people gain competence in the concept via a grasp of the biconditional. But it also contrasts with other stories, in particular with other stories that would also make sense of the traditional view that it is a priori knowable that something is red just in case it is such as to look red in normal circumstances. Here is one example (Johnston 1989). People do not conceive of redness as the property possessed by something that produces red sensations under favourable conditions; they do not access the concept of redness, as under the implausible story considered earlier, via the biconditional linking redness with red-sensations. Rather, so this story says, people gain access to the concept of redness, as they gain access to any concept, through learning a set of platitudes that link redness with other things: a set of platitudes that give the concept its place in the web of belief. But the set of platitudes that supports the concept of redness, the story continues, includes propositions that entail that something is red just in case it looks red to certain observers in certain situations. And so it is a matter of a priori knowledge, for anyone who understands the concept of redness, that that biconditional holds.

Like the manifestly implausible story we considered earlier, the platitudes narrative purports to tell us something about the application conditions of the concept of redness. It does not claim, in the manner of the implausible story, that that biconditional spells out the application conditions that guide those who use the concept. But it does say that the biconditional reflects the conditions in play among such people. My ethocentric story, on the other hand, abstracts from any particular account of the application conditions that guide the users of the concept. It says that whatever the platitudes in play among the users, it is surely the case that they apply the concept on the basis of their sensations and that they correct the cues their sensations give them in order to maintain intertemporal and interpersonal constancy. And that being so, it points out that theorists are in a position to hold it to be a priori that something is red for the participants in a discourse if and only if it looks red to them under conditions that survive negotiation: under conditions that count as normal. The ethocentric genealogy derives the a priori biconditional from reflection on the competence or possession conditions of the concept, not from any particular account of its application conditions.[6]

The comments that we have been making about the concept of redness generalise to all basic concepts. In each of these cases the holist picture means that we think it is a priori that the concept applies to something if and only if that thing engages appropriate extrapolative responses under conditions that survive negotiation across persons and times; the holist picture means that those responses are authorised by the concepts. And in each of these cases the biconditional is a priori, and response-authorisation holds, because the holist picture involves an ethocentric story as to how the users of the concept come to master it. The picture commits us to a story under which the competence or possession conditions of the concept—though not necessarily its application conditions among users—ensure the a priori status of the biconditional.

So much for why we think that a concept like that of redness is response-authorising, being governed by an a priori biconditional that links redness with certain human responses. We may now turn to the question we set out to explore, as to whether this view of the concept of redness commits us to a denial of realist intuitions, in particular to any brand of relativism.[7] That question breaks down into three separate issues. Does the doctrine commit us to reductivism? Or to idealism? Or to anthropocentrism? I will discuss these in turn.

Does response-authorisation mean that talk and thought about redness reduces to discourse about sensations of red: that human beings who employ the concepts are really addressing their own responses? This result might follow, if the biconditional for the concept of redness were thought to define the content of that concept: if the suggestion was, as under the implausible story considered earlier, that people grasp the concept of redness through grasping that biconditional. But the result does not follow if we think in the ethocentric way about how people get the concept going.

The reductivist proposal supposes that we can master the discourse of colour-sensations independently of access to discourse about colour; it supposes that the sensation-discourse is independently given as a mode of thought and talk to which we might think of reducing colour-discourse. And that is a bizarre assumption on the approach we have taken. According to the ethocentric story, we form the concept of redness in things independently of having any concept of redness in sensations: the sensations serve to highlight similarity-classes of colour but without necessarily becoming objects of awareness themselves. A natural extension of the story would be to say that we form the concept of red-sensations derivatively from the concept of red things: red-sensations present themselves as the sensations occasioned by red things, at least under conditions that are not discounted in the course of negotiating about discrepancies. And in such a case there can be no question of reducing colour-discourse to discourse about colour-sensations.

So much for the reductivist form of relativism. The next question is whether the response-authorisation doctrine is likely to support an idealist story. Is it likely to support a picture under which the entities posited in colour-discourse exist but exist only in virtue of the recognition they receive in the discourse? Is it liable to lead us into thinking that properties like redness are the creation of observers: in particular, the communal creation of those observers who acknowledge one another as presumptive authorities in determining what is red?

I do not think that there is a plausible linkage to be found here but there is an argument that suggests such a connection. It goes like this. According to a response-authorisation thesis, people's responses make it the case that the concept of redness applies to something; their presence, under normal conditions, ensures that the concept fits the object. But the concept of redness applies to something if and only if it is red. So, according to the thesis, people's responses make it the case that the thing is red. To say that, however, is to support nothing less than idealism: it is to hold that redness and the other colours are properties that depend for their instantiation, and in that sense for their existence, on the epistemic responses of human beings.[8]

In response to this argument, I distinguish. There are two quite different readings of the claim that people's responses make it the case that the concept of redness applies to certain things. And equally there are different readings of the conclusion derived from it, that people's responses make it the case that certain things are red. The claim may be that people's responses shape those things in such a way that they fall under the preexisting concept of redness. This proposition would certainly involve something like idealism. Or the claim may be that people's responses shape the concept of redness in such a way that it falls upon those preexisting things. And that proposition has nothing of idealism in it. The first proposition represents the things to which the concept of redness applies as being moulded by our responses. The second represents the things as independent, given entities; the role of our responses is merely to determine which of them fall within the extension of the concept of redness.

The second reading is implicit in the ethocentric genealogy of colour-concepts. The story as to how we get the concept of redness going is, quite clearly, not an idealist narrative. We essay thoughts and assertions involving a property that we identify on the basis of certain exemplars. What property do we manage to engage with? What property do we fix upon as the referent of our concept? The genealogy provided directs attention to the red-sensations that we experience under normal conditions: under conditions where there is no obstacle to intertemporal and interpersonal constancy of response. According to the story developed, the property that we fix upon, the property that provides the referent of our concept of redness, is that property whose instances evoke red-sensations in normal observers under normal circumstances. And there is no reason to think of that property as constituted by our recognition of it.

Take the world as populated, independently of us, by a great range of objective properties; we may think of these in any of a variety of ways. With a concept like that of redness, the question arises as to what determines that the concept will hook onto this property rather than that: say, onto this reflectance property, to take a plausible sort of candidate, rather than some other (Hilbert 1987, C. Hardin 1988). In maintaining that the concept is response-authorising, so our genealogy shows, all that we mean is that that question is to be answered in a particular fashion: the concept hooks on to that property, whichever it is, that evokes red-sensations under normal conditions.

There is another way of emphasising the non-idealist character of the sort of response-authorisation claim illustrated in the case of redness. This is to point out that on our approach the assertion that a concept is response-authorising is, precisely, an assertion about the concept, not an assertion about that of which it is a concept: not an assertion about the property or object or operation in question. It is to say that the reference of the concept is determined in such a way that our responses are privileged under certain conditions: they are not capable of leading us into ignorance or error. It is not to say anything about the property or object or operation in itself, and so a fortiori it is not to say that that entity is dependent on us in the fashion envisaged by the idealist (see Peacocke 1992, chap. 1).

If someone who defends response-authorisation is making a point about how our concepts get their referents determined, equally someone who denies that a concept is response-authorising will be making a point in the theory of reference. He will be claiming that the concept has its reference fixed in a manner that leaves open the possibility that even normalised or idealised responses can lead us astray. He may say that the reference of the concept is fixed in a more or less Platonic fashion, by no known naturalistic device. Or he may say that it is fixed in a way that relies only on response-indifferent, natural connections, connections that are not reflected in our dispositions to make judgments. He may say, for example, that the concept of redness is the concept of a property that is causally connected with us in a certain manner: specifically, in a manner that does not affect our sensations of redness.[9] This will allow him to think that the referent of the concept can be fixed in such a way that our red-sensations, even our red-sensations under normal conditions, may lead us astray as to what is red and what isn't.

If we reject reductivism and idealism about something like colour-discourse, then we are committed to thinking that the discourse, if successful, directs us to entities that exist independently of our recognition; we are provisionally committed to the main elements in a realist vision of the discourse. We should note the qualification about the success of the discourse and about the provisional character of the commitment. The ethocentric account of a discourse will never allow us to be anything more than provisional realists, for whether we are to think that the discourse does direct us onto independently existing entities is always subject to the test of negotiable convergence. We are justified at any moment in being realists about a discourse, only on the assumption that participants will continue to succeed in negotiating convergence, filtering out conditions under which they respond alike. The question of whether the discourse deserves to be taken realistically is always *sub judice*: in particular, *sub judice conversationis*. Failing the achievement of convergence, we will be forced to adopt an error theory of the discourse; we will be forced to the conclusion that people are misled in thinking that there is anything there on which they are commonly targetted (Mackie 1977).

We have seen that the response-authorisation doctrine to which our holism commits us does not involve a belief either in reductivism or in idealism. The point has been argued· with the concept of redness but it goes through for relevant concepts generally. The last question is whether response-authorisation means that if human beings are not the subject-matter or the subject-maker of their discourses, they are at least the measure of what is true and false there. The question is whether we are committed to an anthropocentric form of relativism, if not to a reductivist or idealist version.

Here, finally, we have to concede. By our account of response-authorisation, the doctrine means that people in normal conditions cannot be in error or in ignorance about whether something that is presented to them is red. More generally, it means that people in normal or ideal conditions cannot fail to be led, or fail to be led appropriately, about whether something presented to them falls under a basic concept that they employ. In such conditions human beings are the measure of truth-value. They are authoritative about what is what, and so the anthropocentric form of relativism imposes itself upon us. This is scarcely surprising. Any story that puts people's responses and practices at the centre of things, any story that has the referents of concepts fixed ethocentrically, is bound to put people themselves in a central position.

The anthropocentrism forced upon us does not mean that we can ever be certain, individually or communally, that we are not in error or in ignorance; we cannot, because we can never be certain that conditions are normal or ideal: we may always come to revise our view on that matter in the light of later divergence over the same sort of case. Neither does our anthropocentrism mean that we are immune to error or ignorance over a wide range of cases; we may be authorities about whether something in normal conditions is red or not, but we will not be guaranteed against error and ignorance over more complex questions: say, over how many red things there are in the universe. But, all of this notwithstanding, the global anthropocentrism generated by our holist picture of things, generated in particular by the response-authorisation doctrine, is a striking and surprising thesis. It offends against a deeply ingrained tradition of thought, a tradition that has been described as endorsing an absolute conception of what there is (McDowell 1983, and see also Nagel 1986, Wiggins 1980, and Williams 1978 for commentary; the best spokesmen for the tradition may be Quine 1970 and Smart 1963).

The tradition in question marks a distinction between two grades of concept and indeed two grades of knowledge. On the one hand we have concepts like the concepts of secondary qualities that direct us to features of reality that have a title to attention only for creatures with our particular interests, sensibilities, modes of thought, or whatever. On the other hand, so it is alleged, we have concepts that direct us to features of reality that are of enduring significance, features that call for recognition in the most objective, encompassing picture of what there is. The first sort of concept answers to certain human perspectives, and may be important in our aligning ourselves with what there is, but it has no significance in the ultimate, perspective-transcendent vision of reality; it has no tenure in nature. The second sort of concept is made of more enduring stuff and enjoys permanent tenure; it is represented as the sort of concept that must appear in the absolute as distinct from the perspective-relative conception of what there is. If the untenured concepts are those of common sense and the special sciences and the arts, the tenured ones are the concepts that will figure in what is depicted as the only truly detached representation of reality: the austere and comprehensive vision of the ultimate science.

Why does our anthropocentrism run into conflict with this conception of things? The absolute conception motivates the idea that there are at least some distinctively objective areas of discourse—presumably, in mathematics and fundamental science—where human beings never have any immunity to error and ignorance. There are some areas where our concepts track their objects in such a way that the possibilities of human ignorance and error are unbounded.[10] If we take this view, we cannot think that these concepts are such that at a certain limit human ignorance or error about their instantiation is ruled out. We must think that the concepts in question are not response-authorising: that they leave room for our responses to fail to lead us, and to fail to lead us aright. In short, we must reject the global form of anthropocentrism that the holist picture entails.

Anthropocentrism Embraced

The votaries of the absolute conception will have a variety of questions to raise about global anthropocentrism. Doesn't it make all discourse as species-centred as colour-discourse? No, it need not. The responses relevant to some discourses, unlike colour-sensations, may be such that we can hardly conceive of other intellects that lacked them. Alright, the opposition may say. But doesn't global anthropocentrism mean that our basic concepts are all relational and second-rate: each is a concept of that entity, whatever it is in itself, which occasions such and such a response, as the concept of nauseating food is the concept of that sort of food, whatever it is, that occasions nausea? No, for a familiar reason. Anthropocentrism about a concept will have this effect only if the response involved has an intrinsic, introspectible character: only if it is individuated independently of anything it purports to represent. The feeling of nausea is individuated in this way but not, intuitively, the inclination to group these objects with that, as things that are saliently similar.

But this is not the place to respond to these more or less detailed objections. The main issue is whether we can establish the consistency of global anthropocentrism with robust philosophical intuitions. In particular, can we reconcile the anthropocentric vision with our instinctive, realistic view of how we relate to the world: with the view that learning about the world is a matter of discovery, not invention? I believe that we can and in this sub-section I would like to say why. I take two claims that give expression to a realist view of things and I show that we can endorse each of them under our ethocentric brand of anthropocentrism. The first claim is a statement of what I call epistemic servility, the second a statement of cultural openness.

To believe in epistemic servility is to hold that in seeking out knowledge in a given area we have to strive to attune ourselves to an independent reality; we are epistemically servile in our relationship to the world. The most radical form of anthropocentrism would deny this servility. It would represent participants in a discourse as dictators about what is the case and would make them, on that basis, immune to ignorance and error. What they say goes; they do not discover facts, they invent them. I wish to argue that the anthropocentrism to which we are committed is not of this radical variety.

Take the concept of U-ness that used to be in vogue—courtesy of Nancy Mitford—among what we might describe as the Sloane Square set or, for short, Sloanes. To speak of lavatories is U, of bathrooms non-U; to lay cloth napkins at table is U, to lay paper napkins non-U; and so on through a myriad of equally trivial examples. I assume that there is something distinctively collusive in the way Sloanes use the U-concept: that as they individually decide whether something is U or non-U they look over their shoulders to make sure they stay in step—the community is the authority—rather than looking to the thing itself to see what profile it displays. In other words I think that whether something is U or not is a matter of the say-so of those in the appropriate set; the members of that set have an authoritative, dictating role in regard to the concept. That they have this role is borne out by the fact that as the regular bourgeoisie tries to get in on the game, Sloanes are notorious—at least in the oral tradition—for shifting the extension of the U-concept.

The most radical form of anthropocentrism would hold that immunity from ignorance and error comes from the fact that with the relevant concepts, people have the dictatorial role that Sloanes have with the concept of U-ness. It would undermine the idea that getting at the truth in relevant discourse is a matter of discovery, not invention. The first thing I want to argue about the ethocentric admission of response-authorisation is that while it privileges certain responses by participants in a discourse, it does not invest those participants with this sort of dictatorial authority. It leaves untouched the ordinary view that in seeking true beliefs in a relevant area, even true beliefs formed in normal or ideal conditions, we have to try to get in tune with an independent authority: a reality that dictates whether the beliefs we form are in fact true. It leaves epistemic servility in place.

The most striking way of establishing this result would be to establish that under the ethocentric admission of response-authorisation, even normally functioning and normally or ideally positioned subjects have to be seen as getting things right in virtue of their access to an independent realm, not in virtue of their say-so. This would be to say that even with subjects for whom there is an a priori assurance against ignorance and error about certain propositions, getting things right is not a matter of dictating how they shall be. As it happens, I believe that something like this can be established.

There is an intuitive contrast in respect of the dictatorial dimension between U-ness and ordinary response-authorising concepts such as that of redness. The contrast, I maintain, testifies to the fact that we can accept a response-authorising story about ordinary concepts and still think of ourselves, still think even of normalised or idealised subjects, as occupying an epistemically servile position: we can still think of ourselves and of normalised or idealised subjects as having to strive to get in tune with an independent authority. The Sloane Square set, or at least those in normal mode, do not face any task of attuning themselves to an independent authority. What they say goes. Not so with those of us who make judgments of colour and the like; not so, even if it is assumed that we are normally functioning and normally or ideally positioned. Or so I say.

How are we to mark this alleged distinction between U-ness and redness: in particular, redness on the ethocentric sort of account sketched in the last section? It may strike some that here would be a good place to introduce a distinction that came up earlier, in Chapter 2, a distinction that is frequently discussed in the literature (Davies and Humberstone 1981). Although it is a priori that something is red if and only if it looks red in normal conditions, redness may be contingently or necessarily connected with that response. Redness may be identified with the property that the normalised response discriminates in the actual world, so that it can fail to be picked out by that response in other worlds. Or redness may be identified with whatever property the normalised response discriminates, so that it cannot fail to be picked out by that response at any world. The first, contingent reading of redness serves realist intuitions better than the second, for it means that there are more possibilities of error open; it means that had our colour-sensations been different, they would have led us astray. So perhaps the difference between redness and U-ness is that redness invites the contingent reading, U-ness the necessary reading.

I prefer the contingent reading of response-authorising biconditionals, as indicated in Chapter 2; as I put it there, I want to think of the relationship between a rule and the associated extrapolative inclination as contingent. But I do not think that this marks any contrast between the concepts of redness and U-ness. Even with the biconditional linking U-ness with Sloanes, it is possible to offer a contingent reading that expands the possibilities of error. It is possible to understand the biconditional so that what matters for U-ness at any world is what the Sloane Square set says in the actual world, not what it says at the world in question. We can set things up, just as we could with redness, so that even normal Sloanes would have gotten things wrong if they had broken with the responses found in the actual world.

So how then are we to mark the distinction between redness and U-ness? That normal observers judge that something is red establishes that it is red, that normal Sloanes judge that something is U establishes that it is U. So where is the alleged difference between redness and U-ness? Where does the exercise of a dictatorial role show up in the U-ness case, if it is present only there?

Given that red-sensations determine the referent of the redness-concept, U-responses the referent of the concept of U-ness, the only place for a systematic difference between the two cases is in the things that in turn determine those responses. And when we look to what determines the responses then we do indeed find a significant difference. U-responses are determined, under my characterisation of the case, by the efforts of Sloanes to keep in step with one another in their classification of things. But clearly red-sensations do not generally spring from such collusive machinations, even if people sometimes succumb inappropriately to group pressure.

When subjects see something as red, even when normally functioning and normally positioned subjects see something as red, they do so, or so we generally assume, because the thing presents itself—and, if there is no mispresentation, because it is—a certain way. What way is it such that the thing's being that way leads them to see it as red? The only plausible answer is: its being such as to merit the description 'red'—in short, its being red—leads them to see it as red. As the preferred, contingent reading of the biconditional would have us frame the point: the thing's being the particular way that is contingently linked in the actual world—not in all others—with red-sensations, that is what leads people to see it as red. Nothing of the kind can be said in the case of U-ness. It may be because of exposure to an instance of the property that Sloanes judge something to be U but the causally relevant property of the instance in eliciting that response is not the U-ness itself; it is rather the fact that this is a type of case that Sloane rangers generally regard as U.

Someone may baulk at the claim that a property like redness can be causally efficacious in producing sensations. But this would be a mistake. Consistently with the ethocentric genealogy for the colour-concepts, it is possible to think of the colours of things in a variety of ways.[11] But no matter how we think of them, we can make sense of the claim that they are causally relevant to people's sensations (see Campbell forthcoming). For any sensation that a colour produces, it is true that that sensation will be attributable to more basic, microphysical properties of the object and of the light that falls on the object. But we can think of the colour as having a higher-level causal relevance to the sensation, provided that the object's having that colour more or less ensures that no matter how things are disposed at the micro-physical level, they will be disposed so as to produce the sensation. The colour may not produce the sensation in the most basic sense available for that term but it will be causally relevant provided that it programs for that sensation in the sense of programming introduced in the first chapter. An analogy will remind us of the point. A square peg is blocked as I try to push it through a hole. What produces the blocking in the most basic sense is this or that overlapping part. But the squareness is still causally relevant. Given the dimensions of peg and hole, the squareness ensures that there will be some overlapping part—may be this, may be that—that blocks the peg; it programs for the blocking, even if it does not produce it itself.

The contrast between the concepts of U-ness and redness, or between the concepts of U-ness and ordinary response-authorising concepts, is that U-ness fails a certain test that redness passes. In Plato's *Euthyphro* Socrates asks whether something is holy because the gods love it, or whether the gods love it because it is holy. We might ask in parallel whether something evokes the U/red-response in normal subjects because it is U/red or whether it is U/red because it evokes the U/red-response. We know that in one sense it is true that something is red or U because it evokes the U- or the red-response among normal subjects: that it evokes the appropriate response ensures that the property in question is the redness or the U-ness property. But we see now that there is another sense in which the converse is also true for redness—it is true that something evokes the red-response in normal subjects because it is red— whereas there is no sense in which the converse holds for the U-ness case. The sense in which the converse holds for redness is that the property of being red can be thought of as something common to red things that ensures that normal observers will have a certain experience. Ask the Euthyphro question with U-ness and the unambiguous answer is that something is U because it evokes the U-response in suitable subjects. Ask the question with redness, and the answer is less straightforward: in one sense something is red because it looks red to normal observers, in another sense it looks red to normal observers because it is red (Wiggins 1987, p. 106).[12]

There is nothing very anomalous about the claim that the 'because' runs in both directions in these cases. An eraser is elastic and bends. Does it bend because it is elastic, or is it elastic because it bends? In one sense—if you like, a criterial sense—it is elastic because it bends: the capacity to bend is what marks off elastic things. In a parallel sense something is red because it looks red to normal observers: the capacity to look red to such observers is what marks off red things. But in another sense—if you like, a causal sense—the eraser bends because it is elastic: the elasticity is responsible, in part, for the bending. And in a parallel sense something looks red to normal observers because it is red: the redness is responsible, in part, for the thing's looking red. Something is U because it looks U to appropriate subjects, since being U is defined by reference to that U-response. But it is not the case that something looks U to such subjects because it is U; it is not the case that its U-ness is responsible in any part for evoking the U-response: the U-response is driven by different pressures.

I want to make one further comment on the red-U contrast. Those who use the U-concept exercise their will in determining the things to which it will apply, the property to which it will refer. But things would be just as inappropriate were they to be guided by causal pressures that emanated from sources other than the nature of the things and property in question (see Peacocke 1989, 1990). Take the concept of things that are ping as distinct from pong and assume that there is, as psychologists report, a surprising degree of convergence in what people regard as ping-things rather than pong-things: ice-cream is ping, soup is pong, and so on. I presume that what produces the pressure responsible for the convergence is often something as irrelevant as the sound of the name for the thing in question and it should be clear that if that is so, there is no more epistemic servility with the concept of what is ping—though there is a servility of sorts—than there is with the concept of what is U.

So much by way of defending the first claim advertised for this sub-section: that the anthropocentrism to which we are committed allows us to admit epistemic servility in the various discourses to which it applies. The belief in epistemic servility is distinctive of a realist view of our relationship to the world and it is reassuring that our anthropocentrism does not undermine that belief. Now to the second claim we promised to address: not only does our anthropocentrism allow us, and indeed incline us, to believe in epistemic servility; it also allows us to believe in what I describe as cultural openness.

There will be cultural openness, as I understand the idea, just in case divisions between cultures do not create absolute blocks to commonability. Human thought is commonable if it is possible for someone to know which rules of thought another is following, identifying those rules in such a manner that she can follow them herself. Cultural openness obtains if commonability is not necessarily confined within cultural boundaries, however those boundaries are drawn. It is possible, at least in principle, for people to lay hold of the things that the members of other cultures discern and to claim them as a common possession. There are no impermeable divisions in the human community.

Cultural openness is important for realism, again because of the realist stress on discovery rather than invention. If the things people know are indeed a matter of discovery, then it appears that they ought to be commonable across cultural divisions. Anything that one individual or group claims to discover should be something that, short of contingent difficulties, they can invite other individuals or groups to corroborate. Otherwise the claim that it is a matter of discovery begins to look hollow. The anthropocentric relativism to which our holist argument commits us would gain a real sting, therefore, if it meant that there can be impermeable divisions in the community of human knowledge: if it meant that there are really many different communities of knowledge, communities that are hermetically sealed off from one another. Those who proclaim such divisions and such boundaries, for example those who exult in the alleged diversity of mutually incomprehensible conceptual frameworks, are not relativists in any marginal or shrinking sense; they figure among the most prominent challengers to realism (see Davidson 1984, essay 13).

There is nothing we can have against the idea that there might be good evidence that certain individuals follow rules of thought, and know certain things about the world, while no suitable indication is available as to which rules they follow or what things they know. Imagine the situation where we come across a group of people—an exotic human culture—whose members have had no contact with the rest of us. The situation invites observations that parallel earlier lessons. We may be able to tell more or less immediately that the individuals imagined are thinking subjects like us and that they therefore know certain things. If we have not had any interaction with them, however, then we cannot know which rules they are actually following, at least not in the way in which they know this. We may observe them closely and may conjecture that certain of the rules followed—rules that they identify on the basis of inclinations shared among themselves—are this or that or the other rule that we identify on the basis of inclinations of our own. But short of interaction we can only conjecture. We cannot claim the rules as a common possession.

Let it be granted, then, that there may be other human beings, in distant cultures, who know things such that we are not actually in a position to know what they know: who know things that are not actually commonable by us. The question now is whether there is some position we could occupy, at least in principle, which would enable us to achieve commonability and come to know what they know. If there is, cultural openness obtains; if there is not, cultural openness fails. The question in particular is whether our argument for holism and anthropocentrism allows us to countenance such a position and to endorse cultural openness.

Our argument has the implication that there will be commonability in practice, there will be intercultural access, just in case we as interpreters can enter community with those whom we seek to understand: just in case we can be taken successfully within the scope of our subjects' attention, as they begin to assume that constancy between their responses and ours is important, and as they manage to vindicate that assumption, achieving the constancy and consensus required. There will be commonability in principle therefore, so long as it remains possible for this sort of interaction and community to get going. So does our argument give us any reason to deny the possibility of achieving such interaction and community with the members of exotic cultures?

No, it does not. Our argument does not force us to think that there are cases where it is impossible for us as interpreters to be invested by the subjects of another, contemporary culture with the authority of fellow-interactants. More strongly, the argument does not lend the slightest support to the view that commonability will ever be blocked in this way. It shows what commonability properly requires—the practical incorporation of the interpreter in the community of her subjects—but it says nothing on the question of whether there will ever be a bar to such incorporation. To that extent therefore, the argument keeps cultural openness in place. It gives us no reason to expect the closure endorsed by those who allege the existence of mutually incomprehensible conceptual schemes.

But if our argument gives us no reason to expect cultural closure, if it gives us no reason to be pessimistic about cultural openness, does it offer us any reason for optimism? Does it give us any positive reason to think that there is always a possibility of achieving commonability across cultures? Not in itself, it doesn't. But it sits easily with an observation and a conjecture, which together would suggest that there will always be such intercultural access.

First, the observation. In making assertions in any area of discourse, we clearly aspire to be intelligible, and potentially persuasive, for any other human beings who may give us a hearing (Apel 1976, Habermas 1984–89, Pettit 1982). We do not delimit in advance the possible audience that we are addressing; short of some communal idiocy or conspiracy, we do not put restrictions on those whose lack of understanding or agreement can in principle count and weigh with us. This feature of our practice, this universality of communicative intention, shows that even if we cannot provide a theoretical case for the permanent possibility of intercultural access, we are all of us, just in virtue of participating in the practice of assertion, committed to the belief in such access.

Now to my conjecture. What would be likely to happen if some of us systematically failed to understand one another's putative assertions in a given area of discourse, or at least failed to understand them as things worthy of assertion and belief? Would we be likely each to taper our aspirations of intelligibility to some restricted community, a sub-culture of fellow minds, thus introducing divisions in the knowledge community as a whole? My conjecture is that we would not: that we would rather come to treat the assertions as less than assertions, regarding them as utterances that serve to express feelings—or whatever—but not to record anything that others should also be expected to endorse (Price 1988). We would come to see the discourse in the debunking manner in which emotivists claim that we have actually come to see moral discourse; they claim that moral discourse offers an occasion for the venting of feeling, not an occasion for intellectual debate.

The observation on the universality of communicative intentions shows that whether or not cultural openness obtains, we believe in it. The conjecture about how we would respond to breakdowns of communication, if it is correct, shows that our dispositions are such that we actually make the belief in cultural openness true. We are arbiters on what is or is not an assertoric practice—it all depends on how we treat the utterances involved—and the conjecture suggests that for any practice that does not promise the access we expect, we will deny it assertoric status: we will not treat the utterances at issue in the manner of proper assertions. The joint message of the observation and the conjecture, then, is that a practice will get established as assertoric among us only if it proves capable of sustaining the assumption of cultural openness.

But these remarks in defense of the permanent possibility of such inter-cultural access are surfeit to present requirements. The main point to be stressed is that the anthropocentric or ethocentric relativism to which our argument for holism commits us does not undermine the belief in cultural openness any more than it undermines the belief in epistemic servility. And so, it ought not to cause excessive anxieties for the realist who wants to emphasise the fact that human knowledge is a matter of discovery, not invention.

One final matter, in conclusion to this discussion of cultural openness. Our socially interactive view of thought does lead us to expect a certain variation between societies in people's sense of salient similarity. It suggests that in one society, with distinctive paradigms of colour on offer, people may divide up the spectrum differently from how they do in another. It suggests that with differing needs in view, people may trace artefacts through time in different ways: for one group this ship may be the same ship as a vessel of old, despite the replacement of every plank, rope, and sail; for another group it may count as a substitute. And so on. Does the fact that our position leads us to expect such diversity mean that, contrary to what we have just argued, it supports an expectation of cultural closure?

No, it does not. Variations of the kind illustrated do not mean that the members of one culture cannot know what the members of another think and know, say about the colour or temporal identity of something. They will be able to know this sort of thing, if they can cotton on to the extrapolative habits of the other culture and borrow or invent terms to reflect those habits. Of course, it may be that as a matter of human psychology no one can simultane-ously think, so to speak, in the rival terms—say, the rival colour or identity terms—of the two societies; the concepts may not be co-tenable, as we might say. But this does not mean that the concepts are inconsistent with one another or that someone who thinks in the one set of concepts cannot tell what is thought and known by someone who thinks in the other. The variation in question does not give proof of cultural closure.

The upshot of our discussion in this final section is that while the holism explicated and defended in the earlier sections does entail a sort of relativism—an anthropocenrism, as we have called it—this relativism is not something at which we need baulk, even if we are moved by the sensibilities and ambitions of the realist. It allows us to expect epistemic servility and cultural openness with any discourse we practice; it allows us to think that learning about the world is a matter of discovery, not invention. Thus we may return with confidence and enthusiasm to the holist refrain. There is no obvious reason why we cannot relish the thought that human beings depend superveniently on their relations with one another for the realisation of the capacity to think. The lesson of the preceding chapter was that we should join with individualists in rejecting the notion that human beings are compromised as agents by the operation of social forces and laws. The lesson of this is that nevertheless we should join with holists in seeing human beings as essentially social agents, as agents whose ability to think, or at least to think commonable thoughts, is a social property.

NOTES

1. Notice that if rule-following has to be public in some sense, and if public rule-following has to rest on a foundation of interpersonal interaction, then there will be a transcendental argument available to support holism: the only possible foundations for the capacity to think will include the enjoyment of social relations.

2. I am grateful to Christopher Peacocke for some helpful remarks on this matter. Notice that the communicability that Peacocke (1988, p. 488) discusses requires only discriminability, not commonability.

3. The point made in this paragraph is that if we think of you as taking the rule under the description 'the rule that answers to our convergent responses', then the description cannot serve to pick out a rule, in the event that it is false. See Kripke 1980.

4. I conjecture that there is no other possible way in which I, or indeed you, could know which rule is involved. It would be interesting to explore ways of establishing that conjecture.

5. This sort of capacity approach makes the ontology of concepts relatively unproblematic. There will be a certain concept of something just so far as there is a possibility of fixing on the item under consideration in the relevant manner. See Peacocke 1992, chap. 4. For other examples of a capacity approach, see Geach 1957, McGinn 1984.

6. I have benefitted from very useful conversations with Mark Johnston on this point. Christopher Peacocke (1989, 1990, p1992) introduces the notion of the possession conditions of a concept. As mentioned in a footnote to Chapter 2, I find his approach congenial.

7. A more general question is whether the doctrine has non-realist implications, since relativistic implications are a sub-set of non-realist ones. For a discussion of the general question, see Pettit (1991b).

8. Michael Devitt drew my attention to this sort of argument. Related matters are discussed in chapter 13 of a new, forthcoming edition of Devitt 1984. Devitt's own view is that a belief in global response-authorisation, but not in local response-authorisation, involves 'constructivism'. On a global view even the concept of causality is response-authorising, or definable in response-authorising terms, and his thought seems to be that it cannot therefore be used to vindicate the realist intuition that various properties and other entities are at the causal source of our responses. I see no problem. We may think that the type of relation picked out by the concept of causality is at the causal source of the responses with which it is linked, and we may therefore take a realist view

of the relation; we may allow a sort of self-subsumption whereby the concept of causality applies to the relations between the relation it picks out and the responses with which it is associated.

9. This sort of theory would exemplify something akin to what Gareth Evans (1982, chaps. 3 and 4) describes as the photograph model of reference.

10. I ignore the sort of pragmatic paradox which Smart (1987, p. 183) acknowledges, when he says that if my notion of what is physically possible is determined by my core scientific beliefs, it will be paradoxical to envisage the physical possibility of those beliefs being false. In Pettit 1991b, I stipulate that realism admits the unbounded possibility of ignorance and error only with what I describe as substantive propositions; this stipulation is designed to get around the general sort of difficulty involved here.

11. For example: if to be red is to be such as to look red in suitable circumstances, then we may take the redness of a thing to be the higher-level state of having a lower-level state that produces the required effect on observers; we may take it as the lower-level state operative in the thing; or we may take it as the disjunction of the lower-level states that do the job required. In terminology introduced in Chapter 1, we may take the redness of a thing to be a role state, a realiser state, or a disjunction of possible realiser states.

12. The claims made here about the Euthyphro test do not directly conflict, despite appearances, with those made by Mark Johnston (1989, pp. 171–73). He argues, on one interpretation of his remarks, that if the concept of redness is subject to an a priori biconditional then it cannot be that something is *such as* to look red in normal conditions because it is red. But that may be so, consistently with epistemic servility: it may still be that something looks red in normal conditions—looks red, as distinct from being such as to look red—because it is red. If there is a difference between us, under this construal of his views, if it bears on the significance that we attach to our respective claims. I do not see it as a grave compromise of realist commitments, that one should deny that something is such as to look red in normal conditions because it is red. The important thing, in my book, is that one doesn't have to deny that something looks red in normal conditions because it is red. I have benefitted from conversations with Mark Johnston about these matters.

III

MIND, SOCIETY, AND THEORY

Preview

The third and last part of the book looks at the significance of our individualistic and holistic commitments for social and political theory. By social theory I mean the discipline of explaining the doings of individuals in society and the patterns of social events. By political theory I mean the project of evaluating the different social structures that political activity enables us to contemplate as alternatives. Social theory is the common property of the social sciences, ranging from anthropology, history, and political science to more theoretical disciplines like economics and sociology. Political theory is associated in particular with law and philosophy but it also has a part, and indeed a rapidly growing part, in economics and political science.

The fifth chapter is devoted to social theory and is divided into three sections, each concerned with a different variety of explanation. The first section deals with the enterprise of intentional explanation: the explanation of the responses of individuals, singly and in groups, in terms of belief, desire, and the like. I take a causal-information view of such explanation. According to this view, explaining an event, whether well or badly, means presenting information on its causal history: information on causally relevant properties of the antecedent situation. What then are the relevant factors to cite in intentional explanation? Beliefs and desires, of course, as everyone admits. But also, I claim, the sorts of evidential and valuational factors that deliberatively impact on agents, under the picture developed in the first two parts of the book. We must be able to identify the sorts of considerations that make certain beliefs imperative for the agents involved, and the sorts of valued properties which make certain options more or less compelling. We must be able to see things, as it were, from their point of view; we must be able to practise *Verstehen* or interpretation.

Interpretation is generally taken to raise problems as to how we can ever be in a position to know what the contents of another's mind are and, more radically, as to how the contents of a mind can be more or less determinately fixed. But these problems resolve themselves within the ethocentric perspective developed in our discussion of thinking and in our defence of holism. If we are all authorities for the contents of one another's thoughts, authorities to which we each defer in the identification of the rules of thought we seek to honour, then we will collectively introduce a determinacy of content that is unavailable at the purely individual level. And if we fix content by means of this mutual attunement, then we each have a ground on which to expect mutual accessibility, at least in principle: the attunement involved in fixing the content of another's thought also offers us a means of access to that content.

This discussion of intentional explanation or interpretation involves a picture of the human mind that I describe as inference-theoretic. The picture is that we each form our beliefs and desires, not just under any old causal pressures, and not just under pressures that make for rationality; we do so, more specifically, under the sorts of influences that allow us to be represented as reasoning or inferential creatures: as instances of *homo ratiocinans*, not just *homo rationalis*. This inference-theoretic picture allows us to see decision theory as a significant discipline but it contrasts with an image of human beings that is often associated with decision theory: an image under which the pressures that shape beliefs and desires, the pressures that mould the agent into a rational profile, cannot be explicated in a process of reasoning by the agent; they are not fitted to present themselves as premises in theoretical or practical deliberation. I argue that this decision-theoretic picture would make the possibility of thought problematic and that we have to embrace the sort of inference theory described. Decision theory retains a significant place in my conception of things but this associated picture has to be dropped.

Intentional interpretation is the most common sort of explanation to be found in the social sciences, being the stuff of many anthropological, historical, and other approaches. The second and third kinds of explanation that I consider both contrast with intentional interpretation in abstracting from the contents of individual thoughts and even from the identity of the individuals thinking. They offer prescindent, high-level perspectives on the doings of individuals and groups and on the vagaries of social and historical pattern. We saw that our holistic individualism, in particular our holism, enables us to make excellent sense of intentional interpretation, resolving the problems of indeterminacy that often seem to dog the enterprise. But with prescindent explanation it appears to raise a problem. According to holistic individualism it is individuals, in particular intentionally intelligible individuals of a socially embedded and engaged kind, who are responsible for what happens on the social stage. So how can there be a place for explanations that abstract away from individual thought and identity?

The prescindent patterns of explanation considered in the fifth chapter are, on the one side, structural and historicist explanation and, on the other, rational choice explanation. These abstract from individual psychology in quite different ways but in each case I argue for the legitimacy of the abstraction. I defend a sort of explanatory ecumenism, according to which we may believe that an effect is the result of a train of micro-events—and ultimately we may even be committed to the causal fundamentalist view that every effect is the result of micro-physical causes—and still hold that there are a variety of significant macro-style explanations available. In particular I argue that we may embrace the sorts of accounts offered either in structural or historicist explanation, or in rational choice, and do so without resiling from the view that intentional explanation tells us about what are relative micro-causes of the things explained.

Structural explanation is the sort of account that explains something in accordance with what I described in Chapter 3 as a social-structural regularity. An example might be the explanation of a rise in crime by an increase in unemployment. Historicist explanation, which may or may not count independently as structural, is the sort of account that traces an event, not to relatively close antecedents, but to antecedents removed in time: in particular to antecedents such that, whatever occurred in their immediate wake, it is alleged that they were more or less bound to lead to the event explained. An example might be the explanation of the collapse of communism in Eastern Europe by the recognition in the late 1980s that the Soviet Union was not going to intervene to preserve a Communist regime. Structural explanation abstracts in level from intentional explanation. Historicist explanation may or may not abstract in level but at least it abstracts in time, for it treats immediate intentional antecedents as more or less dispensable. It suggests, for example, that even if this or that Communist government fell by the immediate efforts of such and such protestors, the efforts of those individuals were not essential: had those individuals not acted at that particular time, some other individuals would have made up for that inaction at a later moment.

Structural and historicist explanations have been put in the dock by much recent social theory, the main argument being that we get more causal detail as we go to lower levels, and more immediate antecedents, and that the more detail we get, the better the explanation we offer. It may seem that holistic individualism—in particular, individualism—pushes us in this direction, since we are certainly committed to the view that we get more detail on the causal genesis of a social event by going to the intentional deliberations of those immediately involved. But I argue that we can resist the push, on the grounds that more detail does not necessarily make for better explanation.

The crucial element in the argument is suggested by the program model of causal relevance advanced earlier. One of the things that the program model shows is that there is a comparative sort of information on causal micro-process that is available at a macro-level and that is not extractable just from the micro-level explanation itself. The fact that the water is boiling in a closed flask gives us the information that no matter how the molecules are configured, the flask is more or less bound to crack: it gives us the information that the causal process involved will work, not just under the actual-world configuration, but also under configurations associated with various other contingencies, various other possible worlds. The micro-level account gives us contrastive information, information that marks off the actual-world process from the process in those other possible worlds, whereas the macro-level story gives us comparative information about that process: information on a commonality between the actual-world process and other possible processes. The comparative information provided may not be available to someone who possesses just the micro-level story. Someone who knows that the flask broke because of the impact of this particular vibrating molecule on that molecular bond in the surface may be unaware that the water was boiling or, given the water was boiling, that the flask was more or less bound to break. Thus the macro-level story may be informationally indispensable.

Given this point about comparative information, it is relatively easy to establish that, consistently with thinking that there is an intentional explanation available for a certain event, we may acknowledge the interest of a structural or historicist explanation as well. We may have tracked all the individuals involved in this year's crime-statistics and we may be able to pinpoint the novel factors that accounted for the rise over the previous year's level without being aware that there was an increase in unemployment in the relevant period or that this increase made it likely that there would be a rise in crime. Thus we may have a lot to learn from the structural explanation. Equally we may be aware of the exact train of events that led to the fall of the East German regime—at least in principle—without realising that it had generally been recognised that the Soviet Union would not interfere to prop up a Communist regime in the area, or that this recognition made it more or less inevitable that those regimes would topple. Again we may have a lot to learn from a historicist explanation. Thus we should reject the idea that there is a choice to be made between intentional explanation on the one hand, structural or historicist on the other. We should be ecumenical or pluralist about explanation.

The other sort of prescindent explanation considered in Chapter 5 is associated with rational choice theory. Such theory assumes that people are more or less self-regarding in their concerns and that they manage, by whatever deliberative means, to be rational in their response to those concerns. I distinguish between two sorts of concerns highlighted in rational choice theory: a concern with economic gain and a concern with social acceptance. Economic gain stands in for those goods that can be produced, supplied, or made accessible by human action: for short, action-dependent goods. Social acceptance stands in for attitude-dependent goods: for goods that one enjoys in virtue of the attitudes of people, such as status, esteem, and affection. Rational choice theory assumes that a great deal of human behaviour can be explained and predicted by how it serves these two sorts of desire.

As with structural and historicist explanation, it is hard to see at first how our holistic individualism would allow us to find room for rational choice explanations. Our commitments, in particular our holistic commitments, mean that we take human agents to reason their way to action, using the concepts that are available to them in the currency of their culture. But no culture encourages people to reason just in terms of the economic and social gains that different options represent, at least outside certain narrow contexts like that of the market. So how can rational choice theory have anything to say about such behaviour?

In response to this problem I admit that rational choice theory may not serve to explain the occurrence or continuance of anything, at least outside marketlike contexts, since it is people's culturally shaped deliberations that account for such results. I argue that what the theory is typically and usefully invoked to explain is rather the resilience of certain patterns. Consider a ball that rolls on a straight line, where the path it follows is guarded on either side by pinball posts. Suppose that the posts serve the following function: they tend to block any outside force that would disrupt the course of the ball and if such a force gets through, they tend to put the ball back on the straight line. That the ball sets out and continues on the straight path may be fully explained by Newton's laws, without reference to the pinball posts. But that it is going to stick resiliently to that path, that it is going to stick to or tend towards that path under any of a variety of contingencies, is explained only by the presence of the posts. The posts are standby causes in the actual world, causes that are not called upon to play any role, but they are activated in the possible worlds where outside forces interfere; thus they serve to explain the fact that in the actual world the straight rolling of the ball is a resilient phenomenon.

I argue in parallel to this that under some plausible assumptions about people's motivation, we may regard rational choice theory as a potential source of explanations for the resilience, as we take it to be, of certain patterns of human behaviour. Here are some people displaying a collectively desirable pattern of behaviour, and displaying it for a variety of motives and reasons, ranging from the very virtuous to the very cynical. Considerations of economic and social acceptance certainly do not explain why that behaviour appears or continues. But they may explain why it is resilient. They will do so under the following assumption: that were people to begin to depart from that pattern of behaviour, then they would find the departure costly in economic and social terms and those costs would deliberatively induce them to resume the behaviour. The assumption required is that self-interest has at least a virtual presence in people's psychology. As things actually are, self-interest may not move them consciously or unconsciously, explicitly or implicitly, but there are hypothetical conditions under which it would certainly come into play.

We see from this example that even if the self-regarding motives postulated in rational choice theory are only standby causes of actual behaviour, which is all they will generally be in many areas of life, still they may play an interesting theoretical role. We see our way towards a deflationary way of understanding rational choice theory, and one that may help to make the theory more generally acceptable. I try to indicate with the help of some examples that the representation of the theory is not only attractive, but fair to the actual practice of rational choice theorists. Thus I am able to sustain the sort of explanatory ecumenism defended with regard to structural and historicist explanation. People may act on the basis of the considerations identified in intentional interpretation but that does not rule out the legitimacy of structural and historicist theorising on the one hand, or rational choice theorising on the other.

The sixth and final chapter takes us to political as distinct from social theory, to evaluation as distinct from explanation. We argued in Chapter 5 that it is possible to embrace holistic individualism and be extremely ecumenical about explanation. Does the doctrine have a similar impact on evaluation? No. I argue that it forces us into more or less sectarian paths, as we consider how to go about evaluating different institutional arrangements.

There are three main sorts of theorising found in modern political theory. The contract-centred approach asks after the political institutions people would choose if they were placed in a suitable position of choice. The value-centred approach asks after the goods or rights that institutions should realise and investigates the requirements of those values. The institution-centred approach takes different values and asks whether they are institutionally feasible, whether the arrangements they require can be reliably set up and sustained. I discuss each of these kinds of theorising, looking at the impact of holistic individualism, in particular holism, on the different enterprises.

Many different political theorists use contractualist devices in support of their conclusions. What distinguishes the contract-centred approach, properly understood, is the fact that it defines the right by reference to what would be chosen in contract; it invokes contractual eligibility as the essence of rightness, not just as a more or less useful test. The approach goes either of two main ways. It may suggest that what fixes a basic structure as the right one to have is the fact that it would emerge in a process of rational bargaining under conditions that it would be rational of bargainers to endorse. Or it may propose that what fixes the structure as right is the fact that it would attract intellectual consensus among a group of people who are committed to finding the best sort of arrangement: say, the arrangement that answers best to considerations that all would regard as relevant. More briefly, the contract-centred approach may be pursued under an economic or under a political interpretation.

I argue that holistic individualism directs us to problems with both variations on the approach. The main problem with the economic interpretation is that in arranging things so that a rational bargaining solution will be fixed for every relevant predicament, it is forced to pay a cost that is quite intolerable within the inference-theoretic perspective supported by holistic individualism. It is forced to assume that people can find their way to solutions that are deliberatively inaccessible, solutions that no plausible train of reasoning would support. The main problem with the political interpretation stems from the fact that any proposal that contractors would allegedly support will be supported for the valuational properties it presents; this, again, is a lesson of the inference-theoretic perspective that we have adopted. But if the contractors support a proposed structure for its possession of such properties, then it is those properties that make the structure the right one to have, not the fact that the contractors would support it. In other words, it appears that the contractual method is not serving to define the essence of rightness; it is serving, at best, as a useful test for pointing us towards the right choice.

It transpires, then, that holistic individualism impacts very negatively on contract-centred thinking in political theory. I go on to argue that it also impacts quite dramatically on value-centred thought. The general idea defended here is this. Atomists admit the wholly solitary condition as a logically possible option for human beings and when they come to identifying the ultimate values by which different socio-political dispensations should be judged, they will naturally focus on values that are realisable, not just in social life, but under the condition of total isolation; otherwise they will have no basis for ranking the solitary condition against the other options. This means that the ultimate political values that have been articulated in the atomistic tradition are restricted to rights and goods that can be enjoyed out of society as well as within. It means that the tradition has ignored the possibility that such inherently social values as cultural cohesion and the rule of law, or the enjoyment of citizenship and political participation, should count as basic yardsticks of assessment; the tradition has been forced to reject such values or to find some derived reasons for admitting them.

The main impact of holistic individualism on value-centred thinking is to release the enterprise from this artificial restriction to values that are capable of realisation out of society, in the condition of the solitary individual. According to holism, no one who is possessed of the full range of human capacities can live a fully isolated life and so the solitary condition need not be regarded as a serious alternative. Consequently we should feel free to consider the possibility that certain inherently social values may represent the ultimate currency for the assessment of socio-political institutions. This is an abstract lesson, however, and I go on to make it more concrete by considering the likely impact of becoming holistic individualists on the evaluative commitments of the liberal.

There are two main sorts of liberalism, one deontological and the other consequentialist, but both approaches are committed to the value of negative liberty: they hold this value to be the unique or primary basis for assessing political arrangements. I argue that adopting holistic individualism, in particular adopting holism, is liable to impact on any liberal's beliefs, leading him towards a distinctive conception of negative liberty. The conception I have in mind strictly belongs not with the liberal tradition but with the republican tradition of classical Rome. This tradition was revived in the Italian republics of the Renaissance and had a major presence in western European thought down to the time of the American and French revolutions.

The notion of liberty that is shared between the liberal and republican traditions is negative: it is freedom from the interference of others in the exercise of certain basic choices. The liberal and the republican differ, so I argue, on what it is to be free from such interference. For the liberal it is to enjoy the absence or expected absence of interference, on whatever basis, with whatever security. For the republican it is to enjoy the secure or resilient absence of interference: essentially, it is to have such a status in law and custom that one is protected against all sorts of contingencies of interference, not just those that actually materialise or are likely to materialise.

Take an individual who manages by cunning and charm to enjoy a great deal of non-interference in a society where she has few legal rights: she may be the *servus sine domino* of ancient Rome, the slave without a master, as distinct from the *liber* or freeman. The liberal is committed to saying that strictly such an individual may enjoy more liberty than a *liber*, though whatever non-interference she enjoys is relatively fragile: the slave may have solid expectations, given her abilities and the actual dispositions of others, but she is vulnerable to the slightest change of humour or heart on the part of others in her society. The republican, on the other hand, will focus on the insecurity of the non-interference won by the slave, rather than on its quantity or extent, and will deny that the slave enjoys liberty proper. He equates liberty with franchise, where franchise has the old meaning of incorporation within a polity and enjoyment of all the rights and privileges that the polity offers. The *servus* is deprived of franchise, even if she avoids interference better than the *liber*.

Why would our holistic individualism motivate liberals to adopt the republican rather than the liberal conception of negative liberty? Why would it lead them to equate liberty with franchise? The liberal notion of negative liberty is non-social, the republican notion is social, and so atomism is going to motivate the use of the liberal notion. But the republican notion of negative liberty—the notion of franchise—has many attractions, in particular attractions that ought to be felt by any liberal, and so the rejection of atomism is likely to motivate an endorsement of the republican notion instead. Or so at least I suggest.

Would the shift from a liberal conception of liberty to a republican conception make for a radical change in the liberal's commitments? I argue that it would. The liberal is led by his conception of liberty to put in place a presumption against state activity, even activity designed to protect people from interference; every form of state activity is itself a sort of interference, under this approach, and it should only be contemplated when it is absolutely clear that it would reduce the level of interference suffered overall. The republican will take a very different view, for short of the guardians of the law being made so strong as to constitute a threat to others, there is nothing inherently undesirable about state activity. The republican must be actively committed, therefore, to investigating the possibilities of productive state activity, in particular activity designed to enhance the protected status of citizens; he cannot fall back on a general presumption against the involvement of the state in people's social lives.

It is one thing to defend a value like franchise, and to argue for a social structure under which the franchise of individuals is suitably advanced, whether in deontological or consequentialist mode. It is quite another to argue that the sort of structure envisaged is feasible: that we can look to it, not just as a utopian ideal, but as an arrangement or set of arrangements that is capable of institutional realisation. In the final part of Chapter 6 I ask whether our holistic individualism leaves us any room for the exploration of this sort of feasibility question: whether, for example, it offers us any resources with which to argue for the feasibility of the republican ideal. The section deals with the third, institution-centred sort of theorising that is to be found in modern political theory.

We will be able to argue for the feasibility of a proposal if we can show that certain plausible controls on social life would serve to keep the proposed structure in position, under realistic circumstances. I suggest that the controls to which it is most natural to look under the picture developed here are those which rational choice theory invokes. The question with any candidate structure is whether it satisfies people's self-regarding ends in such a measure that we can expect it to survive in a run-of-the-mill population: in a population where many will be virtuous but some will not, and where even the virtuous will have occasional lapses. Can we expect the sort of structure required for the promotion of franchise to survive in a population where rational self-interest may rarely rule but may always strike? Can we expect the structure to be sustainable under a pattern of motivation where self-interest always has at least a virtual presence?

The structure required for the advancement of a value like franchise will involve a relatively rich mix of state agencies, and so the question of feasibility is a pressing one. The richer the state apparatus, the greater the power available to individuals within it, and the greater the power available, the more difficult it is likely to be to keep individuals honest in the pursuit of their official roles. I go along with much contemporary thought in agreeing that fear of the law alone, and more generally a centralised sytem of policing and sanctioning—an iron hand—will hardly be sufficient to guarantee the satisfactory operation of such a structure. So what else might serve to sustain the operation?

It appears that we must be able to identify decentralised sanctions that can be used to keep the structure efficient and in place. But there is a well-known problem, often described as the enforcement dilemma, which seems to represent an obstacle to decentralised sanctioning. The problem is that if people are not virtuous enough to behave well spontaneously in a certain area, then they are hardly likely to be virtuous enough to police and sanction one another into behaving well. I agree that this is a problem for a system of decentralised sanctioning that requires people to sanction one another intentionally: to sanction one another out of a desire to provide such sanctioning. But it need not be a problem for a decentralised system of non-intentional sanctioning. So is there a suitable system in view?

One possibility involves the invisible hand and will be more or less familiar. The competitive market, or so it is alleged, represents a decentralised system of non-intentional sanctioning: people discipline those who try to sell above the competitive price by taking their custom elsewhere, but they do this because it suits them, not because it has a sanctioning effect. There have been recent proposals that a variety of behavioural patterns may be feasible, so far as compliance is rewarded and deviance punished in a parallel fashion: specifically, by the operation of something like a tit-for-tat pattern of interaction among people. Those proposals envisage an invisible-hand pattern of sanctioning on a par with the competitive market. But the proposals involved are sketchy and I argue that none of them looks to be particularly persuasive. If there are invisible hands that may usefully serve in the promotion of civic virtue, they are more likely to be found in traditional devices involving the sharing and the splitting of power: devices, as it happens, that republican theorists have often explored. So much, then, for the invisible hand. Is there any other sort of decentralised, non-intentional sanctioning?

There is. In the discussion of rational choice we saw that people are assumed to be interested, not just in action-dependent goods, but also in attitude-dependent goods: in goods, such as honour or prestige or affection, that are available just from the attitudes of others, and independently of whether they do anything. I argue now that it is possible, at least in principle, that people may police and sanction one another in the pursuit of various activities—say, those required by a republican structure—so far as they are in a position to notice any offences and, noticing them, to develop a poorer opinion of the offender. The possibility envisaged involves people sanctioning one another non-intentionally, but not because their intentional actions have an unintended sanctioning effect, as with the invisible hand; rather, because the sanctioning is effected by the formation of attitudes, which does not involve intentional action at all: here we may speak of the intangible hand as distinct from the invisible hand. Assume that certain activities are commonly taken to be desirable and assume that equally they are commonly taken to be more or less likely to be noticed by others. If the activities do not involve serious countervailing costs, there is good reason to think that they will get more or less reliably established in a society. And so we may be pointed here towards a way of making something like a republican structure feasible. We may be able to rely on intangible hands to support such a structure.

The suggestion fits, as it happens, with the long republican tradition of emphasising that where virtue is not spontaneously forthcoming, it can be elicited by exploiting people's interest in position and distinction, honour and prestige. One example that illustrates it nicely is the system we have in many countries for getting jurors to be conscientious in determining guilt. We vet the jury to make sure that no one has a special interest in the outcome and we shroud the jury in confidentiality to protect jurors from intimidation. We make sure that it is commonly known among jurors that they should be conscientious and, by encouraging discussion among members of the jury, we arrange things so that anyone who is being casual or malicious in forming her judgment will be exposed to ignominy among her peers. In setting up this regime we rely on precisely the sorts of pressures associated with honour and prestige; we rely on the operation of an intangible hand for the disciplining of the parties involved.

The lessons supported by the example of the jury generalise to a variety of other institutions, particularly to institutions of the kind that must function reliably if something like a republican structure is to be feasible. But I am not uncritically optimistic about the possibility of deploying intangible hands in support of such institutions. I argue that there are many ways in which the intangible hand can fail to achieve a desired effect—say, the effect of promoting conscientiousness among jurors—and that it is the business of institutional analysis and design to identify those possible modes of failure and to try and develop remedies for them. The discussion of the intangible hand is not meant to show that there is no problem about designing feasible institutions, only that there is no reason to despair about the project of institutional design. I suggest that this project is of the utmost importance in contemporary political thought, as it was a program of the utmost imortance in the institutional thinking conducted by classical figures like Machiavelli, Montesquieu, and Mill.

5

Social Theory

Our discussions in Part I gave us an insight into two aspects of human competence. It threw light on our capacity as intentional subjects to be guided, more or less rationally, by more or less rational beliefs and desires. And it threw light on our ability as thinking subjects, not just to be more or less rational, but to act intentionally with a view to displaying such rationality. The view that human beings are intentional, thinking subjects amounts to what we have described as intentional psychology; it is the view of ourselves and one another that we adopt and apply in our day-to-day dealings.

In Part II we developed the main lines of a social ontology on the basis of this intentional psychology. We argued in particular for two propositions. First, individualism triumphs over collectivism: whatever structural regularities obtain in social life, they do not compromise our capacity as intentional, thinking subjects; they do not give the lie to intentional psychology. Second, holism triumphs over atomism: our capacity as intentional, thinking subjects—in particular, our capacity for thought—is superveniently dependent on the enjoyment of social relations; it contingently but constitutively requires a community of interacting subjects.

As the second part of the book applies the intentional psychology defended in the first, so this final part will apply the approach—the holistic individualism—defended in the second. Our task is to look at the implications for social explanation and political evaluation, for social theory and political theory, of the combination of individualism and holism that we have been led to espouse. In this chapter we look at the implications for social theory, and in the next at the implications for political.

The discussion in this chapter falls into three sections. In the first we look at the most basic form of social theory: that involved in the intentional interpretation or explanation, particularly the satisfactory interpretation or explanation, of people's attitudes and actions. In this section we develop and sharpen the view of agents that has been emerging in earlier parts of the book; I represent it as an inference-theoretic, as distinct from a decision-theoretic, account of people. We go on in the second and third sections of the chapter to look at more formal styles of social theory: structural and historicist explanation, which we examine in section 2, and rational choice explanation to which we turn in section 3. The one sort of explanation is in apparent conflict with the individualism endorsed in Part II, the other with the holism endorsed there. In each case I argue that the conflict can be resolved.

In arguing this line, I defend a sort of explanatory ecumenism according to which intentional, structural, historicist, and rational choice styles of explanation are complementary enterprises. This explanatory ecumenism may come as a surprise, in view of my defence of causal fundamentalism, but we shall see that the two positions are not in any tension. Causal fundamentalism means that there are no emergent causal forces that appear at non-physical levels: no forces such as Descartes might have associated with the soul or Durkheim, under an uncharitable reading, with social facts. But explanatory ecumenism says that, nonetheless, there are distinctively interesting sorts of explanation to be found at different, non-physical levels. Non-physical sorts of explanation may offer us information not available from physical explanations and different non-physical sorts of explanation may offer us different forms of information.

This explanatory ecumenism means that the ontology that we have presented does not translate into a corresponding methodology for explanation; it means for example that the individualism defended in Chapter 3—ontological individualism—does not entail methodological individualism. But though the ontology does not translate into a corresponding methodology in this way, it does have a methodological impact. The holistic individualism we have developed accounts in good part for the view of social theory we shall be presenting.

Intentional Interpretation

Explanation

Intentional interpretation, as I understand it here, is the procedure of invoking attitudes of belief and desire to make sense of how agents—in particular, human agents—respond. It amounts to nothing more or less than the intentional explanation of such responses. If I describe it as interpretation, that is because my account of intentional explanation gives an important role to what social theorists commonly think of as interpretation. As we shall see, it emphasises themes that are important in the interpretative or hermeneutic tradition of social science (Dallmayr and McCarthy 1977; Macdonald and Pettit 1981, chap. 2).

There are two questions which we must raise as a preliminary to discussing intentional explanation, and indeed social explanation in general. First, what is required for something to count as an explanation, whether in intentional psychology or more generally? Second, what is required for an explanation, in particular an intentional explanation, to count as satisfactory? I will deal briefly with these questions.

I should stress that the sense of explanation I have in mind in raising the questions is that which usually applies—we meet an exception later—when we speak of explaining an event by its ancestry: this is explanation in the sense in which it is generally assumed to direct us to a cause. Such explanation contrasts with the constitutive sense of explanation that applies, as we saw, when we speak of making naturalistic sense of, or naturalistically explaining, something like rule-following. Going along with the general assumption that the sort of explanation I have in mind directs us to a cause, I shall describe it as causal explanation. But here too caution is required. Causal explanation in my sense merely means explanation that directs us to a cause. Such explanation may fail to be causal in another, narrower sense: it may fail to direct us to the sort of formulation of a regularity—the sort of theoretical law or principle, however rough hewn—that would commonly be described as causal. I hold that intentional explanation is causal in my generous sense, and indeed I believe that it bears on causal regularities, but I do not think that it offers us what would ordinarily be called a causal law or principle. The principles of intentional psychology—for example, the principles linking certain belief-desire profiles to appropriate actions—are a priori in a way that familiar causal principles are not. Familiar principles, including the sort of rough-hewn law that links thrown bricks with broken windows, are individually susceptible to discovery, and individually vulnerable to falsification, in a dramatically differ-ent manner (Macdonald and Pettit 1981, chap. 2; Pettit 1986c).

Now to the question of what is required for explanation. In the first part of the book we introduced the notion of a causally relevant property or, as we sometimes called it, a causal power. We stressed that in employing this notion there was no need to make a commitment on a variety of questions raised by causal powers: for example, on whether there are objective causal powers that are primitively efficacious, or on whether a property can be causally relevant through engaging with suitable laws or just through identifying conditions that are necessary, sufficient, or perhaps just probabilifying, relative to certain events. We went on in Part I to give an account of how causal relevance, however it is understood, can be possessed by properties at different orders—for example, neuro-physiological and psychological properties—in relation to the same event; this is what we called the program model. In Part II we showed how this model fits with a general picture, a causally fundamentalist picture, under which causally relevant properties at higher levels are all causally relevant—again, however relevance is understood—in virtue of the causal relevance of properties identified in physics; for practical purposes we take the relevant properties, the relevant physical powers, to be properties like mass and charge and spin but we naturally leave open the possibility that physics may surprise us in various ways.

With the notion of causally relevant properties in play, and with a picture of how properties at different levels can be relevant to one and the same thing, we can deal rather peremptorily with the question of what is required for an explanation, or at least for an explanation in the putatively cause-giving sense that engages our interests. We can say that to explain an event or condition is to provide some information on properties that are causally relevant to its appearance. And we can say that to give an intentional explanation in particu-lar is to provide some information on causally relevant, intentional properties.

I do not intend to offer a defence of this account of explanation; the account may be taken as stipulative. But I would just mention in support of the definition that it is more likely to allow too much to count as explanatory rather than too little. It allows any statement that gives causal information, however slight, however oblique, however unhelpful, to count as explanatory. The statement may not constitute a very satisfactory explanation but it will count as an explanation of sorts. This approach may be unfaithful to some of our intuitions, since we tend to equate the explanatory with the satisfactorily explanatory. But in erring on the generous side, it errs at least in the right way. There will be little harm done if we count too many sorts of information as explanatory. The important thing is not to count too few, and this danger does not loom; if there are explanations of events that do not inform us about properties that are causally relevant in any sense, then they certainly do not bulk large (see D. Lewis 1986, pp. 221–24).

But though I offer no defence of my account of explanation, I should at least link it to the well-known account defended by David Lewis (1986, essay 22; also see Ryan 1970). According to Lewis, to explain an event is to provide some information about its causal history. So how does this account relate to mine? Any information on causal history is certainly going to be information about causally relevant properties. But is any information about causally relevant properties going to be information on causal history? Causally relevant properties, under my approach, may be properties of relatively higher or lower level. If higher-level causally relevant properties are thought to figure in the causal history of something—if causal history is understood generously—then any information about causally relevant properties is going to be information about causal history and my account will be straightforwardly equivalent to Lewis's. But what if causally relevant properties are thought to have a part in causal history only when they figure at a suitably basic level: say, only when they are micro-physical properties? How, in that case, does my account relate to Lewis's? I shall argue that even under this austere account of what causal history is, any information on causally relevant higher-level properties will count as information on causal history.

Consider the molecular process—for our purposes, the micro-physical one—leading to the cracking of a closed flask in which water is boiling; this example will serve us as well here as it has done in many earlier contexts. We have distinguished two sorts of explanation of the cracking. One describes the micro-physical process itself, identifying the exact vibrating molecule that, under our simplified picture, finally cracks a molecular bond in the surface. The other explains the cracking by reference simply to the fact that the water is boiling: to the fact, in effect, that the molecules display such and such a mean motion. Thus it invokes a higher-level property rather than the momentum of the molecule as causally relevant; it invokes a statistical property of the mass of water molecules as a whole, not a property of the molecule that breaks the bond. Can information about this higher-level, causally relevant property be cast as information on the micro-physical process that leads to the cracking: on the causal history of the cracking, where history is understood in an austere sense?

In order to deal with this question, we need to equip ourselves with an important distinction between two sorts of information that can be provided about something like a micro-physical process or, more generally, about the actual world that involves that process. One sort of information helps us to differentiate the actual world from more and more other possible worlds. The other sort helps us not to differentiate the actual world from other possible worlds, but to relate it to them: to show how the actual world runs on patterns found in a variety of possible worlds. The first sort of information is modally contrastive, the second modally comparative; the first focusses on differences between the actual world and other possible worlds, the second focusses on similarities.

The distinction between the two sorts of information applies neatly to the flask example. Learning the explanation of the cracking in terms of this or that molecule clearly increases our contrastive information on the causal process involved; it helps us to differentiate the actual-world cracking from more and more other possible-world counterparts. But, equally clearly, learning the boiling-water explanation increases our comparative information on the causal process. We may already be in possession of the molecular account and be sensitive to what differentiates the actual process at this level. But still, in being made aware of the boiling-water explanation, we learn something new: we learn that in involving the boiling water the relevant causal process is such that in more or less all possible worlds it will lead to the flask's cracking. Thus we have a positive answer to our question about this example. Getting information about the higher-level, causally relevant property means in effect that one gets information, albeit information of a comparative rather than a contrastive effect, about the micro-physical history of the cracking.

Generalising from this example, we can see that no matter how causal history is understood, our account of what it is to explain something is more or less equivalent to Lewis's. We say that to explain is to provide information on causally relevant properties, he says that it is to provide information on causal history, but in effect the two claims come to one and the same thing.

So much for the question of what is required for explanation, in particular for intentional explanation. The other question on which we promised to comment is, what is required for a satisfactory explanation. If explaining means providing information on causally relevant properties, then explaining something satisfactorily will mean providing satisfactory information on such properties. But what goes to make information satisfactory? It appears that there are two main dimensions of assessment, one of accuracy, the other of adequacy (see D. Lewis 1986, pp. 226–28).

It is required of any good explanation that the information it supplies be accurate or, if inaccurate, then at least not inaccurate in ways that affect the information on the score of adequacy. The explanation may idealise certain detail and this will not matter greatly, provided that doing so does not reduce the adequacy of the information; idealisation may even serve to increase the adequacy, if the detail is in some way distracting. The assessment of explanations on the score of accuracy is not altogether a straightforward business, as this shows. But the assessment of explanations on the score of adequacy is even more complex. The accuracy of information is judged by the relation between the information and the world on which it comments, whether in contrastive or comparative mode. The adequacy of information is judged by a different sort of relation: by how well it engages with the mental set of those to whom the explanation is addressed.

There are a variety of respects in which explanatory information may not fit well with that mental set. The information may be misleading through failing to correct—perhaps even through reinforcing—certain errors among the group addressed. The information may be uninstructive through failing to enlighten them on the matters where they are ignorant; it may provide only knowledge they already have. The information may be inaccessible through presupposing a theory or framework that members of the group do not generally understand. Or the information may be irrelevant through failing to engage with their particular cognitive interests. No doubt there are other ways too in which the information may be deficient. But these are the salient possibilities.

These observations enable us to say that a satisfactory explanation of something, at least relative to a given audience, is one that provides relatively accurate information on causally relevant properties, in particular information that is not misleading, uninstructive, inaccessible, or irrelevant. Applied to the interpretative case, we can say that a satisfactory intentional explanation is one that provides that kind of information on causally relevant intentional properties. These remarks leave it vague how exactly to assess a given explanation, for they say nothing on the weights that should be attached to accuracy versus adequacy or to the different dimensions of adequacy. But they should at least give us an idea of how some explanations come to count as more satisfactory than others. They should help to give content to the notion of a satisfactory explanation.

Explanatory Intentional Properties

Let us grant then that satisfactory intentional explanation involves providing satisfactory information on causally relevant intentional properties. So what are these causally relevant intentional properties? What sorts of properties should we expect to have information about when we succeed in explaining people's responses intentionally? What sorts of intentional properties are causally relevant to the emergence of people's attitudes and actions? I outline an answer to that question in this sub-section and I defend it against an important alternative in the next. The answer gives pride of place to what we might describe as inference theory, whereas the alternative privileges decision theory instead. Because of this highlighting of inference theory, the line I defend connects with the hermeneutic tradition.

To be an intentional system, we saw in the first chapter, is to interact with an environment under the control of intentional regularities: regularities that subsume the inputs, outputs, and existing states of the system under characterisations that relate them to propositions. The regularities relevant in the case of a well-designed or a naturally selected system will tend to be at least minimally rational ones. For such a system, at least as we presented things, there will be characterisations of what it registers, what it believes—including what it believes desirable—and how it behaves such that when conditions are favourable, what it believes will tend to be (evidentially) rational in the light of what it registers, and how it behaves (practically) rational in the light of what it believes.

Under this image of the intentional agent, what are the causally relevant intentional properties? It is sometimes said that the only relevant factors, under this image, are beliefs and desires. But that is shorthand. Under the picture we developed, the causally relevant factors include propositionally characterised perceptions and propositionally characterised beliefs, together with the habits of inference involved in moving from perceptions to beliefs, and from beliefs to other beliefs, and together with the desires involved in moving from beliefs to actions. My seeing that p together with my believing that if p then q and my instantiating the *modus ponens* habit of inference are all causally relevant to my coming to believe that q. My believing that I can make it the case that r, that if r then s, and that it is desirable all things considered that s, are all causally relevant, together with appropriate dispositions of inference and desire, to my coming to make it the case that r. And so on.

This image of the intentional and rational agent suggests that in explaining what an agent does we should look to the contents of the perceptions and beliefs that serve as givens and, assuming suitable habits of inference and desire, explain the beliefs and behaviours that appear in the role of outputs as rationally programmed results. Or at least we should do this in the absence of perturbation: for example, in the absence of the sort of practical unreason where the agent's beliefs about what is desirable fail, no doubt because of familiar perturbing factors, to be matched by suitably powerful desires (Pettit and Smith forthcoming). Why did the agent come to believe that q? Because she believed that if p then q and she came upon evidence that p. Why did the agent make it the case that r? Because she believed that she could do so, she believed that if r then s, and she believed that it was desirable all things considered that s. In each case the contents of the inputs and associated states are causally relevant in generating the outputs, or so at least we assume; the inputs and states program for the outputs in virtue of having those contents. The pattern will be familiar.

We are directed, wherever the pattern holds, to an argument that the agent is taken to have endorsed: if p then q, p, therefore q; if r then s, it is desirable that s, I can make it the case that r, so that is the thing for me to do. This style of explanation involves reconstructing the agent's pattern of inference or deliberation, the pattern of explicit or implicit reasoning that she is taken to have followed. It need not represent the agent as justified in what she did, since it may indicate that the premises in the light of which she responded were not true or were not the whole truth, or it may point us to a slip that the agent made in her reasoning. Still, the explanation does introduce us to the viewpoint of the agent herself. Even if it supports a critical stance on her responses, it will have to invoke an intelligible emotion, a common oversight, a standard fallacy, or something of that kind, to make sense of her failure; otherwise it will leave us in the dark about how she, an intentional and thinking system, came to be moved as she was. This is explanation in the sense of what is sometimes called re-enacting the agent's thought, seeing things as she must have seen them, understanding her from within. It involves understanding her via a process that Vico described as *fantasia* (imagination), Herder as *Einfuehlen* (empathy), Dilthey, Weber, and others as *Verstehen* (understanding). This is the process that is commonly described nowadays as interpretation (Macdonald and Pettit 1981, chap. 2, Pettit forthcoming c).

There are a number of ways in which this style of explanation involves understanding from within, reconstruction of how things look from the agent's perspective. In this interpretative sort of explanation we direct attention to factors that came, explicitly or implicitly, to the attention of the agent herself: factors such as the things allegedly observed and believed. Furthermore, in invoking such factors we take it for granted that a rational agent, though perhaps only a rational agent working under certain limitations of evidence, could have observed or believed them and to that extent we give countenance to the observations and beliefs of the agent. And, finally, in invoking those factors we take as given the habits of inference in virtue of which the factors observed or believed give a rational look to the things believed or done. Or if we do not take them as given, we take them as rationally understandable: we can see why someone might embody the habits and we can replicate how such a person might reason. We can readily get in tune with the reasoning of someone who firmly believes in the potency of luck, for example, even if we are sceptical about the notion.

What of reconstructive explanation in those cases where perturbing factors appear to get in the way and we cannot see the agent as rational, in however weak a sense? What of the case where we see her as self-deceived, for example, or as practically irrational: as harbouring desires that do not answer to her beliefs about what is desirable? If the view that we take of the agent in such cases is to be consistent with the attitude that generally goes with reconstructive explanation, then we must make sense of the breakdowns of theoretical or practical reason in terms of factors that we can imagine occasionally disturbing our own responses. Even here it seems that we must try to see things from the inside, from the point of view of the subject herself.

All of this is to say that there is a conspiratorial aspect to the reconstructive style of intentional explanation. We get on side with the agents under investigation; we get to breathe with them, in the etymological sense of 'con-spire'. We adopt the concepts in terms of which the contents of their observations and beliefs are defined; we countenance the presumptions of evidence in virtue of which those contents seem like plausible things to espouse; we give our seal of approval, or at least of rational understanding, to the habits of inference by which accepting the propositions involved rationally leads to the beliefs and actions under explanation; and, finally, we use our own likely responses as the test of whether certain factors are likely to have caused a temporary breakdown of reason.

The reconstructive style of explanation will be familiar from everyday life, since this is how we make sense of one another in our day-to-day dealings. It is also how anthropologists, or at least ethnographic anthropologists, make sense of their subjects. They enter into the conceptual world of the subjects, as most of us are tempted to put it, and they explain how those subjects come to believe new things and come to adopt one or another course of action by reference to the contents of their observations and their existing beliefs, as expressed in the local concepts. Nor is reconstructive explanation confined to the explanation of subjects with whom we can interact, as in everyday accounting and in anthropology. It also extends, though now it is much more tentatively based, to the historical explanation of non-contemporary human subjects, and to the ethological explanation of non-human subjects (Dennett 1987, essay 7). Reconstructive intentional explanation—in a word, interpretation—is the common coin of a variety of disciplines.

We have generated this picture of interpretation from within the perspective provided by our discussion of intentionality in the first chapter. But that perspective leaves in place an indeterminacy about the contents to be ascribed to intentional states, in particular an indeterminacy that threatens to make it possible always to avoid imputations of error. Does the cat believe, falsely, that the doll it pounces on is a desired tit-bit? Or does it have a belief that is not mistaken: say, the belief that the doll is something to play with? This indeterminacy gives rise to two problems that are often raised in the discussion of intentional interpretation. First, an ontological problem. If there is intentional indeterminacy, then how can we assume that only some of the content-properties ascribable to an agent's observations and beliefs are the appropriate ones? And second, an epistemological problem. Even if we can assume this, how can we claim to know which set of ascribable contents is the one that ought to be ascribed?

The answers to both of these questions become salient, as we saw, in light of the distinction between merely intentional systems and intentional systems that can also think: intentional systems that have the capacity to act intentionally so as to promote rationality in their intentional states. Even if we grant that there is an indeterminacy about the contents that belong to the intentional states of the merely intentional system, there can be no such indeterminacy about the contents of the thinking mind. The thinking system not only has beliefs with contents, as we put the matter earlier, it also has beliefs about contents. The thinking system not only displays intentional states that can be paired off—however indeterminately—with contents, it also tries to be faithful to certain contents in the intentional states it forms; it tries to form such states as circumstances make it rational to hold, given the contents by which the states are identified. This means that there has to be a certain determinacy of content with states of thought: that is, with the intentional states that come under the control of thought. Those states must have such contents as the system takes to be normative for the states, such contents as the system tries to honour.

The fact that thought contents have to be determinate does not in itself explain how they come to be determinate: it does not explain what fixes their identity. But our ethocentric solution to the rule-following problem provides the requisite explanation. The rules engaged by the rule-follower, or at least the basic rules engaged by her, are fixed by the sorts of instances that exemplify them and the extrapolative inclinations of the rule-follower and her community under such conditions as are suitable for convergence. The contents of her various intentional states—the propositions she believes, for example—are not of course basic rules. If she tries to be faithful to them in the beliefs she forms, seeking only to hold such beliefs as are true in the light of those contents, she does that through trying to be faithful to the more basic rules represented by the constituents of the propositions: for example, to the rules of belief-formation represented by properties. But the identity of the propositions that serve as contents to her intentional states will still be fixed in the manner described in our story about rule-following. It will be fixed that the person believes this or that, so far as it is fixed in the rule-following way that she is targetted on this or that property, this or that particular, this or that operation, and so on.

So much for the ontological problem of indeterminacy of content. But, given we can assume that there are determinate contents to the intentional states of thinking systems, how can we establish what those contents are? How can we know that such a system is targetted on these rules and on these contents rather than on any others? If the ontological problem is soluble in the light of our approach to rule-following, the epistemological problem is soluble in the light of our account, developed in the last chapter, of how I can be in a position to know which rules you are following, and to know this in a way that allows me to follow them too. It is soluble in the light of our account of the commonability of thought, as we called it: our account of how I can claim the rules you follow as a common possession.

Under that account, I am in a position to know which rules, say which rules of thought, you are following, if we invest one another with authority on what the rules require and if we manage to sustain a critical convergence about those requirements: we converge case by case or we can identify perturbing or limiting factors to make any divergence intelligible. There is no interpretation available, or at least no knowledge-bearing interpretation, without such social interaction. No interpretation, as we put it, without incorporation. And with interaction and incorporation, interpretation is no great problem. We can each get to understand the rules that another is following, and in particular the contents with which her intentional states engage, just by getting in communal step.

Indeed, not only can we understand in this way the rules that another is following. We may understand those rules better than that person herself, having a sharper sense of where this or that ostensively identified rule leads: having a more tutored sense of where it goes in currently unconsidered cases. But this greater capacity need not be surprising. For example, it need not suggest that we know the person's mind better than she does herself; it need not undermine first-person authority. She may have a certain rule and content in mind without being more authoritative on what exactly that rule and content are. But she may still be more authoritative on whether she really has that rule and content in mind: say, on whether she really believes that content. The point comes out in an analogy. We may understand the words that a person utters better than she does herself but she may be more authoritative on whether or not she sincerely asserts those words.

Given our line on the ontological and epistemological problems of indeterminacy, it should be no surprise that reconstructive interpretation involves a conspiratorial aspect. If we reconstructively explain what an agent believes or does, then we assume that we can know what rules she is conforming to in her thought. But if we come to know what rules an agent is conforming to by entering into community with her, as the lesson of the last chapter suggests, then it should indeed be no surprise that in reconstructively explaining her responses we give her countenance in the conspiratorial mode. To treat someone as a fellow in community, as someone with a voice that we have to heed, is precisely to give this sort of countenance.

But if our line on rule-following and on the recognition of rule-following generates answers to the ontological and epistemological problems that content-indeterminacy raises for intentional explanation, we should notice that the answers are straightforwardly helpful only for everyday and ethnographic explanation, where we can interact with our subjects. There remains a problem with the historical interpretation of non-contemporary human subjects and with the ethological interpretation of non-human subjects, whether contemporary or not.

What is there to say on these cases? My comments will be brief. I think that the greater problem arises with ethological interpretation, for if we accept that non-human subjects are not thinkers, and if we countenance the possibility of content-indeterminacy here, then whatever reconstructive explanation we offer with such subjects, we will have to admit that it is no better than a more or less indefinite variety of other possible accounts. What justification can we have for the explanation presented? Presumably only a pragmatic one, that the properties identified program for the responses under investigation—and may predict further responses—and that they do this programming and predictive job in a way that is easily accessible to us: say, because the concepts invoked are familiar ones. The position will be that which someone like Quine takes to obtain for our interpretation, not just of animal subjects, but also of one another; this is the position that he identifies in defending the indeterminacy of translation (Hookway 1978, Quine 1970).

Things are not quite so bad in the historical case, for here it is reasonable to assume that there is the ontological determinacy of content achieved by thinking subjects. The problem that arises is the epistemological one. Given that we cannot get into interaction with our subjects, given that we cannot achieve the exchange of authority involved in each side's addressing rules on which the voice of the other is taken to be relevant, how are we ever to identify the rules and contents engaged by the members of a past culture, particularly a past culture that is discontinuous with our own? One thing that we must clearly avoid is interpreting the subjects of the culture, on the assumption that they are necessarily addressing rules and contents of the kind that are salient with us; we must strive to identify the practices of their own time and milieu they would have taken as constitutive of what they meant to address. This lesson of interpretation Quentin Skinner and others have emphasised in recent times (Tully 1988). But there is always going to be a degree of underdetermination attached to historical interpretation as distinct from the interpretation of contemporaries. We may be able to identify rules such that we can actually explain any difference of response between us in the construal of the rules, but this possibility of explaining divergence cannot come to be a matter of common assumption between us; it cannot come to be the case that the aspiration on the two sides to be following a rule is contingent on the availability of such an explanation for any divergence. We may not be driven to justify the interpretation we prefer on merely pragmatic grounds, as in the ethological case, but any evidence we adduce in its support is likely to leave other possibilities open.

Inference Theory and Decision Theory

Before we leave the discussion of what is involved in intentional interpretation, one important matter remains to be discussed. The account of explanatory intentional properties presented in the last sub-section is not as uncontroversial as I have made it seem. It fits naturally with our earlier account of intentionality—in particular, the emphasis given to deliberation—and our view of the nature of thought, but it is not the only view that might be taken. It constitutes an inference-theoretic view of the properties that are relevant in intentional explanation and it contrasts with an alternative view that is particularly associated with decision theory. I will try to give a sense of that alternative here, and then explain why I believe that the inference-theoretic view is superior. Those who are not particularly interested in decision theory may prefer to skip the section.

The reason why I characterise the view presented as inference-theoretic is that it gives an important place to inference: to explicit or implicit reasoning (Pettit and Smith 1990, 1992). When things go well with a subject, it explains the formation of new beliefs as the product of (evidential) inference from the contents of observations or of other beliefs; and it explains decisions to act as the product of (practical) inference from beliefs about the options, in particular beliefs about the desirable properties of the options and their possible outcomes, and about the probability of those outcomes: it represents them as the products of deliberation. Three features make it appropriate to say that the approach involves a notion of inference (see Chapter 1 and Cummins 1983, chap. 3). The intentional states invoked in explanation are supposed to be causally relevant to the outputs; the contents of those states are meant—in the light of a familiar principle of inference—to make the outputs appropriate, given the contents of the outputs; and it is no accident that the explaining states give a rational gloss to the explained states in this way: it is in some part because of their contents' rationalising the explained states that they are causally relevant to the appearance of those outputs.

Two assumptions play a crucial role in supporting the inference-theoretic picture. The first is that desires are rendered appropriate, at least when the subject functions properly, by beliefs as to the desirability of what is desired, be that a property or an outcome or an option: desires are rationally grounded (see Pettit and Smith forthcoming). I see that an option or outcome would be fun and, believing that the property of being fun is a desirable feature, I desire the option inasmuch as it possesses that property or promises to realise it. I see that otherwise there is not much to distinguish between the options and I conclude that not only is the option desirable inasmuch as it would be fun or would make for fun; it is desirable, all things considered.[1] And, believing that the option is desirable, I do indeed desire it: I prefer it to the alternatives and am disposed to choose it. The property-desire is grounded in the belief that realisation of the property is desirable and the option-desire in the belief that the option is desirable: that it is better than any of the alternatives. The assumption that when things go right with a subject, desire is supported in this way, can be summed up by saying that in the inference-theoretic picture some beliefs are inherently evaluative, some beliefs are inherently suitable for deliberative justification.

A note on the side. Cognitivists and non-cognitivists about motivation will differ on the connection between evaluative beliefs and desire. The cognitivist holds by the view that such a belief is indeed a cognitive state and that just to be in a state of that kind is to desire the object that it holds out as desirable. The non-cognitivist will deny at least one of those claims. We saw in the first chapter that the view adopted here is not committed to cognitivism, despite the role that it gives to deliberation. Consistently with thinking that normally desire is deliberatively grounded in evaluative belief, we might argue that evaluative belief is not a cognitive state in the sense that the cognitivist has in mind or that such belief, even if it is cognitive, does not necessitate desire in the manner required by the cognitivist. The issue between cognitivism and non-cognitivism is orthogonal to our commitments.

The assumption that some beliefs are inherently evaluative, serving to ground desires, is the first of two important assumptions that characterise the inference-theoretic picture. The second assumption is that the beliefs that figure in explanation are all on-or-off states. For any proposition, an agent believes it, disbelieves it, or doesn't have a belief either way; there is no possibility of her half-believing something or of her believing it to any particular degree. This assumption is important, because if we represent an agent as inferring something from the content of a belief, then we have to assume that the belief is an absolute affair; it makes no sense to think of an agent inferring something from the content of anything less than a belief of that kind, as it makes no sense to imagine her inferring something from the content of a desire. That I desire that p will not give me any ground for reasoning from 'p' as a premise and neither will the fact, if it is a fact, that I believe that p to a certain degree: that I believe that p to degree x does not make it suitable for me to take 'p' as a premise in my theoretical reasonings; in order to have a ground for reasoning from 'p' as a premise I must believe—believe, period—that p.[2]

But if my beliefs are all degree-less beliefs, where does uncertainty make an appearance? The answer, on the inference-theoretic picture, is that my beliefs must include beliefs as to the probability of this or that scenario. In fact all my beliefs must be explicitly or implicitly probabilistic on this picture. The explicitly probabilistic ones will assign a probability of less than one to various scenarios; the implicitly probabilistic ones, the ones that seem just to posit non-probabilistic propositions, will assign a probability of one—or a probability approaching one—to those scenarios. I said that the first assumption could be summed up by saying that some of the beliefs allowed in the inference-theoretic picture are inherently evaluative, involving matters of desirability in their content. The second assumption may be summed up, in parallel, by saying that the beliefs allowed must all explicitly or implicitly involve matters of probability in their content.

Decision theory is distinguished in the first place by the fact that it drops these two assumptions. Beliefs are allowed to come in degree, with some being held more strongly, others more weakly. And desires are taken independently of beliefs about underlying valuable properties; they are not assumed to have a deliberative grounding. That beliefs are allowed to come in degree means that the contents of those states need not involve matters of probability. That these degees of belief are paired with independent degrees of desire means that their contents need not involve matters of desirability.[3] The apparatus of inference theory is thoroughly revamped, with the dropping of its two characteristic assumptions. But though decision theory drops the assumptions, we should notice that it need not actually deny them. The decision theorist who tries to work with degrees of belief and desire may admit that consistently with his doing so, and consistently with the theory that he goes on to develop, it may also be reasonable to ascribe the evaluative and probabilistic all-or-nothing beliefs admitted in inference theory. We will return to this point later.

Suppose now that we go along with decision theory in dropping the two assumptions. Our picture of the intentional agent will involve two major posits: degrees of desire and degrees of belief or, in the preferred jargon, subjective utilities and subjective probabilities. What regularities are we going to think of these states sustaining with observation (the input end) and with action (the output end)? We cannot postulate regularities of evidential inference in the one case, or practical inference in the other, even when things are assumed to be normal or ideal. For inferential relations are relevant only for beliefs, not for desires, and only for beliefs in the all-or-nothing sense of belief, not for beliefs in the sense in which they come in degrees. There is no room to think of our subjective probabilities and utilities as being inferred from observation and no room to think of them providing inferential grounds for choice. But how in that case is rationality to be defined for such a system?

Decision theory attempts to provide an answer to this question, though an answer that is only relevant for an intentional system that fits some rather stringent, idealised requirements (Eells 1982, Jeffrey 1983, Pettit 1991a). The system is required to have subjective utilities and probabilities defined over a very complex range of entities or sets of entities, in effect all the propositions that could ever be relevant to its beliefs and desires. Assuming the availability of that range of utilities and probabilities, the theory then goes on to identify conditions expressive of practical rationality on the one hand and evidential rationality on the other.

The condition expressive of practical rationality is that the degree of desire the rational agent has for any option correlates according to an expectation formula with the degrees of desire that she has for different ways the option may work out, different ways the world may be consistently with its being chosen (Eells 1982, chaps. 2 and 3). If the option is represented by O and the different ways it may work out—the different outcomes— by O1, O2 . . . On, then the degree of desire she has for the option will be the sum of her degrees of desire for the different outcomes, with each such degree of desire being multiplied—in effect, discounted—by a probability associated with the outcome in question; the degree of desire for the option will be the expectation of the degrees of desire for the outcomes. We assume that the rational agent will choose the option in any set of alternatives for which she has the highest degree of desire and so this formula is meant to enable us to predict the choice that a rational agent will make, from her existing degrees of desire for the outcomes associated with the different options. The expectation formula is variously interpreted within the tradition of decision theory but we are only concerned with the basic idea (see Eells 1982, chaps. 2 and 3).

The condition expressive of evidential rationality according to decision theory—or strictly, according to the standard extension of decision theory—is usually described as conditionalisation. Here the basic idea, and again it is subject to various developments, is quite simple (Eells 1982, chap. 1). If I discover evidence that p, so that I come to have a revised degree of belief that p, what effect should this rationally have on my degree of belief for other propositions: say, for the arbitrary proposition 'q'? Conditionalisation has it that my new degree of belief that q should be equal to my old degree of belief that q-given-p. As the expectation condition is meant to enable us to predict the choices of the rational agent, so this formula is supposed to enable us to predict the belief-revisions of the rational agent. We predict what her new degrees of belief will be from a knowledge of her prior degrees of belief, in particular her prior degrees of conditional belief, as in the other case we predict her choices from a knowledge of her degrees of desires for relevant outcomes.

I said earlier that the decision theorist need not deny the claims made in inference theory. He attempts to give an account of intentionality, in particular rationality, that works with different ideas: those of degrees of belief and degrees of desire. But consistently with thinking that he has identified certain constraints of rationality among such degrees of belief and desire, he may think that there are also states of the kind postulated within the interpretative stance. He may think that there are degree-less beliefs, in particular degree-less beliefs about the probability of this or that scenario, and he may think that some of those beliefs are inherently evaluative.

There are a number of ways in which someone can reconcile decision theory in this way with the inference-theoretic picture. One congenial line would be to take the degrees of belief postulated in decision theory as being responsive to degree-less beliefs about matters of probability; and to take degrees of desire as being responsive to degree-less beliefs about matters of desirability. If the decision theorist takes this view then, consistently with thinking that the rational agent will revise her beliefs according to Bayesian conditionalisation, he will think that the degrees of belief formed in the course of this revision are responsive to the evidential matters articulated in inference theory: that they are responsive to degree-less beliefs about what is probable. And consistently with holding that the rational agent will form desires for new objects in accordance with the rule of expected utility—will form desires that relate appropriately to other relevant desires—he will believe that the degrees of desire formed are responsive to the valuable properties identified in those objects: that they are responsive to degree-less beliefs about what is desirable.

Does this sort of reconciliationist story reify matters of desirability and probability in a counterintuitive way? Does it require us to think of the desirable and the probable as non-natural realms which we access by some special faculty of intuition? No, it doesn't. Our red-sensations display an object as red—as belonging to the ostensively presented class of red things—when they operate in those conditions that come out as normal in negotiation about discrepancies; this is the ethocentric story told in earlier chapters. The story allows us to say that a red-sensation can be a response to the redness of an object and, for all that the reconciliationist line requires, a similar sort of story may hold in these cases. A desire may display a property as desirable—as belonging to the ostensively fixed class of valuable features—when it operates under conditions that are conducive to negotiated convergence; and so we may think of the desire as a response, when things go right, to the desirability of the object. And, relative to a certain body of information, a degree of belief may display a proposition as probable to a corresponding degree—as belonging to a suitable class of propositions—when it operates under conditions that elicit convergence; and so we may equally think of the degree of belief as a response, when things work well, to the probability of the proposition. There may be nothing to the desirable or the probable that fails to meet the naturalistic eye, then, though whether something is desirable or probable may only be discernible to someone with an appropriate habit of response, embedded in a suitable practice of negotiation; the elements required will be familiar from our ethocentric story (Pettit 1991b).

Being committed to the inference-theoretic picture, I would like to be able to reconcile it with decision theory: perhaps in the manner illustrated, perhaps just by representing ascriptions of probability and desirability as causally redundant expressions of degrees of belief and desire. But the point I should now emphasise is that there is a reading of decision theory—a reading that goes beyond anything asserted in the theory itself—under which it offers an account of explanatory intentional properties that replaces that of inference theory. Under this reading, decision theory denies that values and evidence play the causally relevant role accorded to them in inference theory; it denies that there is any causal role, however redundant, for degree-less beliefs involving matters of probability and desirability. It proposes that the only intentional properties explanatory of belief-revision involve degrees of prior conditional belief, and that the only intentional properties explanatory of choice involve existing degrees of desire for relevant outcomes, together with the associated degrees of belief in those outcomes. It is eliminativist about inference theory; it sees inference theory as a big mistake.

With this reading of decision theory, we come to the challenge we must face if we are to maintain the inference-theoretic picture presented in the preceding sub-section. The question is whether there is any argument for inference theory and against this reading of decision theory. I believe that there is and I want to present that argument now. I assume from all that has gone before that human subjects are thinking systems, not just intentional systems of any old kind. I argue that in order to be a system capable of thought, the human being has to form new beliefs and choices in the manner depicted in inference theory.

To be a thinking subject is to be an intentional system that acts intentionally with a view, among other things, to having rational intentional states. Such a system will be subject to certain causally relevant pressures in the formation of beliefs and intentions and other relevant states. If the agent is generally to be rational, as we assume a naturally selected or well-designed intentional system will be, then she will be subject to pressures that generally make for rationality. And if she is to be capable of thought, capable of intentionally achieving such rationality, then she will be able to identify the sorts of pressures at work, to assess them for their tendency to promote rationality in any given case, and to reinforce or inhibit them as the assessment requires.

Under the inference-theoretic picture, there is no difficulty—free-will problems aside—in seeing how the thinking subject can meet these conditions. The relevant pressures in the case of new beliefs and choices are evidential and valuational respectively; they are pressures associated with the contents of probabilistic and evaluative beliefs. We can see that in a well-equipped system a sensitivity to considerations of evidence and value will tend to produce rationality in a straightforward sense. We can see how a system like one of us can come to be aware of the evidence that inclines us to a particular belief and the values that incline us to a particular choice. We can see, furthermore, how we may come to assess that belief or choice for rationality through paying attention to all the pieces of evidence or all the values that are relevant; we will develop that assessment under the inferential pressure of considering all relevant matters. And we can see, finally, that as we pay attention to all those matters, we will come to inhibit or reinforce the original inclination, depending on what is appropriate. Under the inference-theoretic picture, in short, there is no difficulty about how we can manage, at least sometimes, to act intentionally with a view to promoting the rationality of our intentional states; there is no difficulty about how we can have the capacity to think.

Consider now, in contrast, how things will stand under the reading of decision theory that would present it as a replacement for inference theory. Let us grant, for purposes of argument, that if an intentional system is sensitive to prior degrees of conditonal beliefs and existing degrees of desire for relevant outcomes, in the manner characterised in decision theory, then it will tend to form rational beliefs and rational choices. But how can we expect a person to be able reliably to identify the prior degrees of conditional belief or even the existing degrees of desires that are, allegedly, at the source of the new beliefs she forms and the new choices she makes? After all, the identification of such precisely defined factors would require an extraordinary degree of self-aware-ness and an unusual capacity for introspection. And even if a person could pull off that feat, how can we expect her to be able to assess the rationality of the beliefs and choices in question in the light of such factors? After all, that assessment would require a knowledge of the principles of conditionalisation and expectation presented in decision theory and such knowledge is not widespread.

I conclude that given that people are thinking systems, we cannot plausibly see them as intentionally responding just to the sorts of factors identified in decision theory as constitutive of rationality. Decision theory may offer a useful specification of some aspects of rationality, evidential and practical, a specifica-tion that is not necessarily inconsistent with anything maintained in inference theory. But it does not give an account of the factors that thinking systems like you and me consider as we find our way to revisions of our beliefs and to decisions about what to do. For such an account we do better to rely on some version of the inference-theoretic picture presented earlier. It develops an image of the springs of belief and choice that allows us to make ready sense of how an intentional subject can have the capacity for thought: how she can have the ability to act intentionally with a view to promoting the rationality of her own intentional states.

There is one proposal to consider in conclusion. Assuming that decision theory offers a persuasive specification of evidential and practical rationality, someone might propose that even if we are not actually moved by the factors it invokes, still we ought to sensitise ourselves to such factors if we want to be really rational. The proposal is that we ought to change our inferential habits: that in working out what to think in the light of new information we ought to look at our prior degrees of belief conditional on that information and derive the appropriate degrees of belief we ought to have by the conditionalisation formula; and that equally, in working out what to do in some decision, we ought to look to our degrees of desire for the relevant outcomes, our associated degrees of belief in those outcomes, and derive the appropriate degrees of desire for the different options by the expectation formula. Is there anything to be said for this proposal?

I see nothing to be said in favour of this proposal, since it requires a blind faith in the virtues of formalism to think that we would do better by following the proposed route. But in any case there are things to be said against the proposal. One simple but knockdown consideration is that, as already stressed, we do not have the sort of access to our degrees of belief and desire that the proposal would require (Harman 1986, chap. 9). There is an old quip that goes: how can I tell what I think until I see what I say. In the case of degrees of belief and desire, as these are conceptualised within decision theory, the quip has some resonance. The degrees involved are so finely calibrated that the only way they could conceivably be identified is by reconstruction from behaviour, under the assumption that the behaviour is indeed in accord with decision theory. There is no telling the strength of your different beliefs or desires short of seeing how you behave.

There is also another sort of consideration to be mentioned against the proposal (see Pettit and Smith 1990, 1992, and Pettit 1991a). Someone who gave up working out what to think and do by reference to what she finds probable or desirable, someone who opted instead for deriving her responses from consideration of her existing degrees of belief and desire, would make an important conceptual resource difficult to access. Suppose that I form a belief or perform an action on the grounds that it is probable that p or that it is desirable that q. Should I think that the response would have been rational even in the event that I did not have an appropriate degree of belief that p or desire that q? Of course I should, given that as I actually see things, it is probable that p and desirable that q. My grounds for recommending the response to myself are that it is probable that p or desirable that q, and those grounds remain relevant even in otherwise similar situations where I do not have appropriate degrees of belief and desire and do not hold that it is probable that p or desirable that q. They remain relevant also at otherwise similar times where I do not have the appropriate degrees of belief and desire, but I shall not explore the temporal case.

Consider an example. I believe that I ought to look after myself in old age by contributing to a pension fund on the ground that it is desirable that old people should not have to depend on others. The ground for this belief remains in place even in otherwise similar situations—otherwise similar possible worlds—where I just happen not to have the appropriate degree of desire that old people be independent or the belief that it is desirable they should be independent. And so, as I now see things, I say that even in such worlds the right thing for me to do would be to contribute to a pension fund. Again, I come to judge that New York is larger than London on the ground that it is probable that the statistics I have seen are reliable. The ground for this judgment—that the statistics are probably reliable—remains in place in otherwise possible worlds where I do not have the appropriate degree of belief in the statistics. And so, as I now see things, I say that even in such worlds the right thing for me to believe would be that New York is larger than London. In each of these cases there may be a subjective sense of rightness in which it can be allowed that in the other world it would be right of me not to contribute to the pension fund or not to judge that New York is larger. But the important thing is that there is a straightforward sense in which I must now say that those responses would not be right.

The point I want to make against the proposal under consideration here is that if I worked out what to believe or do on the basis recommended, then I would lose easy access to the possibility of stipulating appropriate responses for myself in worlds where I differ in my degrees of belief and desire. Under the proposal, I would judge that it was right to contribute to a pension fund on the ground that I have a certain degree of desire for people's being independent in old age. Equally I would judge that New York is larger than London on the ground that I have a certain high degree of belief in the statistics that indicate this. But consider now what I ought to say about the right responses for me in worlds where my relevant degree of desire or degree of belief was considerably lower. There is little reason why I should find any sense in saying that in those worlds I still ought to contribute to the fund or judge that New York is larger. Why should I now think that in other worlds I ought to respond on the basis of my degrees of belief and desire in this world rather than in those worlds? If I think here that in view of my actual degree of belief or desire I ought to contribute to the fund or judge New York larger, then surely I ought to think otherwise in worlds where the relevant degree of belief and desire is considerably lower? If I deny this, I appear to give relevance to a particularity—the identity of the world I actually inhabit—in making my prescriptions as to how I should respond and that is not an easy move to defend. It is in tension with the widely accepted requirement of universalisability, that anything that provides a reason for one response rather than another should do so in virtue of universal, non-particular features (Hare 1981, Rabinowicz 1979, Pettit 1987c, 1988a).

The upshot is this. Under our existing ways of reasoning, we have access to the possibility of giving reasoned support to certain responses, support that is stable over counterfactual variations in the degrees of belief and desire that we imagine ourselves having. Under the proposal that we have been considering, we would lose easy access to this possibility. That loss is important, for it would mean that we could not use the standpoint of our actual beliefs and desires, in particular our actual judgments of probability and desirability, in assessing our responses—or indeed the responses of other people—in situations where the relevant degrees of belief and desire were different.

This observation holds up, it should be noticed, no matter what account we give of judgments of probability and desirability. Suppose that we are extreme non-cognitivists about desirability and that we take a similar line about probability. We reject the cognitivist view that the judgment of desirability is a cognitive state that necessitates desire by denying that the judgment of desirability is a cognitive state; and we deny, in parallel, that the judgment of probability is cognitive either. We may adopt an expressivist story about such judgments, casting them as mere expressions of corresponding non-cognitive states: degrees of belief in the one case, degrees of desire in the other. Even under this story, we must see the capacity of the agent to think in terms of the probable and the desirable as an important conceptual resource.

The fact that an agent thinks and reasons in terms of desirability rather than in terms of degrees of desire will mean that she can use the standpoint of her present desires—as expressed in her judgments of desirability—in order to assess her responses over counterfactual variations in the degrees of desire she happens to have (Blackburn 1984a). She will be able to stand with her present desires and, holding firmly to them, prescribe for herself the behaviour they require, even at worlds where she lacks those desires. And a similar point goes through for probability. There will be a difference between the conceptual resources of someone who thinks and reasons in terms of probability and someone who restricts herself to the austere consideration of her own degrees of belief. She will be able to stand with her present degrees of belief and prescribe for herself the conclusions that follow from them, even at worlds where she lacks those degrees of belief.

Structural and Historicist Theory

The Case for Ecumenism

Intentional interpretation bulks large in social science but it is not the only style of explanation that figures there. In this section and the next I will consider quite different explanatory modes that are found in social research. Structural explanation, which I consider in this section, prescinds from the intentional and psychological altogether, finding often quite surprising determinants of social and related facts. Rational choice explanation, which I consider in the next section, stays at the psychological level but restricts the determinants invoked to broadly economic factors: something is psychologically explained under this approach only if it is shown to be in the rational self-interest of the agents involved. In this section I also consider what I describe as historicist explanation; this goes quite naturally with structural, though, as we shall see, it may have a psychological character. I shall argue that we can countenance all these sorts of explanation side by side with the interpretative kind just discussed. It is in virtue of this tolerance of different explanatory styles that I describe my approach as supporting an ecumenism about explanation. I am a causal fundamentalist, as appeared in Chapter 3, but an explanatory ecumenist.

We discussed structural regularities at some length in Chapter 3. As examples, we mentioned regularities such as those that link unemployment with crime; the social benefits of economic stratification with its appearance and stability; the fact that an international initiative is in the interests of a state with its adoption; and the dominance of a Protestant ethic with capitalist activity. A structural regularity has two features, by the account we gave. First, it governs a social event or condition: an event or condition involving the intentional responses of a number of people, as in the rise of crime or the adoption of a state initiative. But, second, it is causally and logically discontinuous from intentional regularities. It does not belong to the same realm as intentional regularities; it does not invoke factors that are related by causal regularities to intentional factors. And neither does it obtain in virtue of the obtaining of intentional regularities: its obtaining is not logically entailed by their obtaining; a structural regularity, as I argued, requires the satisfaction of certain boundary conditions as well as the satisfaction of intentional regularities.

Structural explanation corresponds to structural regularities. Structural explanation explains by invoking a social event or condition that is related to the *explanandum* in accordance with a structural regularity. Why was there a rise in crime this year? Because there was an increase in unemployment. Why do we find economic stratification in Western societies today? Because it is socially functional. Why did France stay out of NATO? Because it was in the state's interest to stay out. Why does capitalism thrive in these countries but not in those? Because a Protestant ethic is abroad in the one sort but not in the other. These may not be good explanations but they serve to illustrate the sort of thing that I have in mind by structural explanation. They show that the explanatory style envisaged is a broad-spectrum variety, stretching from common-or-garden explanations in terms of aggregate factors like unemployment, or corporate factors like the state's interest, to more controversial explanations that invoke functional or ideological antecedents.

So much for structural explanation. What now of historicist? Nothing in our discusssion so far serves to introduce this category of social explanation but it has an important feature in common with the structural kind, and so it is natural to consider the two in tandem.

Structural explanation presupposes that the responses involved in the social consequent explained have intentional antecedents; the structural and intentional levels are in harmony with one another, as causal fundamentalism requires. But the structural explanation invokes an antecedent removed in level from those responses and, unsurprisingly, it does not presuppose any intentional antecedents in particular: it is compatible with any of a variety of intentional configurations. The rise in unemployment that explains the increase in crime presupposes that the offences committed spring from the belief-desire sets of individual offenders, but it does not require any particular intentional profiles. This person may commit an offence for one sort of reason, that other may offend on quite a different ground, and so on.

This feature of structural explanation is, in a word, that the explanation prescinds from the immediate intentional antecedents of the social consequent explained. The best introduction to historicist explanation, as I call it, is that it serves in a parallel, prescindent way to explain social consequents: consequents that involve intentional responses and that spring from a variety of immediate, intentional antecedents. It does not prescind from the immediate, intentional antecedents in favour of a structural antecedent at a different level, or at least it does not necessarily prescind in favour of such an antecedent. What distinguishes this form of explanation, and what earns it the name I use, is that it prescinds in favour of an antecedent that is removed in time—whether or not it is also removed in level—from the immediate intentional springs.[4] It prescinds in favour of a more or less remote historical factor.

Here is a simple illustration of a historicist explanation. It became clear towards the end of the 1980s that the Soviet Union was no longer disposed to use force in order to preserve Communist regimes in its Eastern European satellite states. That turn of events led, we may assume, to the fall of the Communist governments in those states. But many of us who would trace the fall of those governments to that source would be prepared to assert something stronger than that this realisation about the Soviet Union was in the causal ancestry of that political change. We would be prepared to say that the realisation about the Soviet Union was more or less bound to lead to such a change. As a matter of fact, due to this contingency here, and the other contingency there, the realisation led by a particular route to that result. But we feel that even if the associated contingencies had been different, still the political change in the satellite states would have come about. We think that the realisation that the Soviet Union was no longer disposed to use force in order to preserve communism meant that the fall of the Communist regimes was more or less inevitable.

The picture we entertain in holding by this sort of belief is not mysterious in any way. Hold the world fixed at the point where the realisation about the Soviet Union becomes suitably widespread. At that point, as we picture things, future contingencies may go any one of a number of distinct ways. But under the picture entertained, it does not matter which of those contingencies materialise, so far as the fall of the communist regimes is concerned. Under more or less all contingencies the causal paths lead to the fall of those regimes. Thus we say that the realisation about the Soviet Union did not just precipitate the collapse of those governments; it made that collapse more or less necessary.

On this picture the earlier event is causally relevant in an unusual fashion to the later event that it explains. Not only is it relevant to something that is relevant to something that is relevant to something . . . that is relevant to the event explained. It is also relevant to the event explained in a manner that is independent of the particular linkages in operation. It enjoys a double relevance in relation to the event or condition explained.

I said earlier that I wanted to use the notion of causal relevance without presupposing any particular account of what makes for relevance; the only commitment we have made is to a program account of how causal relevance is related across levels. Following this policy, I do not mean to presuppose any particular account of causal relevance in alleging this double relevance in the historicist antecedent. Imagine we think that a property is causally relevant in virtue of engaging with the antecedent of a causal law. In that case the historicist antecedent will be doubly relevant through engaging with a law that links it with the next intervening link in the causal chain and through engaging, at the same time, with a law that links it directly with the sort of event at the end of the chain: the sort of event explained. There is no reason to doubt that we will be able to make a parallel distinction under any plausible account of causal relevance.

Once we are clear about the picture that motivates a historicist explanation, we can generate other examples of the exercise. Suppose that during a demonstration by workers or dissidents or any rebellious group a number of shots are fired by soldiers at the crowd and that as a result there is a riot and, eventually, a full-scale revolt against the authority of the state. We can readily imagine someone arguing that while the shots precipitated the revolt, the social tensions that led to the demonstration in the first place made such a revolt more or less inevitable. Even if that particular demonstration had not taken place, or even if the soldiers had not fired on the crowd, still those tensions would have sparked a revolt by some other sequence of effects. The tensions meant that a revolt was more or less bound to happen.[5]

For another example consider a case mentioned by Richard Miller (1978, p. 410). The availability of stainless steel from the 1920s on, in particular its availability at a relatively reduced cost, led to the replacement of carbon steel as the main material for knives. Now the availability of stainless steel led to this result via a series of particular intentional episodes: episodes involving decisions by the executives of relevant companies. But we may reasonably think that the explanatory relevance of the availability of stainless steel is not exhausted by its contribution to the actual causal history of the replacement of carbon steel in knives. We may hold, with some reason, that the availability of stainless steel made it more or less inevitable that carbon steel would be replaced as the main material in knives: this, because stainless steel keeps an edge better, so that if these particular executives had not noticed, or had not introduced the change, then sooner or later some others would have done so.

For a final example of historicist explanation we might turn to what Tolstoy has to say in *War and Peace* on the burning of Moscow in 1812, at the time of the French occupation. He spurns any of the variety of particular stories about the origin of the conflagration in favour of a distinctively historicist narrative.

> Moscow was sure to be burned, because her inhabitants had gone away, as inevitably as a heap of straw is sure to be burned where sparks are scattered on it for several days in succession. A town of wooden houses, in which when the police and the inhabitants owning the houses are in possession of it, fires are of daily occurrence, cannot escape being burned when its inhabitants are gone and it is filled with soldiers smoking pipes, making fires in Senate-house Square of the Senate-house chairs, and cooking themselves meals twice a day
>
> (Tolstoy 1972, pp. 969–70)

An emphasis on historicist and structural explanation is distinctive of the tradition of social theory that insists on the dispensability of individual contributions to many an important social outcome: the tradition that defends what we described as the dispensability thesis in Chapter 3. The idea is either that no one individual played an essential part in producing the outcome or that if some individuals did, then there were always others who would have taken their places in the event of their not making that contribution. In the turmoil that led to the fall of the Eastern European communist regimes, no one individual played a necessary part: there were more than enough involved in the crucial demonstrations and assaults. Or if there were some individuals whose actions were essential to what happened in this or that country, then there were others who would have acted in a suitably substitute role, had those individuals failed to make their contribution.

We have spent some time in introducing the categories of structural and historicist explanation. The question now is whether we are free to admit the validity of these explanatory styles, given the holistic individualism to which we committed ourselves in the last part of the book. The holistic aspect of that commitment is not directly relevant to the issue. The question is whether we can be individualists in the sense outlined in Chapter 3 while countenancing structural or historicist explanation.

On the face of it, there is no problem. By our account in the last section, to explain an event or condition is to provide information about a property that is causally relevant to its realisation and, on our generous understanding of causal relevance, a structural or historicist antecedent may well enjoy such relevance. In particular, it may enjoy such relevance, consistently with the regular, causal relevance of the intentional properties and states realised in relevant parties: consistently, that is, with the sort of causal relevance on which individualists insist.

We saw in the first chapter that an intentional antecedent may be causally relevant to an action by programming for its appearance and that this causal relevance is entirely consistent with the causal relevance, in a suitable context, of a neural or electronic antecedent; indeed it presupposes that sort of causal relevance, since it programs for the action through raising the probability that there will be a lower-level antecedent sufficient to produce the action. Suppose that the contextual conditions are fulfilled under which a structural antecedent like a rise in unemployment leads to an increase in crime. If there is genuine causal relevance here, then it comes about because the structural antecedent programs for the increase in crime. But if the structural antecedent programs in this way, then its causal relevance will presuppose, and be consistent with, the causal relevance of the intentional factors involved. How can it program for the increase in crime—as it does, given fulfilment of the contextual conditions— without raising the probability of such lower-level, intentional antecedents as will yield the required level of offences? We discussed the point already in Chapter 3.

Is there any extra problem for an individualist in endorsing historicist as distinct from structural explanation? Again, no. The individualist holds that beliefs and desires and such intentional states lead to certain responses according to the patterns charted in our intentional psychology. There is nothing inconsistent with this in the idea that higher-level factors raise the probability of certain intentional antecedents, and program for the effect of such antecedents, as we have just seen. And equally there is nothing inconsistent with the individualist credo in the distinctively historicist notion that a remote cause of the antecedents, and therefore of the effect they produce, may do better than just belong to their causal ancestry. It may make the sort of effect in question more or less inevitable by being such as to have led either to those actual antecedents or, under other contingencies, to variant antecedents that would still have generated that effect. The type of remote overdetermination alleged fits just as easily as higher-level programming with a belief in the soundness of intentional psychology.

The picture emerging from this line of consideration is an attractive one. We hold, as argued earlier, that causal fundamentalism is sound: that causal relevance at all higher levels is determined by the causal relevance of microphysical properties. But consistently with this causal fundamentalism, we are explanatory ecumenists. We acknowledge the distinctive causal relevance, and therefore the explanatory interest, of intentional properties, as the individualist wants to do. And equally we acknowledge the potential causal relevance and explanatory interest of the structural and historicist factors on which collectivists, in particular collectivist social scientists, have wanted to focus.

One of the continuing divides among social theorists marks off those who favour structural and historicist explanations from those who look for social explanations that invoke intentional antecedents or that stay closer to the psychological level. This may be described as the divide between explanatory or methodological versions of individualism and collectivism, as distinct from the divide that concerned us in Chapter 3: that between causal or ontological versions of the doctrines (James 1984, Miller 1978). My position on the ontological divide is individualist but my position on the methodological one is ecumenical. I do not think that there is any need to choose between the two sorts of explanatory strategy which the parties to the debate regard as alternatives. There is nothing in the preference for either sort of explanation that raises a question for the other.[6]

On the one side of the divide we have two groups: first, those who espouse structural or historicist theories such as Marxism or structuralism or functionalism and, second, those who despair of general theory but continue to look for comparative or historical data to underpin structural or historicist claims (Kiser and Hechter, forthcoming). On the other side we also have two groups: first, those who espouse psychological theories such as the rational choice model discussed in the next section and, second, those who reject such general theory but insist that the explanations to seek in social science must refer us to individuals and their psychological workings (see, for example, Veyne 1984). There will be detailed questions between adherents of the two sides, and among the adherents of each, as to what sort of explanation makes best sense of a given phenomenon. But my position is that there is no general issue over which the sides need to divide. There is no need to choose between the different styles of explanation that they champion; there is no reason to think that the validity of either style undercuts the validity of the other.

In taking this ecumenical line, I should stress that I am not endorsing all the actual explanations or theories put forward on the two sides. Thus I have already offered reasons in Chapter 3 for thinking that the selectional model presupposed by at least some functionalist theories is false (see also Macdonald and Pettit 1981, chap. 3). And in the next section I will be offering serious reservations about rational choice explanation, as it is mostly understood. My point here is that there can be no objection to a style of social explanation just on the grounds that it is structural or historicist or psychological in character. The picture represented by our holistic individualism allows us to countenance explanations in different styles. It allows us to avoid sectarian, explanatory commitments.

Beyond the Fine-Grain Preference

This ecumenical line on social explanation will not go unopposed. I would like to discuss one instructive sort of opposition that has commanded a good deal of attention among self-styled individualists: among those who would go along with the ontological individualism defended in Chapter 3 but argue also for a methodological or explanatory form of the doctrine. This opposition derives from a principle relating causal structure to explanatory strategy, a principle I have described elsewhere as the fine-grain preference in explanation (Jackson and Pettit 1992b). Understanding and undermining this form of opposition will develop our sense of the explanatory ecumenism to which we are committed.

The fine-grain preference, as I understand it, assumes that there is a grain in causal structure such that for any event or condition we may get a more or less fine-grained view as to its exact causal antecedents. The preference is a preference for any explanation that is more fine-grained over any that is less so. Thus Jon Elster, who also thinks there are instrumental reasons for wanting more fine-grained explanation, assumes that in any case just the fact of getting nearer the detail of production makes such an explanation intrinsically superior: 'a more detailed explanation is also an end in itself' (Elster 1985, p. 5). Michael Taylor (1988, p. 96) agrees: 'A good explanation should be, amongst other things, as *fine-grained* as possible'.

The fine-grain preference can be invoked against both structural and historicist explanation. It makes a case against structural in combination with the natural assumption that as we go to lower levels of causal relevance, we discover more and more basic causal factors; this assumption is entailed by our causal fundamentalism. If we are supposed to prefer finer-grained explanation, and if going to lower levels reveals finer grain, then we should always prefer a lower-level explanation to a higher-level one; among other things, therefore, we should prefer any intentional or psychological explanation of a social consequent to an explanation of a structural sort. 'The *search for micro-foundations*, to use a fashionable term from recent controversies in economics, is in reality a pervasive and omnipresent feature of science. . . . To explain is to provide a mechanism, to open up the black box and show the nuts and bolts, the cogs and wheels of the internal machinery' (Elster 1983a, pp. 24–25).

The fine-grain preference combines with an assumption that parallels that of causal fundamentalism in order to make a similar case against historicist explanation. The assumption is one of causal localism and it asserts that as a higher-level factor is always causally relevant in virtue of the relevance of a lower-level—this is causal fundamentalism—so a more remote factor is always causally relevant in virtue of the relevance of a more proximate. This assumption is explicitly endorsed by the opposition and I am happy to go along with it (Elster 1983a, pp. 28 ff; 1976).[7] The assumption means that the fine-grain preference tells against more remote explanation as well as against higher-level. If we are supposed to prefer finer-grained explanation, and if going to closer stages reveals finer grain, then we should prefer any more proximate explanation to any more remote one. In particular, we should prefer an explanation in terms of proximate intentional antecedents to one that invokes a historically removed factor, even if it does so in historicist fashion. 'The goal of research should be to substitute for past causes the traces left in the present by the operation of those causes, but this we are not always able to achieve'

(Elster 1983a, p. 33). Paul Veyne (1984, p. 109) puts the point even more strongly. 'For a historical explanation to be admissible, it must not present any interruption of continuity in causal relations that connect the agents involved in the plot'.

The fine-grain principle would not have us despair altogether of structural and historicist explanation. Presumably we should stick with the lowest-level, nearest-stage explanation available for any phenomenon, pending the discovery of something better; and that explanation may be a structural or historicist account. But the fine-grain preference does mean that we cannot be complacent about any explanation of a structural or historicist kind. We will have to see such an explanation as a holding exercise, an account that suffices for the moment but that we should discard at the first opportunity. Thus the fine-grain preference gives the lie to the sort of explanatory ecumenism that I have been pressing.

There is something strange about relying on the fine-grain preference to defend explanatory individualism. Such an individualism would privilege the psychological explanation of social consequents. But while the fine-grain preference would certainly favour such explanations over explanations of a structural kind, it would equally favour neurophysiological explanations over psychological, and so on down the hierarchy. It would support an explanatory fundamentalism, not an explanatory individualism.[8] Where I want to combine causal fundamentalism and explanatory ecumenism, it would argue for fundamentalism in both dimensions. But we should not be distracted by the issue of whether the fine-grain preference would serve the purposes of the explanatory or methodological individualist. No matter how unsatisfactory it may be in that regard, the fine-grain preference directly challenges the explanatory ecumenism that I wish to defend. Thus we must be able to find reasons to resist it.

The fine-grain preference is motivated by a view of explanation similar to the one presented in the last section. Under this view to explain something is to give information on causally relevant properties or, in David Lewis's equivalent formulation, to provide information on causal history. Let us work with Lewis's formulation, as this is closer to Elster's way of suggesting the causal-information line; in his preferred formula, the job of explanation is to depict causal structure (Elster 1983a, p. 34).[9]

The causal-information view of explanation would support the fine-grain preference, provided one further premise is granted: that information about finer causal grain is always better causal information. If explaining is providing information on causal history, and if finer-grain causal information is always better, then it is better to provide a finer-grain explanation of anything. The fine-grain preference goes through. The causal-information view is a theory of what makes an explanation an explanation. The extra premise invoked in the argument is a theory on what makes, or helps to make, an explanation satisfactory. In the preceding section I said that a satisfactory explanation is one that presents accurate and adequate information on causally relevant properties or, equivalently, on causal history. This premise states that finer-grain information is always more adequate information and always makes for a better explanation.

But the premise is false. We saw in the last section that there are two grossly different sorts of information that may be offered on any causal history, or more generally on the actual world. The one sort is contrastive information that enables us to narrow down the class of possible worlds to which the actual world belongs. The other sort is comparative information which directs us to features that belong in common to the actual world and to more and more other possible worlds. The molecular explanation of why the closed flask breaks directs us to a specific process involving a specific molecule and helps us to contrast what actually happens with other ways the flask might have broken. The explanation that mentions the boiling of the water contained in the flask abstracts from that detail and directs us to constancies that bind the actual world to more and more other possible worlds: it reveals that in practically all worlds where the water boils, however they differ at the more specific level, the flask cracks.

This distinction lets us see why it is false to hold, as in the premise we are examining, that information on finer causal grain is always better causal information. Information on finer causal grain, like the information provided by the molecular explanation of the cracking, is always better contrastive information. But it will not be better comparative information, for it will direct us only to contrasts between the actual world and other possible worlds, not to constancies that obtain in common between them. Comparative information is clearly a distinct kind of information from contrastive and, equally clearly, it has an autonomous interest. Suppose we know that the flask cracks under the impact of this or that molecule. We still learn something important on receiving the information that it cracks because the water is boiling. We learn something we may not have noticed before: that the molecules had such and such a mean motion and that their having that level of motion would have been sufficient for the flask to crack, even if the molecule that was actually responsible did not do the job.

I see no hope of making a case for the fine-grain preference other than by arguing in the manner described, from the causal-information view of explanation. But that argument goes wrong in assuming that all information, or at least all interesting information, is contrastive rather than comparative in character. Thus I see no prospect for a compelling defence of the fine-grain preference. We may return with renewed commitment to our explanatory ecumenism. With any social event or condition we will get better contrastive information on its origins by looking to the immediate intentional antecedents than by identifying a structural or historicist determinant. But that is not to say that we should prefer such intentional explanations in general. For equally we may get better comparative information on the intentional origins of the event or condition by looking to structural or historicist antecedents.

Consider structural explanation; in particular, consider structural explanation in relation to a corresponding form of intentional explanation. In learning the structural account, we learn that in more or less all possible worlds where the structural antecedent is fulfilled—and the required contextual conditions are realised too—there is some intentional configuration sufficient to bring about the social consequent explained. In the actual world such and such a configuration does the job but the information supplied in the structural explanation is that even if that configuration had been absent, still there would almost certainly have been a suitable array of attitudes present: at least there would have been such an array, provided the structural antecedent—and the required contextual conditions—continued to be satisfied. Thus the structural explanation gives us comparative information on the causal history at the intentional level, as indeed it gives us such information on the causal history at any lower level.

A similar claim goes through for historicist explanation. The structural explanation gives us the information that for more or less any world that is similar in regard to the structural antecedent—and the required contextual conditions—the sequence of intentional episodes will be sufficient to lead to the social consequent explained. The historicist explanation gives us the information that for more or less any world that is similar in regard to the temporally removed determinant in question—whether this is structural or perhaps intentional—the sequence of intentional episodes will again be sufficient to lead to that consequent. In each case we are offered potentially useful information of a comparative kind. We are not introduced to the points of contrast that mark off that intentional history from more and more other possibilities. We are directed to features that make for a constancy in the overall result across a variety of possible worlds.

Residual Opposition

It must surely be agreed that structural and historicist explanation looks capable of providing comparative information on causal history, in particular on intentional causal history. So is there anything further which enthusiasts of the fine-grain preference, or at least of explanatory individualism, can say? All that they can hope to argue, it would seem, is that the comparative information provided in structural or historicist explanation is not information on causal history or that, if it is, it is only second-grade information in some way.

How could the information provided in structural or historicist explanation not be information on intentional causal history? The answer is that it might be information about the presence of a standby cause, a cause whose influence is preempted by the actually relevant factors and which is not therefore part of the causal history (Elster 1989, p. 6). Standby information is not about anything to do with the actual causal history but it does give us information about certain possible worlds. A person dies early in 1990 because of a car accident but it was more or less inevitable that she would die in that year because she had an advanced form of cancer. The information that the person had cancer tells us that in a range of possible worlds, including the actual world, she dies in 1990. But such information is not information on the actual causal history of the event, it is only information on a standby cause, and so it does not constitute an explanation of what happened. The suggestion under consideration is that the information provided by structural or historicist explanation is of the same kind.

But the suggestion comes to nothing. The information provided in historicist explanation, or at least historicist explanation that is not also structural, is information about a temporally removed factor in the intentional causal chain leading to the event explained, and so it is clearly going to be information on causal history; it contrasts directly with the non-causal, standby information provided in the cancer story. The information provided in structural explanation is not information of that historical kind—it is not information about an earlier causal link—but it is also information of a causal character. If a state programs for a result, as the state invoked in a structural explanation is meant to program for a social result, then it must ensure that a lower-level state of a certain kind is present, or at least it must raise the probability of the presence of such a state, and a state of that kind must actually be causally responsible at the lower level for the production of the result. Thus the structural explanation provides causal information on the genesis of the social result explained; it provides the modally comparative information, as we may wish to put it, that the actual causal process at the intentional level is such that even if it is varied within the limits consistent with the structural factor remaining in place, it will still tend to produce the required result. The information about the motorist's cancer is not of this kind, because while the presence of the cancer may raise the probability that there will be some lower-level state—some physical pathology—present that is sufficient to kill the person in 1990, the physical pathology that actually accounts for the death of the person is not made more probable by the cancer; it occurs as an independent result of the motor accident.

So much for the first possible response that may be made by defenders of the fine-grain principle: the response according to which structural and historicist explanation do not provide suitably causal information. The second possible response is for them to argue that while such explanation provides causal information, it only provides causal information of a second-grade kind. I conclude this discussion with a consideration of this response.

The comparative information provided in structural or historicist explanation will be second-grade under either of two scenarios. One is that the information is going to be available to anyone who is possessed of the intentional explanation—in particular, the more or less immediate intentional explanation—of the social consequent in question. This would mean that on getting a suitable intentional explanation one could dispense with the structural or historicist account. The other scenario is that the information provided will be available—the proposition in question will be rationally credible—only for someone who already knows what the relevant, intentional explanation of the social consequent is. This would mean that while the structural or historicist explanation might provide new information, it would be parasitic in an important way on that intentional explanation.

The first scenario is illustrated, as we noted there, by an example introduced in the first chapter. The elasticity and the molecular structure of an eraser provide higher and lower explanations of the eraser's bending. But the elasticity explanation is not of any great interest—it is distinctly second-grade—since anyone who learns the explanation in terms of the molecular structure is in a position to know the truth of the higher explanation too. Will structural and historicist explanation do better than this sort of explanation? Will they generally provide comparative information on intentional history that is not available just from a knowledge of how things go at the intentional level?

They will, if they are sound. The elasticity case is special. To be elastic is just to have such a molecular structure as will lead to the sort of event explained: to bending under such and such pressure. The cases where structural and historicist explanations appear to be available are quite different. The increase in unemployment that programs for an intentional configuration sufficient to generate a rise in crime is not defined in such a higher-order way. Thus it is quite possible to know that such and such offenders made for the rise in crime without knowing either that there was an increase in unemployment or that it more or less ensured the rise in crime. Nor is the urbanisation that programs for a decline in religious practice defined in a higher-order way. Thus it is equally possible to know that the paucity of churches, the break with a seasonal liturgy, or the absence of peer pressure led to a decline in religious practice without knowing that urbanisation had occurred or that it more or less ensured the religious decline. And so on in other structural cases.

What of historicist explanations? Here the point is equally compelling. We may know that a particular demonstration or series of demonstrations led to the fall of an Eastern European Communist regime without knowing either that it had been widely realised earlier that the Soviet Union would not intervene to save that regime, or that this realisation made it more or less inevitable that the regime would fall. And equally we may know that the decisions of such and such executives led to the use of stainless steel in the production of knives without knowing either that stainless steel had only recently become available at a price comparable to carbon steel, or that its availability made the switch more or less inevitable. In these cases, as in the structural ones, the information available is just not available from the more or less immediate, intentional explanations alone.

The other scenario under which the information would be second-grade needs a little more introduction. Consider the relation between the information provided by the intentional explanation of someone's behaviour on the one hand and on the other the information provided by a corresponding neural—or neural-cum-contextual—explanation. All of us readily accept the intentional explanation, with the comparative information that it carries about the neural level, without knowing much about how things are neurally produced. Moreover, most of us would not be deeply disturbed if neural scientists could not explain at the neural level how the regularities postulated in the intentional explanation come to be sustained; we would assume that there was something they were overlooking. All of this is to say that we put credence in the intentional explanation more or less independently of information or understanding at the lower, neural level—or indeed at any lower level—as to how the event explained comes about. We have no image of how an intentionally explicable response must be neurally produced, such that we only endorse the intentional explanation if we know that there is a suitable neural explanation available. We give the intentional explanation autonomous credence, as we might call it.

The scenario projected now for structural and historicist explanation is that the relation between such explanation and the intentional level of explanation is quite different from that just described for the relation between intentional explanation and the neural level. The idea is that far from having reason to endorse a structural or historicist explanation independently of information or understanding about the more or less immediate intentional origins of the event or condition explained, it would never be rational of us to accept such an explanation without having such an intentional account also: in particular, without having an intentional account that fitted in some required way with the structural or historicist one. If the parallel scenario obtained in the intentional-neural case, then we should only be prepared to accept a particular intentional explanation of a response, given access to a suitable neural explanation of how the response comes about.

Is there any reason to think that such a scenario obtains with structural and historicist explanation relative to intentional? I do not believe there is but the question requires some consideration. Discussing it will direct us to some significant aspects of such explanation that we have not noticed before now.

There are two distinct reasons why we might give autonomous credence to intentional explanations of human responses. One is that with such explanations we can generally test them, or at least look to see if they are borne out. We are told that someone did something because of such and such a belief. We can test that claim to some extent by seeing what she goes on to do in other circumstances and by seeing how she responds to certain overtures of our own. Thus we can find some grounds of a non-neural kind for endorsing the explanation. The other ground for endorsing intentional explanations independently of neural knowledge is that we can see reason, or so at least I have alleged, why evolution should have shaped us neurally so as to be rational and intentional creatures. Thus we can see non-neural reason why in general there will be a neural account of a human response that fits with a persuasive intentional explanation.

These sorts of reasons for giving autonomous credence to intentional explanations are not available, or at least not in the same measure, for structural and historicist explanations. The extent to which we can test a given explanation in these categories is limited, as social scientists routinely complain. And if we think that social events and conditions lend themselves to such explanations, that is not generally because of believing that societies have been shaped by evolution to be susceptible to such accounting; the point was discussed in the critique of the outflanking thesis in Chapter 3.

In view of these points of difference, it should be no surprise to find that there is indeed a contrast in the extent to which structural and historicist explanations invite autonomous credence and the extent to which intentional explanations do so. We may give an intentional explanation credence without any thought about the neural. But if we give a structural or historicist explanation credence, that will always be because we can see at least roughly why it might be intentionally borne out. If we think that urbanisation programs for a decline in religious practice, that is because we can see that with urbanisation there is likely to be a shortage of churches, a break with the seasonal basis of liturgy, and a lessening of peer pressure to attend church. If we think that the realisation that the Soviet Union would not intervene made the fall of the Eastern European Communist regimes more or less inevitable, that is because we can see that given such a realisation, hitherto unthinkable challenges to unpopular governments would occur and would lead sooner or later to success.

The fact that the credibility of structural and historicist explanation is dependent on the ability to tell an intentionally plausible story of this kind is a familiar theme among social scientists. It amounts to something like a *Verstehen* constraint: the requirement that if we accept various causal stories about why things happen as they do in the social world, we do so only so far as we can see how individual people can plausibly behave in a manner consistent with those stories (Weber 1949). We accept such stories only to the extent that we can find the stories credible in the light of our intentional psychology. The idea is that even if there were a fair amount of evidence for a certain structural or historicist explanation, we would reserve judgment about the linkage alleged in the absence of a sense of why it might be sustained at the intentional level. Part of confirming such an explanation then will involve this resort to *Verstehen* (Putnam 1978; contrast Abel 1948; Cohen 1979, chaps. 9 and 10).

It should be no surprise that structural and historicist explanation are subject to this sort of confirmation constraint, by contrast with intentional explanation, in particular more or less proximate intentional explanation. Not only is it less well supported by independent evidence than such intentional explanation, as already noted. Intentional explanation represents one of our most firmly entrenched schemas of understanding and it is perfectly natural that we should require other relevant patterns of explanation to establish their consistency with it. But the question we must face is not whether the *Verstehen* constraint is reasonable. The question is whether the existence of the constraint, and in general the lack of autonomy in the credence we give to structural and historicist explanation, means that such explanation is second-grade. Does it mean, as in the second of our two scenarios, that we should not be prepared to endorse such an explanation without already being in possession of an intentional explanation—a suitable intentional explanation—of the matter to be accounted for?

No, it does not. The contrasts noted between the position of structural and historicist explanation on the one hand, and the position of proximate, intentional explanation on the other, are important and worthy of attention. But they do not mean that we cannot rationally endorse a structural or historicist explanation without having access to a suitable intentional account of what is to be explained. We may endorse such an explanation while being in considerable ignorance as to what the intentional genesis of the matter in question is.

With structural or historicist explanation, we do not have powerful indirect evidence that things are so organised at the immediate, intentional level—or at any lower level—that the linkage alleged in the explanation is reliable. Here there is a contrast with the situation that prevails with intentional versus neural explanation: observational and evolutionary considerations both provide indirect evidence that things are so organised at the neural level that intentional regularities obtain. It is not surprising, then, that we should look for direct evidence of the linkages postulated in structural and historicist explanation. It is not surprising that we should look directly to the intentional level to see that, plausibly, things are so organised that any structural or historicist explanation we endorse is suitably underpinned. People are such that no matter how urbanisation goes, for example, it is likely to lead to a decline in religious practice. And so on in the other cases.

But the fact that we need such direct intentional evidence to support any structural or historicist explanation does not mean that we have to be in possession of the intentional explanation of the event or condition explained. That is why the second scenario mentioned does not obtain. Looking to the intentional level, we see that the urbanisation-secularisation link is plausible, because of the factors mentioned: the paucity of churches, the break with a seasonal liturgy, the absence of peer pressure, and so on. But in seeing that this is so, we do not identify the intentional explanation of any particular process of secularisation that we use the structural factor to explain. We do not identify an intentional explanation that mentions particular individuals, of course. And neither do we identify an intentional explanation that resorts to quantificational talk of unspecified individuals or statistical talk of averages and the like. We may have no idea how the intentional story goes, being ignorant as to how far the lack of churches, the liturgical break or the decline in peer pressure plays the crucial part.

It is true that the structural and historicist case contrasts with the intentional. But it is not dramatically different from other cases of program explanation. If we believe that boiling comes to a certain mean level of molecular motion—if we have become physicists—then it is plausible that we can rationally give high credence to the boiling-water explanation of why the flask in our earlier example breaks, only if we can see at the molecular level why such a linkage should reliably obtain. After all, the linkage may be robust under testing—more so than any structural or historicist connection—but we will have no evolutionary or other reasons for thinking that there must be a suitable molecular mode of organisation. We get direct molecular evidence as to why the linkage should obtain when we realise that if the molecules in the closed flask have the mean level of motion required for boiling, then it is very likely that there will soon be a molecule that has a position and momentum sufficient to crack a molecular bond in the surface. Having found this evidence, we will be relaxed about invoking the boiling in any instance to explain the cracking. But in most such instances we will lack a molecular account—at least an account in terms of specific molecules—of why the cracking takes place.

The case with structural and historicist explanation is much closer to this than to the case of intentional explanation. In order to be prepared to endorse any particular sort of structural or historicist account we must be able to see, in our intentional psychology of people, why the type of linkage involved is likely to be reliable. But the capacity to see this does not mean that for any structural or historicist explanation we offer, we will be in a position to tell a proximate intentional story, even an intentional story of a quantificational or statistical kind. There may be any number of intentional mechanisms that combine to persuade us that the linkage is reliable and in a given case we may be quite ignorant as to which of those mechanisms, or which mix of those mechanisms, is actually in place.

This completes my defence of explanatory ecumenism, in particular ecumenism as between the intentional or psychological explanations favoured by explanatory individualists and those structural and historicist accounts that attract explanatory collectivists. In conclusion I would make just one further point. This is that while this ecumenism puts me out of step with both of the methodological schools mentioned, and out of step therefore with the majority of social theorists, it puts me in line with the tradition of social thinking represented by the likes of Montesquieu, Tocqueville, and perhaps Weber. This is the tradition of distinguishing, in Montesquieu's terms, between the general and the particular causes of an event, allowing that both may be relevant in different contexts of explanation. The particular causes refer to the immediate antecedents at a given level. The general causes may refer to programming factors at a higher level or predisposing factors—factors of the kind that will appeal to the historicist—at an earlier point in time (Montesquieu 1965).

Take the explanation of the rise in crime that traces it to the increase in unemployment. In Montesquieu's terms, the increase in unemployment would count as a general cause of the rise in crime, a cause that made it more or less inevitable that there would be a rise in crime. Take on the other hand the particular changes in motivation and opportunity that led to more crimes being committed. In his terms, they would have been the particular causes responsible for the phenomenon to be explained. Montesquieu argued that general-cause and particular-cause accounts complement one another, and similarly we say that accounts at different levels, and at different removes, may be complementary.

Tocqueville gives a nice statement of the explanatory ecumenism that we endorse, again in the language of general and particular causes.

> I have come across men of letters who have written history without taking part in public affairs, and politicians who have only concerned themselves with producing events without thinking of describing them. I have observed that the first are always inclined to find general causes, whereas the latter, living in the midst of disconnected daily facts, are prone to imagine that everything is attributable to

particular incidents, and the wires which they pull are the same that move the world. Both should be considered mistaken.

(Tocqueville 1980, p. 262)

This sentiment is exactly that which I have been defending.

Rational Choice Theory

The Rational Choice Approach

Apart from structural and historicist theorising, the most dominant style of explanation found in contemporary social science is probably the so-called rational choice approach. If social theorists tend to be either explanatory individualists or explanatory collectivists—rather than ecumenists in the style I favour—then rational choice explanation is the approach most explanatory individualists espouse; it is the style of explanation with which defenders of the fine-grain principle are generally associated, for example (Elster 1983a; M. Taylor 1988).

The rational choice approach, like interpretation, is an instance of what we may describe as the intentional or psychological approach to social theory. The structural approach, as we have seen, explains social events and conditions by factors that are related via structural regularities to the matters explained. They are related to the matters explained via regularities that are causally and superveniently discontinuous with intentional regularities. The psychological approach, including that of rational choice theory, explains social events and conditions by factors that are related via non-structural regularities to the things explained. Non-structural regularities differ from structural, as might be expected, by the fact that they are continuous, causally or superveniently, with intentional regularities. They count as social-intentional regularities, to use a phrase from Chapter 3, as distinct from social-structural regularities.

A question in passing. Where does historicist explanation fit relative to this divide between structural and psychological varieties of social theory? We have implicitly given the answer to that question already. We noted in the preceding section that a historicist regularity may or may not be structural: it may or may not fail to be continuous with intentional regularities. We considered historicist explanation in company with structural for two reasons: first, that it goes traditionally with structural explanation; and, second, that there is an analogy between the different sorts of comparative information that they each provide. But the regularity involved in a historicist explanation need not be a structural one; for all we have said, it might also be psychological, even rational choice, in character. For example, the regularity governing the use of stainless steel in knives is surely a psychological rather than a structural regularity. Since we have already shown how to make room for historicist explanations, I shall only be concerned in this discussion with psychological explanations—rational choice explanations—that are not also historicist in character. If historicist explanations come up in discussion, I shall not be paying any attention to their historicist character.

We have seen that rational choice explanation is a psychological sort of social theory, invoking factors that figure in social-intentional rather than social-structural laws. But if rational choice explanation is psychological in character, then what distinguishes it within the category of the psychological? What distinguishes it, for example, from ordinary intentional explanation?

The rational choice strategy is the sort of explanatory approach associated with the discipline of economics, where the structural and historicist approaches are associated more with anthropology and sociology (Elster 1986b). There are two features by which it is distinguished and by which it is distinguished in particular from other psychological strategies of explanation. The first is an assumption that agents are to a good extent, if not predominantly, self-regarding in their desires; this is not often made explicit but it dominates almost all rational choice theorising. The second is a complementary assumption that in acting on self-regarding desires, agents will generally conform to the sort of formal model of rationality represented by decision theory. These features can come apart, and they do in some interesting explanatory projects, but I shall take the rational choice approach to be distinguished by the fact of combining them.

Together these assumptions have an immediate lesson for the sort of psychological approach that rational choice theory represents. If we assume that the theory is not meant to be revisionary of our intentional conception of agents, they mean that it introduces factors—people's self-concerns—that figure in laws which, allegedly, are causally continuous with intentional regularities. Rational choice theory may extend to laws that go statistical, for example, and are superveniently continuous with the causally extended set of intentional regularities. But its distinctive innovation is to posit lawlike antecedents of intentional attitudes that our received intentional psychology does not recognise. The theory may allow that we human beings are the intentional sorts of agents we appear to be, agents who form our attitudes more or less rationally and who act more or less rationally in the light of those attitudes. But it invites us to look more closely at what we actually do and to recognise that there is a pattern of rational self-interest present in our responses that may escape the casual eye.

Each of the assumptions that distinguish rational choice theory needs some further comment. The assumption of self-regarding desires is usually understood to postulate that agents are concerned with what is described as their economic gain but the category of economic gain is not often defined. It certainly includes the procurement of all forms of money and all forms of tradable goods but it extends further too. It must also include the enjoyment of the services provided by others, even when these are not tradable; the enjoyment of public goods, which everyone enjoys if anyone does; and the enjoyment of material goods that no one provides: goods like a nice climate and a beautiful environment. What do all of these have in common? The supply or accessibility of the goods depends on intentional action, one's own or that of others. The goods are distinctively action-dependent.

The main question about the first assumption that distinguishes rational choice—the assumption that people are self-regarding—is whether this means just that they desire such action-dependent goods for themselves. I think that within the rational choice tradition it is commonly assumed that there are no other goods to desire for oneself, so that the assumption that people are self-regarding is often equated with the assumption that they seek their own economic gain (Holmes 1990). But action-dependent goods are not the only sorts of goods that a person may want for herself and, if people are self-regarding, then it is reasonable to assume that they desire other sorts of goods for themselves as well.

Apart from action-dependent goods, we should also recognise a category of goods that may be described as attitude-dependent. Action-dependent goods I get by grace of what I or others do. Attitude-dependent goods I get by grace of what I or others think. They include the good of self-esteem, which I enjoy so far as I come to think well of myself. And, more saliently still, they include goods like the esteem, or gratitude, or affection of others, which I enjoy so far as those others come to think well or fondly of me. I may do things intentionally with a view to securing these goods but the goods will actually be conferred, not in virtue of what anybody intentionally does, but in virtue of non-intentionally produced attitudes. There may be value in my being induced by the evidence to think well of myself, or in others' being induced in a similar way to think well of me. But even if it is possible for someone to think well intentionally—that will depend on her resources of self-deception—there is no value to me in her intentionally thinking well of me: in her thinking well of me as a result, at least in some necessary part, of a desire to do so. Even if I can sensibly enjoin a person to think well of me, I cannot savour the good opinion procured on the basis of such an injunction.

Although attitude-dependent goods are not generally countenanced within the rational choice tradition, there is no reason why they should not be recognised (Pettit 1990c). It is encouraging to find that there are some leading figures in the tradition who agree. John Harsanyi (1969, p. 524) formulates the assumption of self-regard in the following terms: 'people's behaviour can be largely explained in terms of two dominant interests: economic gain and social acceptance' (see also Becker 1976, chap. 1). If economic gain refers us to action-dependent goods, social acceptance refers us to attitude-dependent.

If we take rational choice theory to hold that people are concerned with social acceptance as well as economic gain, there is still a question about the relative standing of these concerns. The current orthodoxy holds that economic gain is a concern in its own right, an end in itself, and we need not reject this. The question is whether social acceptance should be treated as a concern of equal, intrinsic appeal or whether it should be seen as of only instrumental value relative to economic gain. Under the instrumental reading, it would still be a good. It might even be a good that people are in the habit of prizing in this or that instance, without explicitly linking its enjoyment there with the prospect of any economic gain. But it would be a good of purely derivative standing. People would be able to justify it in reflection as something worth attaining, only because of its alleged link with economic gain: only because of its capacity to increase the expectation of economic gain.

There is a long tradition of regarding social acceptance as an intrinsic and indeed compelling concern (Lovejoy 1961, lecture 5). Adam Smith, despite his association with the economic and rational choice approach, is one of the most outspoken representatives; indeed he sometimes suggests that it is because of a desire for social acceptance that people often seek economic gain. 'Nature, when she formed man for society, endowed him with an original desire to please, and an original aversion to offend his brethren. She taught him to feel pleasure in their favourable, and pain in their unfavourable regard. She rendered their approbation most flattering and most agreeable to him for its own sake; and their disapprobation most mortifying and most offensive' (Smith 1982, p. 115). This tradition of taking social acceptance to be an intrinsic concern of people is an argument in itself against the instrumental reading. It amounts to testimonial evidence that such acceptance is an intrinsic concern.

Another line of reflection supports the same conclusion. This is that we often seem to care about the acceptance and approval of others, even when it is not a token of any prospective economic gain. The most dramatic case of this is the concern that people often display for their posthumous reputations. Why care about what people will think of us after we die, if our only interest in their approval is tied up with a concern for the economic gain they may bring our way? But short of this dramatic case, there are other examples too where we reveal a concern for social acceptance that is unlinked to economic gain. Consider how we smart and flinch beneath the gaze of strangers, on realising that we have been noticed doing something that is culturally compromising or demeaning: it may be as humdrum an activity as browsing over another person's mail, breaking a red light, or picking your nose.

A third line of thought also gives credence to the intrinsic reading of the concern for the approval of others. Assume that the concern for economic gain is an intrinsic concern, as the rational choice orthodoxy has it. In evolutionary terms, this assumption is plausibly read as postulating that members of our species have been selected for a sensitivity to their individual economic gain. Suppose now that social acceptance is indeed helpful in the achievement of economic ends, as the instrumental reading suggests and as it seems reasonable to admit. If mother nature has implanted an intrinsic concern for economic gain, then it is plausible under this supposition that it will also have implanted such a concern for social acceptance. It is plausible, in other words, that nature will have preempted our individual reasoning about the linkage between acceptance and gain and will have made sure that we do not neglect acceptance.

In view of these considerations I am prepared to assume that social acceptance should rank in rational choice theory as a concern that is of equal standing with economic gain. It may often be that we value someone's acceptance, in whole or in part, because of the economic benefits it promises. But the point is that such acceptance is also the sort of thing we can cherish in itself. Indeed it may even be, as Smith suggests, that it is the sort of thing for the procurement of which we sometimes cherish economic benefits. There is no reason why there should not be a two-way connection between the goods: no reason why each should not be valued for producing the other, as well as being valued in its own right.

Before passing to other issues, it is worth deflecting one argument against the instrinsic reading of social acceptance. This is the argument that in general we seem to care more for the social acceptance of those who are in a better position to confer economic benefits on us. We care more if the person who sees us crashing a red light is someone we know, for example, rather than a total stranger. I do not think that this argument is compelling, though it may persuade some. The reason I tend to discount it is that while those whose social acceptance we seek most keenly are often in the best position to confer economic benefits on us, they are also the ones with whom we have most of our dealings, or at least most of the dealings that matter to us. They are the ones who are likely to think most often about us and therefore the ones whom we are naturally most anxious to have think well.

So much for the first assumption that distinguishes the rational choice approach among psychological ways of doing social theory. The second distinctive assumption is that in pursuing the ends which interest them—these will include self-regarding ends such as economic gain and social acceptance but may also include other goals—people conform to formal models of rationality like decision theory. I do not say that rational choice theorists, even technically minded economists, work with the Bayesian decision theory sketched in the first section. They do not. Mostly they avoid the complications of such theory by postulating perfect information, or information of some more or less fixed kind. I do not say either that rational choice theorists invariably use technical tools of a kind with decision theory. Sometimes they do, sometimes they do not.

When I use decision theory as an exemplar, I mean only to draw attention to two facts about rational choice theory. First, the models of rationality preferred in the approach put aside the question of how agents are supposed to think and deliberate as they find their way to action; in this, as we have seen, they resemble decision theory. Second, the preferred models tend to represent people as instrumentally rational in relation to the ends postulated: they represent people as taking those ends for granted and as pursuing the best means of realising them. In a phrase, the models used in rational choice depict human agents as means-end automatons: as black boxes designed, we know not how, for the instrumental pursuit of certain goals (Hindess 1988, Pettit 1991a).

The sort of model used will take as given the ends that the agents in any situation are supposed to care about; although other ends are allowed in principle, the ends usually postulated boil down to self-regarding concerns. The model will then give a derivation of what it is rational for the agents to do in the light of those ends, a derivation that may involve some sophisticated deduction. For example, it may identify rational behaviour with the behaviour that would maximise the relevant expectation of whatever good is in question. And then the model will postulate that the agents in question may be expected do the rational thing. It will not tell us how the agents themselves are supposed to work out their decision. And it will not allow that they might be as concerned about the ends they should be pursuing, as they allegedly are about the means that are best suited to the given ends.

We have seen that rational choice explanation is a strategy of psychological explanation, engaging non-structural regularities, and that it is distinguished among psychological approaches by two assumptions: that people are predominantly concerned with self-regarding ends like economic gain and social acceptance; and that in pursuing those concerns they conform to formal models of rationality of a kind illustrated by decision theory. There is one more thing to say in introducing the strategy. This is that most interesting rational choice explanations, in particular most of those that bear on large-scale social phenomena, involve an appeal not just to factors that explain people's responses, but also to factors that explain the collective effects of such responses. Rational choice explanations appeal to aggregative mechanisms, as we may put it, as well as to motivational ones.

The notion of an aggregative mechanism is nicely illustrated with the help of an example that we used in Chapter 3 to illustrate a regularity that is logically continuous with intentional regularities. This is the example in Schelling (1969) of a mechanism whereby a mixed neighbourhood will go over to total residential segregation if each household comes to prefer not to be in a minority among its immediate neighbours. Each household has up to eight neighbours, as each square on a chess board has at most eight adjacent squares, and the aggregative mechanism to which Schelling draws attention is this: that the first wave of removals that is prompted by the desire not to be in a minority will leave previously happy households dissatisfied, and that this will trigger another wave of removals and another wave of dissatisfaction, up to the point where more or less total segregation ensues.

The aggegative mechanism in the segregation case means that what the different parties individually do has a collective result that is not necessarily visible to them. They are individually motivated to move house and in being so motivated they are not necessarily aware of the collective upshot of their actions. When an aggregative mechanism is not necessarily visible in this way, it is common to speak of it as an invisible hand.[10] The idea of the invisible hand probably goes back to Bernard Mandeville's fable of the bees (Hayek 1978, essay 15). But the phrase itself was introduced by Adam Smith in his somewhat sanguine account of how the rich distribute the necessities of life in the course of employing others for 'the gratification of their own vain and insatiable desires'. 'They are led by an invisible hand to make nearly the same distribution of the necessaries of life, which would have been made, had the earth been divided into equal proportions among all its inhabitants' (A. Smith 1982, p. 185; see too Rosenberg 1988, chap. 6; Ullmann-Margalit 1978).

Many of the aggregative mechanisms invoked in rational choice explanations can be cast as invisible hands of roughly the same kind. Some of the results associated with such hands will be benign, as in Smith's own example. But others will generally be regarded as malign, as indeed in the segregation example. In these cases rational choice theorists sometimes speak of an invisible backhand, or an invisible foot, as distinct from an invisible hand proper.

A useful way of understanding the invisible hands and backhands often invoked in rational choice theory is via the prisoner's dilemma to which game theory has drawn attention (Resnik 1987). A number of parties are in a dilemma of this kind if they each face two options, one saliently cooperative, the other saliently an act of defection; if defection is better for each, no matter whether the others cooperate or defect: defection is a dominant option, as game theorists say; and if each is better off under joint cooperation than under joint defection.[11] Suppose you and I meet on a desert island and we each have the choice of giving the other information on our source of food: you on the oysters you have found, I on the edible nuts I have discovered. We are in a prisoner's dilemma if misleading the other is a dominant option for each of us but if nevertheless we are each worse off under joint deception than we would be under joint disclosure of our food supplies. We would be in such a dilemma if the best thing for each of us was to get the other's information without giving our own away; if the worst was to be induced to give information without getting any in return; and if, as between the other possible outcomes—joint disclosure and joint deception—we were each better off under joint disclosure. Such a prisoner's dilemma will constitute an invisible backhand so far as it leads each of us rationally to deceive the other, with the result that we are each worse off than we would have been had we only got our act together and revealed our individual sources of food.[12]

The prisoner's dilemma bulks large in the rational choice literature because generalisations and analogues of the dilemma feature in many of the invisible hands and backhands that the literature invokes. The most important generalisation is to the many-party case and the most important analogue is the free-rider problem; in fact this predicament is an analogue of the generalised, many-party dilemma (Pettit 1986b). The free-rider problem arises with the littered park, the congested road, the overfished river, the polluted environment, the underfunded community effort, and the like (Barry and Hardin 1982; R. Hardin 1982, Olson 1965, Ullmann-Margalit 1977). Everyone is better off if everyone contributes to the relevant cause—say, keeping the local park clean—but the costs and benefits are stacked in such a way that the rational thing for each to do is not to contribute, hoping that the good in question—a good from whose enjoyment she cannot be excluded—will be produced by others. Defecting may not be strictly a dominant option. But defecting is still rational, for it offers the best expectation of satisfied desire. The common feature in free-rider cases is that it is very unlikely that any individual's contribution will be essential. Thus each may think that it is fixed, independently of what they do, whether the good is attained or not. And thinking this, each may reckon that in either scenario an individual is better off not contributing: if the good is attained, contributing would be a redundant effort; if the good is not attained, it would be an effort in vain.

Free-rider problems are clear cases of invisible backhands. But the prisoner's dilemma and its analogues can also figure in invisible hands proper, in invisible mechanisms that have benign effects. Consider the part they play in perhaps the most famous example—or at least alleged example—of an invisible hand: that of the competitive market. In the competitive market producers are drawn, so it is alleged, to sell at the competitive price—roughly, the lowest price at which they can stay in business—and thereby to generate an important social benefit. The assumption is that producers are each self-interested and rational and that in any area of production new producers are free to enter at any time. This means that in each area of production everyone will be better off if all cooperate in fixing a price than if all defect, but that defection will always be better for each than sticking with any agreed non-competitive price: defection will promise to increase that individual's share of the market more than is sufficient to compensate for the lower price charged. Each area of production will involve a free-rider problem and as the individual purchasers seek to free ride on the fidelity of others to any agreed price, they will force the price down to the lowest at which they can stay in business. Thus everyone will be better off in the long run: the loss that an individual may suffer *qua* producer in some area will be outweighed by the gain that all enjoy *qua* consumers in different areas.

I have said enough, I hope, to indicate what I have in mind as the rational choice strategy of social explanation. The strategy is the very stuff of economic theory but it has also been used in a great variety of contexts by other social scientists. (For a short survey, see Friedman and Hechter 1988.) Successfully or not, it has been used to explain the platforms adopted by political parties in democratic elections, the greater failure of collective action in larger than in smaller groups, the levels and varieties of taxation that different states have imposed, the posture of participants in the arms race, and the pattern of aiming to miss that evolved in trench warfare. Indeed the rational choice strategy can even be used, as we shall see in the next chapter, to provide explanations for the appearance and ubiquity of moral norms.

An Ecumenical Interpretation

I do not intend to provide an overview of rational choice explanations here, or an analysis of the extent to which they are compelling. Having introduced the idea of rational choice explanation, I mean only to address the question of its general legitimacy. In particular, I mean to address the question of whether the explanatory ecumenism we have embraced can extend to provide a place for rational choice theory. I shall argue that it can.

In the last section we addressed the question of the legitimacy of structural and historicist explanation. The problem, given our ecumenical aspiration, was to show how such explanation could be countenanced consistently with the holistic individualism we have endorsed. The problem here is to show how rational choice explanation can also be countenanced consistently with that holistic individualism. But where it was the individualism that raised a question about the legitimacy of structural and historicist explanation, it is holism that raises a question about the rational choice strategy.

Under the rational choice approach, as we have seen, people's responses are traced to self-regarding concerns with economic gain and social acceptance.

This approach is in tension with the sort of holism that we adopted earlier. The central strand in our case for holism is the claim that people avail themselves of socially resourced concepts in order to think and, among other things, in order to reason their way to action. This strand is made explicit in the deliberative or inference-theoretic view of agents that we articulated in the first section of the present chapter. Under that view, people find their way to action in response to properties that they register in the options before them, properties that are valued in common with others and that can be invoked to provide at least some justification of their choices. If I help out someone financially, it is because I think, implicitly or explicitly, that she is in urgent need, or that she is a good person who is worthy of assistance or, simply, that she is a friend. And these properties in virtue of which I help her out are sufficient to make it clear why the option is desirable; they serve, if only weakly, to justify what I do.

The rational choice strategy of explanation appears to conflict with this inference-theoretic view of agents. It suggests that with most things that people do, there are self-regarding motives to be identified. But how can we trace people's actions to such motives if we believe, as the inference-theoretic view has it, that people act on the basis of socially justifying reasons, in particular reasons that need not have a self-regarding character? If I go to the assistance of someone on the basis that she is a friend, how can I be held to perform such an action out of a self-regarding motive: out of a concern for economic gain or social acceptance? The rational choice approach appears to run into direct conflict with the view of agents that is projected in our case for holistic individualism, in particular our case for holism. Thus it is hard to see how we can be ecumenically tolerant of it.

We mentioned two assumptions that distinguish the rational choice strategy from other psychological approaches. The first assumption is that people are motivated, in good part, by a concern for their own economic gain and social acceptance. The second assumption is that, however they deliberate—the matter is put aside—they are instrumentally rational in the pursuit of their goals. The problem that arises for rational choice theory within our perspective is how we can make such assumptions. How can we think that they are sound, given that people reach their decisions on the basis of implicit or explicit deliberation, and given that the deliberation they practise is not particularly self-regarding in character?

There are two obvious ways in which rational choice and the inference view might be reconciled but neither is satisfying and this may account for some of the opposition to the approach (see, for example, Mansbridge 1990). The one way would be to hold that, contrary to first appearances, the properties in the inferential light of which people act are all self-regarding: the reasons that move people are all reasons of self-interest. Call this the calculative reconciliation, since it represents people as self-interested calculators.[13] The other mode of reconciliation would admit that on the face of it, people do attend to reasons other than self-regarding ones. It would temper this admission, however, with the claim that such reasons only carry weight when they are more or less unconsciously recognised as counselling self-interested behaviour. Call this the cynical reconciliation, since it represents a rather cynical view of human reasoning.

The calculative reconciliation may be appropriate for limited ranges of human behaviour where it is taken as socially acceptable to attend only to self-regarding matters in the making of decisions. It is acceptable to do this in market behaviour, at least under cultures that the science of economics has influenced. But outside the market it is rarely taken as acceptable for agents to reach their decisions on the basis of self-regarding considerations alone. The friend, adviser, or politician who defends her initiatives on the grounds that they are in her own interest loses all claim to affection, attention, or respect. And it is unlikely in areas where it is socially unacceptable to choose on the basis of self-regarding considerations that people, nonetheless, make their choices on such grounds. Of course we all recognise that people may sometimes act on considerations that are not socially acceptable: say, that politicians may act with a view to their re-election even when they invoke more high-minded grounds for their policies. But if we are not to take an entirely jaundiced view of human beings, we must think that in many cases people are really attending to the socially acceptable grounds that they invoke in justification of what they do.

The cynical mode of reconciliation is likely to be found more attractive than the calculative, since it does not fly in the face of appearances in the same way; it does not suggest that contrary to how things seem, people actually deliberate with an exclusive or dominant eye to their own interests. The cynical mode of reconciliation admits that people deliberate on the basis of a more varied range of considerations but postulates that nevertheless they unconsciously tend to act only in ways that satisfy their self-interest. Gary Becker (1976, p. 7) suggests that he is sympathetic with this point of view: 'the economic approach does not assume that decision units are necessarily conscious of their own efforts to maximize or can verbalize or otherwise describe in an informative way reasons for the systematic patterns in their behaviour. Thus it is consistent with the emphasis on the subconscious in modern psychology' (see too McCullagh 1991).

If the only possible way of reconciling the rational choice approach with the inference theory that we have espoused were by means of this cynical postulate, then I think that we should be extremely loath to endorse the approach. The reconciliation may not fly in the face of appearances in the manner of the calculative line, for it does not say that we attend only to self-regarding considerations. But it requires us to accept a very controversial story about why the considerations to which we attend actually move us, a story that runs counter to our immediate sense of ourselves and one another. No doubt that story sometimes applies; no doubt the best of us are subject to occasional self-deception about the things that move us. But the idea that the story applies to most of us most of the time is extravagant and implausible. Under this mode of reconciliation, the epistemic cost of coming to accept rational choice theory would surely be too high; it would require too deep a revision of the view we spontaneously take of most human beings. However ecumenical we wish to be, we should still baulk at the proposal.

But we should not despair at the failure of the calculative and cynical readings of rational choice. For there is a third, more satisfactory way in which the approach can be reconciled with inference theory and, more generally, with the holistic individualism defended in this book. On this third mode of reconciliation the self-regarding concerns postulated in rational choice do not generally enter into deliberation, contrary to the calculative rendering, and do not unconsciously mould our deliberation, contrary to the cynical. Rather they serve as constraining concerns that agents will not in general flout: they will not act in a manner that reduces satisfaction of the concerns below a given aspiration-level. We may call this the constraining mode of reconciliation.

The constraining mode of reconciliation has not received any attention in the literature but it has a great deal to be said in its favour. There are two lines of argument that I would like to present in its defence. First, I argue that, combined with some plausible assumptions, the inference-theoretic view defended here implies that self-regarding concerns will figure as constraints in the manner envisaged. Second, I argue that on the picture involved in the constraining mode of reconciliation, rational choice theory serves a useful and plausible explanatory role.

People deliberate their way to action, on the inference-theoretic view. If they do not deliberate explicitly, they do so implicitly: they are responsive to such properties—in particular, such valuable properties—as make the action reasonable, at least from their standpoint. The first argument for my preferred reconciliation is that this view combines with two plausible assumptions to imply that people will generally not flout their self-regarding interests. The extra assumptions required are these. First, if the course of behaviour adopted by an agent flouts her self-regarding interests, plunging her below a relevant aspiration-level, then she is likely to become aware of the fact and to begin to deliberate in terms of those interests. Second, if the agent does take account of those interests in deliberation, then she is likely to give them considerable weight: she may not let them outweigh other considerations altogether but she will certainly put them in the balance with such considerations.

The first assumption deploys the notion of a relevant aspiration-level. So how is the appropriate level determined? This is a question that requires an interpretative understanding of the agent involved, including an appreciation of the circles in which she moves. If a general theory of aspiration-levels is required, then perhaps the best available suggestion is that people's levels of aspiration are dictated by their normative reference group: by the group with which they mainly identify, the group which they see as setting the norms that govern how they should act (Runciman 1972, chap. 2). A person's interests in economic gain and social acceptance will be flouted—she will be plunged below the relevant aspiration-level—at the point where their frustration impacts on her capacity to maintain an identification with that group. On this story the manual worker's economic interests will not be flouted just because she is less well off than the company director but they will be flouted if her mode of behaviour means that she cannot maintain the life-style of the colleagues with whom she identifies. Rational self-interest dictates behaviour according to social context, not in a vacuum (Hollis 1987).

Back to the main line of argument. Are the two assumptions I presented plausible? I think they are. The first assumption will certainly be plausible under the interpretation just given of the relevant aspiration-level. It is widely accepted that we judge our own lot by our capacity to keep pace with our fellows, in particular our fellows in the reference group with which we identify (Frank 1985). If any agent sinks below the level that is standard for those in that group, then it is very plausible that she will notice the fact and reflect explicitly on how her behaviour is serving her economic and social interests. The second assumption goes on to posit that if an agent begins to reflect in this way, then she is very likely to give some weight to those interests in deliberating on how she should act. It posits that, at the least, the agent will be likely to ensure that her behaviour does not put her below the relevant aspiration-level. This assumption does not represent agents as particularly self-concerned, for it allows that they may fail to optimise on their own interests for the sake of furthering some other goals; it allows that they may satisfice (Slote 1989). It merely postulates that they are not likely to sacrifice their own interests beyond a certain culturally given point. And that is surely just a common-sense reflection.

Let us admit these extra assumptions, side by side with the inference theory defended earlier. The upshot is that things will conform to the model projected in the constraining reconciliation between rational choice theory and the inference-theoretic view of agents. People will find their way to action on the basis of the deliberative considerations that are socially available to them, considerations that need not be particularly self-regarding. Thus it will always be possible to make sense of their doings in the interpretative manner described in the first section. But people will look to self-regarding considerations, and give them weight in their decisions, if the actions to which they are otherwise led begin to flout their self-interest: their interest in economic gain and social acceptance. Self-interest may not have much of a conscious or even unconscious role to play in motivating their actual behaviour but it has a virtual presence that puts constraints on how their actual behaviour is likely to go. We can be sure that in general no pattern of behaviour displayed by human beings will flout their interest in economic and social benefits.

So much for our first argument in favour of the constraining mode of reconciliation. The reconciliation posits a distinctive model of people's responsiveness to their own interests but the model involved should be attractive. It is more or less bound to obtain under the combination of our inference theory with the two extra assumptions. The second argument for the constraining mode of reconciling rational choice with inference theory is that the reconciliation makes good sense of the explanations offered in the rational choice approach. It shows how those explanations can be plausible and useful.

The rational choice explanations that need vindication are those that address behaviour outside the market and related arenas: those that address behaviour that, unlike market behaviour, we cannot plausibly take to be deliberatively generated by self-regarding considerations. But how can we invoke self-regarding concerns to explain behaviour that issues from a pattern of deliberation that does not take self-regarding matters into account? It is traditional in rational choice circles to distinguish two different *explananda*: on the one hand, the emergence of a certain pattern of behaviour on the part of an individual or group, or something associated with its emergence; on the other, the continuance of that behaviour on the part of the individual or group, or something associated with its continuance. But if the behaviour is generated by deliberation over matters that are not self-regarding, how can the emergence or continuance of the behaviour be explained by the agent's self-regarding concerns? Things would be alright under the calculative or cynical reconciliations, which would give a causal role to the self-regarding concerns. But under our reconciliation the concerns need have no causal role. Hence the problem.

Under our reconciliation, the self-regarding concerns may have the role only of standby causes. The following scenario may obtain. The behaviour is actually produced by deliberation over considerations as to what is socially fitting, what is just, or whatever. But did those considerations fail to produce that sort of behaviour, did they support choices that flout the self-interest of the agents, then those agents would begin to think in self-regarding terms and their self-regarding concerns would take the place of the actual causes of the behaviour. The self-regarding concerns are not the actual causes of anything; they are preempted in that role by the agents' deliberations over other matters. And not being actual causes, the concerns cannot be invoked to explain the emergence or continuance of the behaviour. Information about the presence of such standby causes is not information on the causal history of the behaviour; it does not direct us to properties that are causally relevant to the emergence or continuance of the behaviour.

The solution to this problem is to recognise that there is a third sort of *explanandum* relevant in social theory apart from the emergence or continuance of a certain sort of behaviour or response. We are often concerned to explain not why a certain sort of behaviour emerges or continues, but why it is resilient. We are interested in showing that, and explaining why, the behaviour can be relied on to remain in place—more or less—under a variety of contingencies, including contingencies that remove its actual causes. And in seeking to explain such resilience we may rightly be directed to standby as distinct from actual causes.

Consider the following set-up. A ball rolls along a plane on a straight line, say in acccordance with Newton's laws. Suppose the path of the ball is marked out with little pinball posts on either side and that these posts are equipped to serve a double role: they will tend to deflect any force that would push the ball off its path; and they will tend to return the ball to the straight path in the event that a perturbing force is effective.[14] In such a case not only does the ball actually roll on a straight path; it rolls on such a path resiliently. Not only does it roll on a straight path in the actual world; it rolls on such a path, or tends towards such a path, in various possible worlds where certain perturbing forces are present: in particular, forces with which the pinball posts can cope. Suppose now that we want to explain the resilience of the ball's rolling on a straight path. In this case we will need to mention the pinball posts, although those posts play no role in actually producing the straight-path behaviour; they are merely standby causes.

We agreed earlier that to explain something like an event usually means providing information on causally relevant properties. Now we meet an exception. How does a reference to standby causes, as in the ball-rolling case, provide relevant information? The resilience of the phenomenon under explanation is a feature that it possesses at the actual world but a feature that it possesses in virtue of how the phenomenon fares in a range of possible worlds: those in which the relevant contingencies—say, those associated with perturbing forces—are realised. If we are to speak of a certain property as being relevant to such a phenomenon then it must be a feature of the actual world but a feature that determines that the phenomenon fares appropriately at the associated possible worlds: a feature that is causally relevant in each of those worlds to the appearance of the phenomenon there. The property of the actual world that is relevant in the ball-rolling case is precisely that there are suitable pinball posts at either side of the ball's path. Those posts are not causally relevant to the ball's actually holding to a straight path—they have no actual effects—but they are still relevant to the ball's being a resilient straight-roller. They constitute it as a resilient straight-roller.

That there are pinball posts in position can be seen as programming, if only in a constitutive way, for the resilience of the straight roll. The case fits the various conditions associated with the program model. The presence of the pinball posts is relevant to the resilience of the straight roll only so far as the configuration of possible physical effects—the configuration of possible micro-physical sequences—that is consistent with the presence of those posts is relevant to the resilience. The presence of the pinball posts ensures or makes it more probable that there will be a configuration of possible effects, a configuration of possible micro-physical sequences, which is sufficient to guarantee the resilience. And that configuration of possible effects or sequences is actually what accounts for the resilience of the straight roll; the resilience is not produced by something independent.

The lesson should be clear. While self-regarding concerns may be merely standby causes of the behaviour addressed in rational choice explanations, still they can be suitably explanatory. The presence of such concerns can program for the resilience of suitable behaviour, as the pinball posts program for the resilience of the ball's rolling on a straight path. But is it fair to say that social theory, in particular rational choice theory, is concerned with explaining the resilience of certain forms of behaviour as distinct from their emergence or continuance?

A number of points suggest that this is fair. One is that by the account of the last section structural and historicist explanations serve to explain not just the emergence and continuance of certain phenomena but also their resilience.

Both sorts of explanation abstract from the immediate intentional antecedents of the responses associated with those phenomena, identifying a higher-level or temporally removed factor whose presence means that even if those antecedents had been different, still the phenomenon would have occurred. If one sort of social theory is concerned with explaining resilience as well as explaining emergence and continuance, it should not be very surprising to find that another is concerned with explaining resilience alone.

A second point worth mentioning is that there is a possible sort of structural explanation that would explain resilience without explaining emergence or continuance. Consider the functional sort of explanation that explains a certain pattern of behaviour in terms of the social benefit that it has. Consider in particular the sort of explanation that would lend itself to a gloss that we mentioned in the third chapter. The gloss would legitimate the explanation by these three claims: first, although no departure has actually occurred, any departure from the pattern would put the relevant benefit at risk; second, if the benefit were jeopardised by a departure from the pattern, then the previous beneficiaries would act to reverse the departure: there are no collective action predicaments, for example no free-rider problems, to inhibit them; and third, any departure would therefore be quickly followed by a return to the pattern. Such an explanation cannot invoke the social benefit of the pattern to explain its emergence or continuance if, as is quite possible, the benefit plays no causal part in generating the behaviour. In that case then, all that the functional observation can be represented as explaining is the resilience of the pattern. Thus there is a plausible sort of structural analogue for rational choice explanation, under our representation of rational choice explanation.

There is a third point to be made in support of the view that rational choice theory may reasonably be cast as explaining resilience. The patterns that rational choice theory—and social theory in general—is concerned to explain are generally resilient in character. History may seek to explain why this or that contingent event occurred. But rational choice theory is more usually concerned to address questions about patterns that are seen as robust across a variety of possible differences. Thus it is plausible that outside the market and related areas, what the theory is designed to illuminate is the resilience of certain phenomena rather than their actual genesis or maintenance.

So much by way of defending the constraining mode of reconciling rational choice theory with our inference-theoretic view of agents. In conclusion, a query. We said earlier that rational choice theory is plausibly taken to be an extension of intentional psychology, identifying laws that are causally continuous with intentional regularities. Can we continue to maintain this view if we say that the factors to which rational choice theory directs us often serve only as standby rather than actual causes? Yes, we can. The factors involved as antecedents in causal regularities need not serve as actual causes in every case where the regularity obtains and is satisfied. The regularity whereby certain forms of cancer lead to death would remain a causal regularity, even if victims routinely put an end to their lives by other means. Similarly, the regularity whereby certain requirements of self-interest lead to their satisfaction remains a causal regularity, even if those requirements are rarely called upon to play a causal role in the process leading to their satisfaction.

The Interpretation Illustrated

We have argued that under the constraining reconciliation between rational choice and the inference-theoretic view of agents, it is possible to take rational choice explanation as accounting for the resilience of certain phenomena. This gets rid of the problem raised by the fact that the concerns such explanation invokes as explanatory may often figure just as standby causes. It allows us to maintain our ecumenical stance. But we need to substantiate the emerging view of rational choice explanation with reference to some examples. We need to show with some representative examples, that it is reasonable to take rational choice explanations, at least outside the market arena, as explanations of the resilience of certain phenomena. Here are three examples for discussion.

First example: David Lewis (1969) offers an explication of conventions that, among other things, suggests a rational choice explanation as to why certain conventions should emerge and continue. Take the sort of convention governing how cars should pass when they meet on the road: say, that they should pass on the left. It is going to be in the self-interest of drivers to drive on the same side under any plausible scenario and it is not going to matter to them which side is chosen. This means that they will tend to drive on the same side, provided that it is possible for a preferred side to be indicated in advance. But this will be possible, Lewis argues, under uncontroversial assumptions about how one side rather than the other can emerge as the salient choice: assumptions about the role that can be played in this regard by temporal precedents, cultural analogues, explicit agreement, and the like. Thus we may expect a convention like that of driving on one side of the road to emerge and persist in any society. More generally, as Lewis puts it, we may expect conventions that solve various coordination predicaments to emerge and persist. These are the conventions that resolve predicaments in which it is in everyone's interest to do the same as others; they extend from the road-passing convention to examples like dressing to the same standard for social occasions, assembling at the same locations for certain purposes, and using words to mean the same things.

For a second example of a rational choice explanation consider the more or less standard, free-rider account of why individuals often fail to combine in opposition to some force that oppresses them. The explanation is that while all would be better off if the oppressor were removed, each is in a position where it is not attractive in self-interested terms to do anything to resist. Douglass North (1981, pp. 31–32) gives a representative statement of this sort of explanation: 'the free rider accounts for the stability of states throughout history. The costs to the individual of opposing the coercive forces of the state have traditionally resulted in apathy and acceptance of the state's rules, no matter how oppressive. An historical counterpart of the low voter turnout in many current democracies is the failure of individuals to act as classes and of large groups to overthrow society in the past'.

The third example of a rational choice account takes us to more specific matters. It is the now familiar explanation as to why slavery remained firmly in place in the south of the United States prior to the Civil War: it was an economic arrangement that suitably rewarded the plantation holders. The explanation is suggested, for example, in the classic text by Fogel and Engerman (1974, p. 4). 'Slavery was not a system irrationally kept in existence by plantation owners who failed to perceive or were indifferent to their best economic interests. The purchase of a slave was generally a highly profitable investment which yielded rates of return that compared favourably with the most outstanding investment opportunities in manufacturing'.

The question now is whether it is fair to say with examples like these that the rational choice explanation provided explains the resilience of the phenomenon in question but not its emergence or continuance. In other words, does the constraining way of reconciling rational choice and the inference-theoretic view of agents enable us to give a fair showing to such explanations?

The first thing to say about the explanations is that none of them presupposes that the agents calculate in terms of the self-regarding concerns that are adduced as explanatory. Lewis is explicit on the point, when he admits that conformity to a convention may be actually produced not by reflection on its self-regarding benefit, but by mere habit (D. Lewis 1983, p. 181). But in the other cases too it is surely important for the plausibility of the explanation offered that it does not assume self-regarding calculation. If North is to offer an attractive explanation of the stability of states over history, he cannot be taken to suggest that those who fail to resist an oppressive state must work out in a sequence of explicit reasoning that they are better off doing nothing themselves, though they may hope that others will do something on their behalf. Again, if Fogel and Engerman are to claim an interesting account of the slavery system in the southern United States, then they must surely admit that the plantation holders probably did not think in the self-regarding language of a modern, economically literate manager. They must admit, as is often mentioned, that there was a distinctive culture surrounding the slavery system, a culture that would have encouraged the plantation holders to make sense of their commitments in quasi-religious terms.

If the explanations do not lend themselves to the calculative construal, neither can they be plausibly reconstructed in the mould of the cynical reconciliation. It would be outlandish to take any of the authors involved to be suggesting that though the subjects of their explanation think in other ways, they are subject to the unconscious pressures of self-interest and that it is for this reason that their behaviour can be explained by rational choice. Nothing in what they say, nothing indeed in what rational choice theorists generally say, points to this interpretation of the enterprise.

Can we fairly represent the explanations in the image projected by the constraining reconciliation? Can we see them as explanations that cast people as virtually self-interested, if not actually so, and which focus on the resilience of the behaviour under investigation rather than its emergence or continuance?

Take Lewis's explanation of conventions first of all. In acknowleging that a more or less unthinking habit may be at the origin of actual conventional behaviour, he makes a point that fits the resilience reading of the explanation. 'If that habit ever ceased to serve the agent's desires according to his beliefs, it would at once be overridden and corrected by conscious reasoning' (D. Lewis 1983, p. 181). The picture Lewis seems to have is that although deliberation in self-regarding terms is not at the actual source of conventional behaviour, if whatever actually is at its origin failed to produce that behaviour, then self-regarding deliberation would step in to do the job. That is to say that he considers the self-regarding concerns invoked as standby rather than actual causes of the conventional behaviour. And that in turn means that he can think of such concerns as explaining not the emergence or continuance of the behaviour but only its resilience: the fact that it can be relied upon to persist across a wide variety of contingencies.

A similar reading goes through quite naturally in the other two cases. Take the free-rider account of why people do not resist oppression. For all that the account says, the psychological states at the origin of people's passivity may be any of an indefinite variety of kinds. People may not resist oppression for any of a variety of reasons: because of a lack of imagination or a lack of precedent, because they accept an ideology of subservience, or because they think resistance would be immoral. The possibilities are endless. The point that the rational account emphasises is that non-resistance is in the self-interest of the agents involved, whether they realise or not. This fact cannot explain the emergence or continuance of the non-resistance, given that the actual sources do not connect with thoughts about self-interest. But it can explain the resilience of the phenomenon. All we have to admit is that if an individual began to consider resisting, say because the actual pressures towards passivity were neutralised, then she would soon recognise that her self-interest would be jeopardised by resistance; and that being so, she would almost certainly resume her passive posture. Admit this and we can immediately see that the free-rider account is a natural explanation of the resilience of the passivity.

Consider, finally, the Fogel and Engerman account of why the slavery system was entrenched in the southern United States. Here is the way that our approach would cast it. The plantation holders may have stuck by the system for any of a variety of non-economic reasons; some may have stuck by it out of mere habit, some out of a sense of moral commitment, some out of a yen for playing master. But if the point made by Fogel and Engerman is correct, then had any of them moved out of slave-holding, they would have found themselves no better off; moreover, had any large number of them moved out of the system then they would probably have found themselves individually worse off than they would be as slave-holders: slave-holding activities would have become more profitable with the lessening of competition, other activities would have become less profitable with the increase of competition. The suggestion is that any departure from slave-holding in the period before the Civil War would have been short-lived. In virtue of making good economic sense for the plantation owners, the slavery system enjoyed an important resilience. The fact that it made economic sense may not explain why the system emerged, and almost certainly does not explain why it continued. But that fact does explain why the system was a resilient feature of the antebellum South.

I hope that these remarks are sufficient to vindicate the constraining way of reconciling rational choice accounts and the inference-theoretic view of agents. We have seen, first, that if it is supplemented with two plausible assumptions then the inference-theoretic view implies that people will be responsive to their self-regarding interests in the manner projected in that mode of reconciliation. And we have seen, second, that under this mode of reconciliation rational choice accounts are useful and plausible sorts of explanation: outside the arena where people may be expected to deliberate in a self-regarding way, they are explanations of the resilience of certain phenomena. Thus our explanatory ecumenism extends beyond structural and historicist explanation to encompass explanations of the rational choice variety.

In conclusion to this discussion I would stress just one point further. This is that on the account presented here, rational choice explanation is really rather a humdrum enterprise. It only applies to a limited range of patterns: those where any alternative to the pattern explained would flout the interest of the agents in economic gain or social acceptance. It can go through without any self-regarding calculation on the part of the agents involved. And it can concede that even if agents were to calculate in a self-regarding way, they might not maximise on the fulfilment of their self-regarding concerns; they might be content to attain some lesser level of satisfaction.

Presented under such a deflationary profile, rational choice theory may raise some questions. Where has the distinctive exactitude and technicality gone? Where has the customary mathematics disappeared? Here I would make two remarks. First, rational choice theory should be distinguished from the abstract theory of rationality pursued in decision theory, game theory, and the like: it should be distinguished from the sort of theory that aims, often with great mathematical acuity, to identify what is ideally rational for agents with given preferences and probabilities. Second, we have been concerned with rational choice theory outside the market arena, and in general outside contexts where it is plausible to assume that agents calculate, and perhaps calculate with great sophistication, in self-regarding terms. We have been concerned with rational choice theory not as a theory of abstract rationality, and not as a theory of rationality for calculative agents, but in its role as a potential schema of explanation for uncalculated human activities. It should be no surprise that what the theory offers in this area is non-technical and inexact. It should be no surprise if what it offers is close to common sense.

Notes

1. Having considered all things, I believe it desirable; I believe in an all-out way that it is desirable. This judgment should be distinguished from the judgment that the option is desirable relative to all considerations: that it is desirable-all-things-considered, as we might put it. See Davidson 1980, essay 2.

2. Such 'flat-out' belief, as Bratman (1987) calls it, need not be mysterious. Even simple theories can explain its place. Consider the expressivist view that the belief to degree x that p is expressed in assent to the proposition that it is probable to degree x that p. Flat-out belief amounts to assent on that theory.

3. We thereby avoid problems of the sort raised in Lewis 1988 and Collins 1988. For an independent response to those problems, see Price 1989.

4. My usage of the term 'historicist' connects most closely with that of Popper 1960. There are other usages, as for example in Kiser and Hechter (forthcoming), where historicist explanation is taken to be focussed distinctively on individual events, without any concern for generalities.

5. This is a real-life example discussed by Eduard Meyer and by Max Weber. See Richard Miller 1978, pp. 409–10, for a commentary.

6. In Macdonald and Pettit 1981, chap. 3, I defend a doctrine described as methodological individualism but what is in question there bears on laws rather than explanations: the position is that there are unlikely to be social-structural laws that are inexplicable in psychological terms.

7. In endorsing causal localism, I abstract from complications that quantum mechanics may force us to recognise.

8. This line of argument is an example of what Ned Block (1990b) describes as the Reductionist Cruncher!

9. Elster distinguishes between causal and other forms of explanation—for example, intentional—but he makes clear that all explanations direct us to causal history; the distinction invoked is really one between merely causal explanation on the one side and causal-intentional on the other. Unlike me, Elster seems to restrict the notion of causal explanation to explanation that directs us to what would commonly be taken as a causal principle; as mentioned in the preceding section, I describe any cause-giving explanation as causal (Elster 1983a, pp. 70–71).

10. It is unfortunate that this phrase suggests that the effect is necessarily not visible rather than not necessarily visible.

11. Alternatively, and more weakly, a number of people are in a prisoner's dilemma if, no matter what the others do, none can do better than she does by defecting—defecting is an equilibrium strategy—but if some are better off and no one is worse off under universal cooperation than under universal defection: universal cooperation, as it is put, is a Pareto-superior outcome.

12. It should not be thought that only those who display the self-regard that rational choice postulates have to learn to live with prisoner's dilemmas. Such dilemmas can arise also between parties who are not self-regarding in this way. See Parfit 1984, part 1, and Pettit 1985.

13. This approach may not be fetching but many writers come at least close to endorsing it. 'By rational choice approach, tradition, or framework', Elinor Ostrom (1991, p. 243) writes, 'I mean all work that is based on methodological individualism and assumes that individuals compare expected benefits and costs of actions prior to adopting strategies for action. By rational choice theories I mean the more specific assumptions made by a scholar about the type of information, valuation, and calculation involved in individual choice.'

14. This is a variation on a set-up suggested by Jon Elster in conversation.

6

Political Theory

In the preceding chapter we looked at the significance of the argument of this book for the project of social theory. We examined the impact of the holistic individualism defended in Part II on the enterprise of social explanation. In this chapter we turn from social explanation to political evaluation: to the evaluation of the politically variable arrangements under which we live. We are concerned with the significance of the holistic individualism we have embraced, for the project of political theory.

The results of the preceding chapter were encouraging. They support an attitude, as I described it, of explanatory ecumenism. Not only can we find room for the intentional interpretation of social agents. We can also make a place for structural and historicist explanation on the one side, rational choice explanation on the other. Under our perspective, the many divisions among social theorists about the best way to pursue social explanation cease to have the significance that is ordinarily given to them. They reflect differences in explanatory aspirations and styles, not differences of substance. The results of this chapter will not support a similar tolerance. If we are allowed to be ecumenists on the explanatory front, it turns out that we are forced to become relatively sectarian on matters evaluative. The holistic individualism we have embraced raises serious questions for a number of approaches to political theory, and pushes us in a distinctive direction. So at least I shall be suggesting.

'By political theory', John Plamenatz wrote, 'I do not mean explanations of how governments function; I mean systematic thinking about the purposes of government' (1960, p. 37). Political theory is a normative discipline, designed to let us evaluate rather than explain; in this it resembles moral or ethical theory. What distinguishes it is that it is designed to facilitate in particular the evaluation of government or, if that is something more general, the state (Hamlin and Pettit 1989). We are to identify the purposes of government—more strictly, the proper purposes of government—so that we can decide on which arrangements it is best for a government to foster in a society: which basic constitution it is best to establish and which procedures or outcomes it is best to prescribe in the day-to-day operation of the society (Rawls 1978).

There are three broadly distinct sorts of theorising found in contemporary political theory and this account of political theory lets us see why that should

284

be so (Pettit 1990d). One sort of theorising concerns the question of what arrangements are eligible, what arrangements would properly be chosen if people were contracting into society. A second sort bears on the issue of what arrangements are valuable, what arrangements best answer to currently recognised political values. A third sort explores the matter of what political values can be reliably institutionalised by government, what values are feasible ideals.

Take theorising about the eligible. If political theory is to identify the proper purposes of government, then it is easy to see why some might think of exploring the matter by asking after the sorts of arrangements that the founders of a society would agree to put in place. Those who take this approach to political theory are often described as contractualists. The contract-centred approach has seventeenth- and eighteenth-century antecedents but it has only come into full flower in recent political thinking.

But suppose, for whatever reason, that we do not go the contractualist path. In that case the obvious approach to political theory must be to identify the values that ought to guide political arrangements and to investigate the question of which arrangements would be required by different values. We might identify liberty or equality or community as goods that ought to be maximised by the politically variable institutions of our society and might then look to see which institutions would promote them. Or we might identify certain associated rights that we think that people possess against each other, and against the government, and might then investigate the institutional requirements of those rights.

This value-centred approach to political theory is essentially incomplete. For if it is alleged that a certain value is the appropriate one by which to judge our social institutions, and that the value requires such and such arrangements, then we may always ask whether the arrangements required could really be instituted or, if instituted, whether they could be realistically expected to remain in place. There is no point in establishing that something is valuable unless we can also show that it is feasible. And so we see that it is only to be expected that political theory should involve institution-centred thinking as well as thinking that is centred on matters of contract and value.

In looking at the impact of our holistic individualism on political theory, I shall take each of these three sorts of theory in turn and pick some of the reverberations that the doctrine is likely to have on those areas. I argue in the first section that holistic individualism must lead us to think ill of the contract-centred approach, at least if the approach is offered as an independent way of doing political theory. I then go on to argue in the second section that it is also bound to have an impact on the way we think about political values. I make the point by showing that if we start out with a liberal view of politics—the sort of view that is dominant in contemporary Western culture—then the espousal of holistic individualism facilitates a move towards a perspective that I describe as republican; this involves a radicalisation of liberalism rather than an outright rejection. Finally I argue in the third section that holistic individualism leaves room for a distinctive brand of institution-centred political thought, that it channels us towards an interesting way of exploring whether various values are institutionally feasible. This final section connects up closely with the discussion of rational choice theory in the preceding chapter.

It may surprise some readers that in tracing the import of holistic individualism for political theory, I do not make a connection with communitarianism (see Kymlicka 1990 for a survey). The communitarian school of thought asserts the truth of social holism and argues that once we understand just how socially embedded human beings are, then we must find fault with any politics—in particular, any liberal politics—that claims that a neutral state can be devised to satisfy the demands of people who belong to different cultures and traditions. The failure to make a connection with this school of thought will be especially surprising to those who recognise that communitarianism is intimately associated with the romantic movement, a movement that I mentioned as a source of the social holism defended in chapter 4 (Holmes 1989; Larmore 1987; Pettit 1993).

I do not make a connection with communitarianism, because I do not think that social holism entails an opposition to the ideal of the neutral, or at least the relatively neutral, state. I argue that if we are social holists, then we are likely to radicalise our liberal commitments in a republican fashion. But republicanism in the sense in which I find it attractive is a very different doctrine from communitarianism, despite the fact that communitarians sometimes identify themselves with what they see as the republican tradition (Sandel 1988). Far from raising questions about the neutral state, republicanism hails as the main political value an ideal of liberty that ought to prove attractive across a broad band of religious and other affiliations. Communitarians make many congenial points, but none of the points they make support the rejection of the neutral state.

The lessons I derive for political theory are not meant to provide a full-blown philosophy. Although holistic individualism impacts very negatively on the contract-centred approach—it suggests that this cannot be an independent way of doing political theory—I do not think that it provides a determinate view of what is valuable, for example, or what is feasible. The point I want to stress is not that the doctrine generates a substantive political theory, all on its own, but that it has an important impact on political theory as a discipline. It has an impact in this area that parallels the impact it has on how we think about social explanation.

Before moving on, I should comment on assumptions that I share with most contemporaries in political theory. Political theory, by the account given, addresses the question of which politically variable institutions are normatively satisfactory: in particular, normatively satisfactory from the point of view of some constituency (Barry 1965, chap. 1). Modern political theory, at least in the Western tradition, makes two assumptions in giving further specification to the enterprise. The first is the personalist assumption, as I shall call it, that the relevant constituency is human beings, usually the human beings who are to live under the institutions (Hamlin and Pettit 1989; see too Broome 1990 and 1991, chap. 8, Raz 1986, p. 194, and West 1990, p. 143). And the second is the universalist assumption, that the arrangements should be normatively satisfactory from the point of view, equally, of all relevant individuals (Dworkin 1978, p. 180; Kymlicka 1988; D. Miller 1990).

These assumptions are almost constitutive of Western political theory and, unsurprisingly, I shall be going along with them here. Bentham (1843, p. 321) gives nice expression to the first assumption. 'Individual interests are the only real interests. Take care of individuals; never injure them, or suffer them to be injured, and you will have done well enough by the public'. The second is also summed up well in a slogan attributed to Bentham. 'Everybody to count for one, nobody for more than one'.[1] The first assumption has meant that political theory is not concerned in the modern Western tradition with the interests of corporate or aggregate entities, with the interests of God, or, perhaps less plausibly, with the interests of non-human nature. The second has meant that the discipline is not concerned, at least in theory, with the interests of some human beings more than others.

Personalism is the more controversial of these assumptions. If persons are all that matter, they surely matter equally, so universalism is hard to resist. But the question is whether persons are all that matter, or all that should matter, in politics. One line in criticism of personalism will be that it is excessively focussed on individuals. Some will point out that many of the goods cherished by political theorists do not exist in individuals alone: that they are interpersonal goods like friendship and participation, not private goods like pleasure. That is so, of course, but personalism is consistent with acknowledging such interpersonal goods; it need not be committed to thinking that only private states matter.[2] Others will say that many of the goods countenanced are not properties of individuals at all but rather of corporate or aggregate entities that individuals constitute: they are goods like the solidarity of a community or the continuity of a culture. That is so too but, again, it must be stressed that personalism need not be blind to such goods. While the goods are not properties of individuals, they are still properties that make an impact on individuals; persons will fare differently, depending on whether they belong to a fractured or solidaristic community.

A second line of criticism that will be brought against personalism is more challenging. It is that if political arrangements are devised with a view just to the interests of human beings, then they are quite likely to be damaging to the rest of nature. I cannot discuss this objection here, for it would take us into matters of some detail. All I would point out is that the interests of human beings are tied up in great part with those of non-human nature, as has become clearer and clearer in ecological discussions. Thus we may hope that in approaching political matters from a personalist standpoint we will not be insensitive to the needs of non-human species and of the habitat that we share with those species.

The main role played by personalism in political theory is connected with the individualistic commitments outlined in Chapter 3. It is to emphasise that supra-individual entities like communities and cultures, nations and states, are not independent of persons in such a way that they might represent distinct interests for social arrangements to try and satisfy. The political impact of the individualist component in our holistic individualism may be confined to this assertion of personalism. In looking further at the impact of holistic individualism on political theory, as we do in this chapter, we shall be charting effects that are due to the holist aspect of the doctrine.

Contract-Centred Thought

Introduction

The direct method in political argument would be to argue for the greater value of one or another basic structure: to argue that one or another structure is the most desirable from the point of view of individuals. That direct, value-centred method is exemplified in two influential approaches. One would identify certain rights that people allegedly enjoy and would investigate the institutional requirements of those rights. The other would identify this or that allegedly compelling good—say, the good of liberty or equality—and would look at how those goods are institutionally best promoted. The first approach would hail the preferred institutional arrangements for the property of satisfying relevant rights, the other approach would hail them for the property of promoting relevant goods.

The direct method in political argument may well be found unfetching. Each of the approaches mentioned presents a certain categorical property of the arrangements preferred as the property that makes it right to have those institutions in place. But neither, it appears, has many resources to deploy in the event of counterassertion: in the event that someone denies that natural rights matter and proposes net human happiness instead, or in the event that someone rejects human happiness as an appropriate goal and prefers to support the goal of maximum liberty for all. There are many different stories about the rights that arrangements should allegedly honour and about the goods that they should allegedly promote. How to find a way through such a maze of possibilities?

Perhaps the best way to present the contract-centred approach to political theory is as an indirect alternative to the direct, value-centred method of argument. The direct method leads us to ask what sort of basic structure is best for individuals; the indirect method represented by the contract-centred approach would have us ask what structure those individuals would identify as best if they were suitably positioned. The idea is that we may be able to avoid having to immerse ourselves in questions to do with what is desirable in a basic structure if we only look carefully enough into the question of what structure is suitably eligible: what structure would be chosen by individuals in an appropriate context of choice.

The contract-centred approach harks back to the social contract tradition in political theory, particularly as that tradition emerged in the seventeenth and eighteenth centuries: for example, in the work of Hobbes, Locke, and Rousseau (Barker 1947, Kymlicka 1991, Lessnoff 1986). Under a more or less standard reading, this is the tradition of thinking that society and government exist, or properly exist, only in virtue of a contract—explicit or tacit—among the parties involved. It is the tradition within which existing institutions are vindicated, if they are vindicated, through being presented as the product of tacit contract. They are vindicated not for their current, categorical properties—not for their justice or efficiency or whatever—but for the historical property of being the product of contract. They are vindicated, as we might put it, not for the desirability of their content, but rather for the legitimacy of their origins.

The social contract tradition, under this reading, runs into great difficulties with the notion of tacit contract that it is forced to invoke. For that reason it tended to disappear in the nineteenth century. But in this century adherents of the contract-centred approach to political argument have taken up some elements in the tradition, often also suggesting that the standard reading is mistaken. They have taken up in particular the idea that what matters in a basic structure is not its categorical properties. What matters, as they put it, is the fact that the structure would command the allegiance of individuals were they in a position to choose between structures. What matters on this approach is not a categorical property, and not even a historical one, but rather a hypothetical property: the property of being such as to be chosen under suitable, hypothetical conditions.

Modern adherents of the contract-centred approach say that if we are concerned to assess basic structures, then we should not look to the categorical, potentially valuable properties of the structures, as those who follow the direct method do. Nor should we look to properties related to the actual history of the structures, if indeed they have a history. We should rather look to see which of the structures, or which family of structures, would emerge as the object of choice in a suitable hypothetical contract. That a structure is normatively most satisfactory from the point of view of the individuals involved means, under this approach, that it is the structure on which they would converge in a suitable context of choice. The structure is the right one to have in virtue of its possessing this hypothetical property, not in virtue of any categorical or historical features.

It is important to notice that contractualists not only appeal to what people would choose under certain conditions; they take that appeal to be definitive of what is right. The contractualist employs the contractual method of argument and then interprets the result of its employment in a particular way. He uses the method to identify the suitably eligible basic structure and he then interprets the significance of that finding as follows: that the structure in question, in virtue of being eligible, in virtue of being such that it would be chosen, is normatively the most satisfactory. Being such as to be chosen in a suitable situation, he thinks, is nothing more or less than being normatively satisfactory. The property of eligibility constitutes what it is to be satisfactory in normative regards; it is the ultimate criterion of rightness.

This point is important to notice, because it shows that just the fact of using the contractual method of argument does not make someone a contractualist. Political theorists may use the contractual method without believing that being contractually eligible is what it means to be normatively satisfactory. They may believe that to be satisfactory is to have a certain categorical property—say, to be fair—and they may hold that a good sign that a basic structure is fair is the fact that it would be chosen, as the contractual method reveals, in appropriate circumstances. Political theorists, in other words, may use the contractual method for strategic rather than principled reasons. They may use it in support of an approach that belongs ultimately to the direct method, an approach that aims at identifying plausible constraints or consequences for a basic structure to be concerned with.

John Rawls may be read as using the contractual method in this strategic way, although there are other strands in his thought as well (Kukathas and Pettit 1990, chaps. 2 and 4; Scanlon 1982). He wants to identify the structure that counts as most fair, in an independently given sense of fairness, and he uses the contractual device as a heuristic way of picking out that structure. A fair outcome, he assumes, is one selected by a fair procedure. He reasons then that, other things equal, we should take as fair any basic structure that we would each prefer if we were invited to make a choice under a veil of ignorance about our own characteristics: about our physical, intellectual, and social prowess, about our gender, race, and religion, and so on. The thought is compelling and to anyone concerned with fairness there is great interest in Rawls's contractual deduction. But for all the interest of the exercise, Rawls as presented does not count as someone adopting the contract-centred approach. The property of being contractually eligible serves for him as a useful pointer to the property of being fair and being right: being normatively satisfactory. Contractual eligibility does not constitute rightness on this approach; it merely serves as a symptom and a surrogate.

In Plato's *Euthyphro*, Socrates asks whether something is holy because the gods love it, or whether the gods love it because it is holy. Take this to be an inquiry as to whether the criterion of holiness is the love of the gods; ignore the causal reading mentioned earlier (p. 202). The question suggests a parallel query by which to distinguish contractualism proper—constitutive contractualism—from the heuristic counterpart (Pettit 1982). Is a basic structure right or just because it is contractually eligible? Or is it contractually eligible because it is right or just: say, because it has a right-making property like fairness? The contractualist proper holds that it is right or just because it is eligible. He takes contractual eligibility to be of the essence of rightness. The theorist who uses the contractual method in a heuristic way holds that the structure chosen is contractually eligible because it is right or just. He thinks that if it would be chosen in appropriate circumstances, that is, because it is fair or whatever.

These comments serve to introduce our notion of the contract-centred approach to political argument. The question now is whether such an approach has any appeal or merit. I believe that some very interesting work has been done in pursuit of the approach but my judgment on it is ultimately negative. I reject the idea that this is the way that political theory ought to go in seeking to assess basic social structures. I reject this idea, in good part, because of the holistic commitments that we have espoused; the point will become clear, as we proceed.

The Economic Version

There are two main variations possible on the contract-centred approach. The approach may be interpreted, as I will put it, in an economic or a properly political fashion (Kukathas and Pettit 1990, p. 32; see too Barry 1989, p. 371, and Hamlin 1989). Under the economic interpretation, the upshot of the contract is presented as an attractive bargain: as something that is to the mutual advantage of all participants. Under the political interpretation, the upshot is presented as a compelling conclusion: as something that each is led to endorse, on the basis of the same reasoning. I will discuss each of these in turn but I will devote considerably more time to the economic interpretation.

The economic approach hails as satisfactory the sort of basic structure that is to the mutual advantage of the parties involved. Why does it look to mutual advantage rather than individual rationality? The prisoner's dilemma serves to explain why. An outcome can be individually rational for all of a number of parties—it can be an equilibrium outcome from which no one can unilaterally depart with benefit—while still being, for example, a Pareto-inferior result: that is, while still being dispreferred to another possible outcome by some, and not preferred to that outcome by any. It may be individually rational for each to leave care about the litter in the park to others but all may be worse off with a littered park than they would be if each took care not to cause any litter herself.

But if the economic approach is to try and identify the basic structure that does best by the mutual advantage of parties, how is it to proceed? It may assume some static criterion of what maximises mutual advantage, some criterion that assumes no bargaining, in particular no offers of concessions between parties. Pareto-superiority is an example of such a criterion. It would have the contractualist argue that the structure to be preferred is that which belongs to the family of Pareto-optimal alternatives: the alternatives such that no other does better by some and worse by none; the alternatives such that it should be unanimously preferable that one of them be implemented (Brennan 1987; Buchanan and Tullock 1962; but see Brennan and Lomasky 1984).

The alternative way of proceeding for the economic approach is to assume a dynamic rather than a static criterion of mutual advantage. Under this interpretation, the contractualist will assume some criterion to the effect that the most advantageous alternative for any group of individuals, in particular for any participants in a social contract about basic structure, is going to be that which bargaining between rational individuals would produce. In this context rational individuals are individuals, roughly, who conform to decision theory and maximise expected utility. David Gauthier (1986, pp. 29–38), who is the champion of this approach, also specifies that the preferences maximised should be considered preferences but this restriction need not concern us.

As between static and dynamic criteria, the dynamic must win out. There is a problem with any static criterion like Pareto-superiority. A number of outcomes can satisfy it, but they are not intuitively of equal attraction. One may do so well by some individuals that they are capable, by offering concessions, of winning the cooperation of those who otherwise would do worse under it than under other outcomes; it may do so well by them that they can suitably compensate those others (see Sen 1970, chap. 2). Intuitively, some outcome involving compensation is most likely to maximise the mutual advantage of the parties, not the Pareto-superior outcome. That intuition suggests that the economic interpretation of the contract-centred approach should use a dynamic, bargaining criterion of advantage, not a static one.

What are the prospects for a bargaining theory that will identify a certain basic structure as uniquely rational? We can certainly specify certain initial bargaining positions such that it would be uniquely rational for parties in such a position to choose one or another basic structure. But the specification of initial bargaining positions is not a matter of free choice. Someone following the economic version of a contract-centred approach will have to allow only such an initial bargaining position as he can show that it would be rational— rational, not moral—for people to acknowledge. Thus David Gauthier (1986, chap. 7) argues that a starting point in which people equally enjoy certain minimal rights against others is the appropriate baseline to acknowledge, on the grounds that rational individuals will comply only with agreements reached in such a context. The question then is whether an austere baseline of this kind will enable us to identify a particular basic structure as the uniquely rational, bargaining outcome.

The question is particularly challenging as there are a variety of bargaining theories and they offer conflicting advice on the proper solution of various predicaments (Barry 1989, part 1). The variety on offer leads John Rawls to describe the bargaining problem that has to be solved on the economic approach as 'hopelessly complicated'. 'Even if theoretically a solution were to exist, we would not, at present anyway, be able to determine it' (Rawls 1971, p. 140). But not all contractualists baulk at the challenge. Gauthier (1986) faces it fairly and squarely and develops his own preferred bargaining theory. I shall assume that differences among bargaining theories can be resolved in some such manner and will not make anything of the diversity to be found in the literature.

I do this because, even if such differences can be resolved, there are still problems that arise for the economic version of the contract-centred approach, in particular for any approach that helps itself to bargaining theory. I shall raise three problems, all of them problems that are particularly salient in the light of the holistic individualism we have endorsed. The first two problems are inspired by points made by Robert Sugden (1990) in criticism of Gauthier's project. (See also Pettit and Sugden 1989 on related matters.)

The first problem is this. Every bargaining theory capable of doing the job required by the economic approach has to maintain that there is a unique rational solution to typical bargaining problems. But bargaining theory defends this claim only at a cost that we must regard as intolerable. It develops the concept of a rational solution in such a way that for a given, supposedly rational solution to an interaction between certain parties, there may be no train of reasoning that we can describe that would give those parties reason to adopt the required strategies. In other words the outcomes that bargaining theory countenances as rational include outcomes that are deliberatively inaccessible to the individuals involved.

In the preceding chapter we argued that the intentional interpretation of human agents involves identifying the evidential and valuational factors to which they are responsive; this theme was a development of the holism defended in Part II. Intentional interpretation involves understanding how those agents might reason their way to the changes of mind and the behavioural initiatives that they adopt. In short, it means being able to see the agents as instances of *homo ratiocinans*, not just *homo rationalis*: as ratiocinative or deliberative creatures, not just rational ones. This view of intentional interpretation means that in running the contractual method, it is necessary to be able to see how participants in the contract would individually reason, and perhaps reason with one another, about the right choice to make. It means that any contract-centred approach should identify a train of argument, and perhaps debate, whereby contractors would converge on the choice of the allegedly optimal basic structure. The first problem with the economic approach, so I allege, is that it is incapable of meeting this requirement. There are some predicaments where the sort of bargaining outcome it presents as rational is not an outcome that is deliberatively accessible by participants.

The best example of such a predicament is exemplified by the so-called battle-of-the-sexes game in which a man and woman want to spend an evening together but the man wants to spend it at a prize fight, the woman at a ballet (Luce and Raiffa 1957, pp. 90–91). The game is of interest, despite involving only two people and despite referring us to a somewhat quaint example. It epitomises the many bargaining situations where parties will benefit by cooperation but where different modes of cooperation are differentially rewarding.

In such a situation it appears that whatever the outcome of the bargaining, it should rationally be an equilibrium: it should be an outcome from which neither can unilaterally withdraw with benefit. This seems to mean that both should go to the fight or both to the ballet, since each of those outcomes is an equilibrium. If the parties are at the fight or at the ballet neither can leave on their own without thereby becoming worse off: after all, they each prefer being together to being separated. But now bargaining theory faces a problem, for neither of these equilibria looks to be any more rational than the other. Contrary to what the theory has to claim, there appears to be no unique rational solution to the predicament illustrated.

In order to get over this type of problem, bargaining theory introduces the notion of a mixed strategy and a mixed strategy equilibrium. The introduction of such notions, as Gauthier (1986, p. 64) acknowledges, 'is essential to the very possibility of rational choice in strategic situations'. Mixed strategies are defined by contrast with pure. In our example each of the parties faces two pure strategies: going to the fight or going to the ballet. Given such a set of pure strategies, we can introduce the notion of a mixed strategy: it will be the strategy of allowing one's choice to be determined by a lottery over the pure strategies. Allowing one's choice to be determined by a lottery that selects the fight with probability x and the ballet with probability 1 minus x will be a mixed strategy.

With the notion of mixed strategies introduced, we can go on to define a mixed strategy equilibrium for a predicament like the battle of the sexes. Let the pure strategy payoffs be represented in the following matrix, where the woman's payoff is represented first in each box, the man's second, and where the higher the number the better the payoff.

	Man: Ballet	Man: Fight
Woman: Ballet	3,2	0,0
Woman: Fight	0,0	2,3

We can represent the payoffs of a mixed strategy as an expectation, in the familiar decision-theoretic way, of the pure strategy payoffs. Thus the payoff of a mixed strategy that gives the woman an x probability of going to the ballet, and a 1 minus x probability of going to the fight, will be x times the ballet payoff—this payoff will itself be an expectation, since it depends on what the man does—plus 1 minus x times the fight payoff. Once we have taken this step, we can define the notion of a mixed strategy equilibrium. Such an equilibrium, as might have been expected, will be constituted by any combination of mixed strategies on the part of the players that is such that neither can unilaterally depart from it with benefit.

We must now see how bargaining theory uses the notion of a mixed strategy equilibrium. Looking at the battle of the sexes predicament, we asked before what equilibrium a bargaining theory can select as uniquely rational in this game and we raised the problem that neither pure strategy equilibrium looks superior to the other. Now it turns out that in this game there is also a mixed strategy equilibrium. This is the outcome that is associated with the woman following a lottery that gives her a .6 chance of the ballet, a .4 chance of the fight, and the man following a lottery that gives him a .6 chance of the fight, a .4 chance of the ballet. The response of bargaining theory to the problem we raised is to hail the mixed strategy equilibrium as the uniquely rational solution. By appealing to that sort of equilibrium, it gets out of the difficulty that neither pure strategy equilibrium looks superior to the other.

What is supposed to establish the superiority of the mixed strategy equilibrium over the others? The fact, so it is alleged, that it is the only symmetrical solution, the only solution that deals in the same way with each of the two parties. And what is supposed to be so attractive about symmetry? Not of course the fairness it involves, for we cannot appeal to moral considerations in a theory of rational bargaining. Rather the fact, universally alleged within the relevant tradition of bargaining theory, that symmetrical predicaments, predicaments in which the parties occupy symmetrical positions, should have symmetrical solutions (Sugden 1990, pp. 774–75, comments on this).

We can now, at last, illustrate the problem that arises with a bargaining theory that appeals to such a solution. The problem is that there is no train of reasoning that would compel either party to adopt the mixed strategy involved in the solution. The solution, as we said, is deliberatively inaccessible. Suppose that you are the woman, I the man, and that we are each persuaded—perhaps because we are familiar with bargaining theory—to play our respective lotteries. Imagine that I am directed to go to the ballet, you to the fight. I may or may not know how your lottery has gone but in neither case can I necessarily find a deliberative reason to abide by the result of my lottery. If we assume that the payoff figures pick up everything that matters to me, then it turns out that I am likely to have deliberative reason to ignore the directive of the lottery.

Suppose I know that you are directed to go to the fight by your lottery and that I am certain you will abide by the ruling. In that case the rational thing for me to do will be to depart from the mixed strategy and to go to the fight too. Going to the fight guarantees me a payoff of 3, whereas running the lottery offers an expected payoff of 1.8: that is, a .6 chance of 3 plus a .4 chance of 0. I will not be able to give myself any reason for remaining faithful to the lottery. Suppose on the other hand that I do not know how your lottery has gone but that I am sure you will abide by its ruling; suppose, in short, that I assign a .6 probability to your going to the ballet and a .4 to your going to the fight. In that case, as is easily checked, the expected utility of going to the ballet and going to the fight is the same—in each case, it is 1.2—and so the options are equally rational; going to the ballet, as directed, is not uniquely rational for me. Again I will not be able to give myself any deliberative reason for sticking by the mixed strategy. Other possibilities besides those considered here will call for attention once we allow that I may not be certain you will abide by the ruling of your lottery. The same point will be borne out with them, however, as with those considered. In many cases I may not be able to provide myself with any reason for abiding by the mixed strategy that is supposed in bargaining theory to be my uniquely rational response; in many cases abiding by that strategy will not make deliberative sense.

The problem that I have raised for the economic version of the contract-centred approach is often recognised, though it is not cast in terms of the deliberative inaccessibility of mixed strategy equilibria (Luce and Raiffa 1957, pp. 74–76; Sugden 1990, p. 777). There have been a number of attempts to solve the problem, arguing that it will be rational to stick with mixed strategies after the lottery has been played (Gauthier 1986, p. 70). But none of these attempts points us towards a plausible train of reasoning that the parties are expected to follow. None of them introduces deliberative accessibility. And that failure is a fatal flaw, at least in the perspective developed here.

I promised to raise three problems for the economic version of the contract-centred approach and I should turn to the other two. Happily, they can be stated more briefly than the first. The second problem arises even if the first is solved, the third even if the second can be ignored.

Suppose that the first problem does not arise: suppose that bargaining theory points us towards a plausible train of reasoning whereby parties in a battle-of-the-sexes encounter can get to follow the mixed equilibrium strategies. It might be that reasoning in terms of symmetry offered at least coherent grounds for abiding by the mixed strategy. We would each think as follows. 'Here we are in a symmetrical bargaining situation. Here are three possible equilibrium outcomes. But only one of those outcomes, the mixed strategy equilibrium, is symmetrical. So we should each do whatever is necessary to secure that outcome. Contrary to the points made previously, we should each play a lottery and we should more or less blindly stick by its ruling'.

The second problem is that even if it is granted that the parties in the encounter can coherently reason their way to action in this manner, their capacity to be moved by that reasoning is vulnerable to cultural and historical influence. There are cultural forms, and historical patterns, under which it would be crazy for the participants in the game to endorse the assumptions in this reasoning (Sugden 1990, pp. 778–83).

Suppose that you and I, the man and woman involved in the battle of the sexes, live in a chivalrous culture where it is standard for men to give into women in the sort of predicament in question. If we are each rational creatures interested solely in maximising our expected utility, then you will surely have good reason to stake a claim to the ballet and I will be hard put to it to find any reason why we should not go along with that outcome. I may introduce you to my preferred bargaining theory and show that the mixed strategy equilibrium, unlike the outcome you claim, is a symmetrical solution to a symmetrical bargaining problem. But you can surely laugh at such nonsense, calmly reiterating that you are going to the ballet. If we then have to make our decisions independently, I will be very silly if I follow a mixed strategy, for I should surely be persuaded that you will be going to the ballet.

The upshot of these reflections is that the outcome that it is the rational to expect in a bargaining situation may be a function of culture. The salience that our chivalrous culture gives to my going along with you to the ballet is far too potent a force to be resisted by the abstract considerations of symmetry that might be marshalled in bargaining theory. In particular it is too potent a force to be resisted by rational agents who are anxious to maximise expected utility: agents who want above all to go to the same venue, avoiding the utility loss involved in attending different events.

The first problem I raised is that bargaining theory fails to specify a train of reasoning by which we are supposed to reach preferred solutions. The second problem is that even if it offers a train of reasoning by which we should allegedly be persuaded, that train of reasoning will not be found compelling in all cultures. The final problem I wish to mention is that even if both the other problems are solved, even if bargaining theory can specify a culturally invulnerable train of reasoning for a bargainer to follow, it is not clear that we should here and now be attracted to a basic structure on the grounds that we would be driven to that structure in the hypothetical bargaining envisaged.

Here and now, as we know from earlier discussions, we find our way to individual action and political allegiance on grounds that are presented to us in culturally familiar concepts. We stick by this path because it presents itself as a kindly way of dealing with others; we attach ourselves to these institutions because they give a fair hearing to all individuals; and so on. Why should we care if we are told that the sorts of institutions that we cherish for being fair are not those that we would converge on were we involved in the appropriate bargaining game?

Bargaining is a distinctive sort of enterprise. The parties involved take their reasons for preferring one or another outcome as given and they are impervious to any reasons that others may offer for rethinking their preferences. They each stick by their own reasons and their own preferences and they try to engineer an outcome that is as near as possible to what they want. They will have no care for what other parties want and will be disposed, if they are let, to impose their own wishes on others (Gauthier 1986, p. 311). If they agree to something less than their preferred outcome, that is not because of being intellectually responsive to the ideas of others or affectively responsive to their desires. It is only because they believe that nothing less than that concession will get others to play ball.

The distinctive thing about bargaining, as appears in this characterisation, is that it is close-minded and hard-hearted. It involves an enterprise of reciprocal manipulation, without any pretence at openness to what others think or want. The parties may not be entirely self-regarding in their preferences—in theory they may even have one another's good in view—but they take no account of one another's preferences or one another's perceptions in pressing their own line. They are intellectual and emotional solipsists (Schick 1984).

The third problem that I raise for the economic version of the contract-centred approach is that it is unclear why we should generally care about the fact that such solipsists would converge on the choice of this or that basic structure in a suitable bargaining position. If there are intellectual and emotional solipsists among us, particularly if there are solipsists who are inclined to reject all moral constraints, then they may be impressed by the derivation on offer. But most of us are not solipsists of this kind and most of us accept one or another set of constraints, one or another basic structure, on grounds that we express in familiar moral concepts, concepts that we expect to have a general hold. So why should we be moved by the bargaining argument? Why, for example, should we be alienated from a basic structure that we espouse for the help it gives the poor, on the grounds that it would not be compelling to a group of bargainers with whom we cannot identify?

The Political Version

This is enough on the first of the two interpretations of the contract-centred approach, the economic as distinct from the political interpretation. The last problem raised for the economic approach offers a useful way of introducing the political. Under the economic interpretation, the best proposal is the one that would represent the best bargain for individually rational bargainers. Under the political interpretation, the best proposal is the one that would be selected by the individuals involved, not in bargaining with one another, but in reasoning—on their own or together—about the right structure to create: in particular, to create for people like themselves who presumably wish to form a society and establish a state. It is the proposal that would represent the compelling conclusion for politically attuned reasoners.

Perhaps the best-known representative of the political verson of the contract-centred approach is Jürgen Habermas (1973). According to Habermas's theory, the best basic structure is that which would be supported by people involved in collective debate, under ideal speech conditions. Ideal speech conditions are, roughly, conditions of equality, where each has the chance to speak and where each can call anyone else's assertions into question (Habermas 1973, p. 255; Pettit 1982). Habermas supposes that in such collective, publicly oriented discussions it will be pragmatically impossible for parties to introduce any grounds that refer to their own good alone, or to that of one particular sub-group; if they introduce such grounds, the idea is that they will be laughed out of court, they will not be taken seriously (Elster 1986a). He is confident then that the basic structure that will be chosen in the discussions is not going to be selected for the wrong reasons, as it were. It is going to be the structure that is most in the common interest, not a structure that has the support of the strongest coalition.

Bruce Ackerman (1980) also appears to be committed to a political version of the contract-centred approach. He suggests that the best sort of state, the best basic structure, is by definition the kind of arrangement that will be supported in a dialogue that satisfies a certain neutrality condition.[3] This is the condition that no one is allowed to assert either that her conception of the good is better than that asserted by any of her fellows, or that she is intrinsically superior to any of her fellows (Ackerman 1980, p. 11). In Habermas the idea is that the best structure is the one that would be chosen in ideal, collective debate. Here the idea is that the best structure is the one that would best survive the tests of dialogue under the neutrality constraint.

Finally, it is worth mentioning that in an influential article, T. M. Scanlon (1982) also seems to commit himself to a political form of the contract-centred approach. For him the best basic structure will be characterised, at least in part, by 'rules for the general regulation of behaviour which no one could reasonably reject as a basis for informed, unforced general agreement' (Scanlon 1982, p. 110). The non-rejectability test can be given an interpretation that would put Scanlon in the economic camp with Gauthier; being non-rejectable, on this reading, would mean being an irresistible bargain. But Scanlon writes as if the test is to be understood in a political, rather than an economic, spirit: as if being non-rejectable means being so well supported by reasons as to be intellectually compelling. We shall take his proposal in this way. Thus, where Habermas directs us to ideal, collective debate, and Ackerman to dialogue under the condition of neutrality, Scanlon has us look to reasoning—individual or collective—of a kind that no one can reject. The best basic structure is the structure that will be supported, at least in part, not by ideal, nor by neutral, but by non-rejectable reasoning.

It is important to distinguish the political notion of contract—the notion deployed in the political version of contractualism—from two other notions that sometimes surface in discussion. The political conception represents the contract as a universally compelling conclusion; the other two notions represent it, respectively, as a compromise and as an accommodation. Under the political conception, the contractors are induced separately or collectively to approve of certain principles of organisation, on the basis that the principles have certain commonly recognised, inherent attractions. Under the compromise conception, the parties are induced to approve of the principles, not on the basis of any commonly recognised attractions, but on the basis of varying sets of considerations: one person approves for these reasons, another for those reasons, and so on.[4] Under the accommodation conception, the parties are induced to approve of the principles, not on the basis of any inherent attractions at all, but on the basis that the principles represent the best or only chance of securing agreement.

Under the compromise or accommodation conceptions of contract, the hypothetical property that a certain structure would be chosen in appropriate conditions of contract is not a plausible criterion of right. Whether a given structure is to be eligible as a compromise or accommodation will be affected by whether or not various crazy views are adopted by anyone in the society. Does the society contain anyone of this or that intuitively outlandish mentality? If so, a given compromise or accommodation will be out of the question. The structure in question would be right for the society, if there were no one of that mentality present; it is not right, given that there is. This sensitivity to the presence of outlying mentalities would make it difficult to conceive of the hypothetical, contractual property as the property that determines whether certain principles for the organisation of society are right or wrong. It would mean that there is little or no generality in the matter of what sorts of principles are to be recommended by political theorists. Thus it is not surprising that contractualists do not endorse the compromise or accommodation conception of contract.[5]

Back, then, to the main line of argument. What are we to say about the political interpretation of the contract-centred approach? Where I mentioned three problems confronting the economic interpretation, I have only one line of query to raise about the political. This line of query, like the problems raised with the other interpretation, derives in good part from the holistic commitments that we have espoused. In particular, it derives from the inference-theoretic view of human agents we have been led to defend. This is the view that people are responsive to communally established, evidential and valuational pressures in the responses they display, that they always respond in such a way that we can identify a pattern of reasoning, implicit or explicit, in those responses.

The inference-theoretic view means that when contractors come, individually or collectively, to desire a particular sort of basic structure, then they will desire it for a certain non-hypothetical property—or properties—that they see it instantiating. The property in play may be that a structure best promotes liberty, that it best answers to the culture-relative interests and customs of the individuals involved, or whatever. The presence of that property will supply each of the contractors with a reason for concluding, other things being equal, that the structure that exhibits it is the one to have. Each will think: this is the structure that best promotes liberty (or whatever); and that is the supreme consideration of relevance; so let us have it. As contract-centred thinkers we will be moved then to hail the structure selected, believing that the reasoning in question satisfies the appropriate tests of ideality, neutrality, or non-rejectability. The structure selected is the right one to have, we will say, because it is the one that would be supported in a suitable form of contractual reasoning.

But the inference-theoretic view raises a serious question about this line of thought. We admit, under that view, that the contractors would defend a particular structure on the grounds of its having a certain non-hypothetical—in effect, categorical—property. But we then go on to say that the criterion of the right structure is the hypothetical property that, in virtue of having that categorical property, it would be selected by suitable contractors. But how can we pass over the categorical property in this way? How can we treat the hypothetical property as the criterion of rightness when that property is grounded, as it were, in the categorical property? How can we avoid the conclusion that the real, right-making property is the categorical one and that what the contractual connection does is draw attention to the appeal of that property? In other words, how can we avoid the conclusion that under the political reading, the use of the contractual device can only be heuristic, not constitutive, in significance?

Suppose that the contractors argue for a basic structure on the grounds that it best promotes liberty. In that case, so this argument goes, the real right-making property is the categorical one of maximising the expectation of liberty. Suppose that they argue for a structure on the grounds that it best corresponds to these or those particular interests and customs. In that case the right-making property is the categorical one, either of best answering to those particular constraints or, more plausibly, of best answering to whatever local interests and customs prevail (Pettit 1988a).[6]

If we go along with this argument, then we deny that the criterion of rightness is contractual eligibility; after all, we have identified a more basic property, a categorical property at the source of contractual agreement, as right-making. Running the contractual argument may be a useful heuristic for picking out categorical, right-making properties. But contractual eligibility no longer serves as constitutive of rightness. The *Euthyphro*, criterial test goes the wrong way. We will have to say that a basic structure is contractually eligible because it is right—for example, because it has the categorical right-making property of best promoting liberty—not that it is right because it is contractually eligible.

Is there a response that can be made to this line of objection? One that Scanlon (1982, p. 118) suggests is that the objection presupposes a cognitivist view of value: it presupposes that the property of promoting liberty, or whatever value is in question, is a property such that to understand and ascribe it is necessarily to have some degree of desire for the bearer of the property. According to this response, if we say that the reason for which the parties endorse a given structure must constitute the basic right-making property—

this, rather than the fact that they endorse the structure—then we are assuming a cognitivist stance; we are assuming that the property is an objective, cognisable value. And, so Scanlon claims, the contract-centred approach is designed precisely as an alternative to such cognitivism. 'Thus, while there are morally relevant properties "in the world" which are independent of the contractualist notion of agreement, these do not constitute instances of intrinsic "to-be-doneness" and "not-to-be-doneness": their moral relevance—their force in justifications as well as their link to motivation—is to be explained on contractualist grounds' (Scanlon 1982, p. 118).

Scanlon's line of thought lends itself to the following explication. Suppose a contractor is disposed to choose some arrangement for its possession of a certain property—say, that of promoting liberty—and that she recommends it to her fellows on that basis. Allow, in non-cognitivist spirit, that the property is not an objective value: allow that the property supplies the person with a reason for choice, and a ground of recommendation, only contingently on the presence of a connected desire. It follows then that it is not really the property itself that supplies a reason for choice or a ground for recommendation. Rather it is the joint consideration that the thing in question has that property and that she, the chooser, and they, her fellows, have suitably connected desires. If we are non-cognitivists therefore, as we must be if we adopt the contract-centred approach, then we will not think that the contractors choose their preferred structure just for the property of best promoting liberty or whatever. And so we will not be tempted to think that it is that property which is right-making.

This line of argument is not going to help the contractualist. Suppose the conclusion is granted: suppose it is granted that really the contractor desires and recommends the arrangement for its having the complex property of liberty-given-that-that-simple-property-connects-with-desire. The same line of argument applies also to this property, for as it was granted in the spirit of non-cognitivism that liberty weighs with people only contingently on their desire, so it must be granted that liberty-given-that-it-connects-with-desire will weigh with people only contingently on their desire: say, only contingently on their desiring to satisfy their desire for liberty. Applied to that property, the conclusion would then be that the contractor desires and recommends the arrangement for its having the doubly complex property of liberty-given-that-that-simple-property-connects-with-desire-given-that-that-complex-property-connects-with-desire. The line of argument is self-defeating, for if it is allowed to apply at any stage, then it is going to apply in an infinite regress.

Our discussion of inference theory in the last chapter shows what goes wrong in the contractualist's line of reasoning. We argued there that even if non-cognitivism is true, even if a property can be reason-giving only contingently on the possession of a suitable desire, still it does not follow that the real reason for a choice based on that property is the sort of joint one mentioned: that it is a consideration to the effect that the option has the property and that you, the chooser, have a connected desire. It is one thing to want something for the property of being F, it is another to want it for the property of being F-given-that-you-have-a-connected-desire. If I want to do my duty, then I want that I do my duty even in the event of a connected desire failing me. There is a difference between my wanting to do my duty and my wanting to-do-my-duty-given-a-suitable-desire. And so we must allow, even if we are non-cognitivists, that run-of-the-mill properties, properties that are not objectively valuable in the cognitivist sense, may still supply grounds for desire and choice.

The lesson applies directly to the imagined, contractual case. The contractor I have in mind desires a certain arrangement for its capacity to promote liberty and, in doing so, desires it for a variety of contingencies, including the scenario in which her own feelings for liberty are different. What matters to her is the fact that the arrangement is suitably liberal, not that it is liberal-and-thereby-connects-with-her-current-desires. Equally when the contractor recommends the arrangement to her fellows, she recommends it on the ground that it is liberal, recognising that they are likely to be moved by that property of the proposal. She does not recommend it on the ground that it is liberal-and-therefore-likely-to-connect-with-their-current-desires. This being the case, it appears that if the contractors do finally agree to the arrangement on the grounds that it promotes liberty, then the right-making property must be that it promotes liberty.

We may return then to our earlier query. Just so far as we endorse the inference-theoretic view of human agents, and independently of how we decide between cognitivism and non-cognitivism, we will naturally envisage political contractors as making their choice on the grounds that the preferred structure has this or that property. But if we envisage the contractors going on in this fashion, it is hard to see why we do not take the right-making property of the structure to be the non-hypothetical property which recommends it to the contractors. It is hard to see why we should ignore that property and focus instead on the hypothetical property of the structure, that it would indeed be chosen by parties to the contract. On the economic interpretation, the contract-centred approach fails in its own terms. On the political, it collapses into a more direct approach which picks out certain categorical properties of structures as the appropriate right-makers.

Value-Centred Thought

We have found reason to reject the view that the criteria of right basic structures are hypothetical in nature: that they boil down to one or another version of contractual eligibility. This means that in answering the question of what makes a basic structure normatively satisfactory—in particular, normatively satisfactory from the point of view of all relevant individuals—we should look to the categorical properties that structures may have. We should do political theory in a value-centred and institution-centred way. We may make use of the contractual method of argument, as someone like Rawls does, but we should not think of this exercise as having more than heuristic significance.

We turn now to value-centred political thinking, to see what impact our holistic individualism, and in particular our holism, is likely to have in this area. Value-centred thinking comes in two broad forms, one centred on the rights of individuals that allegedly require to be honoured, the other on the goods that political arrangements should allegedly promote. The first approach is deontological, as it is usually put, the second consequentialist.

The difference between the approaches can be formulated on the following lines (Pettit 1991c). The deontological approach takes a certain universal value—say, the respecting of such and such rights—as given and argues that institutions should be shaped in a way that bears witness to this value, honouring it punctiliously in the treatment they give to different human beings.[7] In particular, institutions should be shaped to honour the value even if this means, as a result of various side-effects, that there is less of the value realised overall: even if it means that in general other agents respect people's rights less than they might have been brought to do.

The consequentialist holds that with any universal value, the rational thing to prescribe for a set of institutions, and more generally for agents, is that, other things being equal, they should promote that value by whatever means available. There is some ambiguity about what it means to promote something but I ignore that issue here (Pettit 1991c). The important point to stress is that the consequentialist does not think it is always right to honour a value, to keep one's hands clean in respect of the value. He thinks that institutions and agents ought not to honour a value on certain occasions, if not honouring it is the best way of promoting it.

Consider an example. Every pacifist thinks that peace is a very important value, in particular an important value for states to recognise. But a pacifist may be deontological or consequentialist about the value in queston; these are the salient possibilities, if not strictly the only ones. The pacifist who thinks that peace is a value that should always be honoured by a state, even if this means that there is less peace thereby realised, is a deontologist. He will say that an individual state should never go to war, even if the war would do better than appeasement in promoting peace. On the other hand the consequentialist pacifist will say that the state should go to war, if this is the best way in the long run of promoting peace: if this is truly, for example, the war to end all wars.

I said that a certain sort of political deontologist would hold that political institutions—if you like, the state—should honour the value of respecting such and such alleged rights. It is worth noticing that even if a consequentialist became convinced that the respecting of those rights was indeed a value, he would still take a different view of what the state should do. He would say that the state should be designed to promote that value, and not necessarily to honour it: that the state should not honour the value in certain sorts of cases—other things being equal, of course—if the overall realisation of the value is thereby increased. He would say that the guiding philosophy should be a 'utilitarianism of rights' (Nozick 1974, p. 28; Pettit 1988b and 1989c).

The distinction between consequentialism and deontology is of the greatest importance in political and moral theory, marking two different ways in which any categorical property may be hailed as a value. The distinction even extends beyond categorical properties, for it may be that a historical or hypothetical property can be treated in the consequentialist or deontological fashion; I did not mention this before now, as the distinction is not often applicable in practice to such properties. But we need not concern ourselves further with the difference between consequentialism and deontology. The adoption of a holist as distinct from an atomist perspective is likely to have a major impact on how we think about political values, regardless of which strategy we endorse in regard to those values. I will discuss the impact of holism, without commenting on whether the values are to be treated as consequentialist targets or deontological constraints.

I will first sketch the general line of reasoning that supports the view that holism is likely to affect our thinking about values and then I will try to sheet home the point by looking at the transformative impact I think that the espousal of holism should have on liberal commitments. The focus on liberalism is justified by the centrality of the doctrine in current Western practice and thinking. I argue that a holist perspective should facilitate the replacement of the liberal conception of negative liberty with a conception that is republican in character. I see the republicanism that is thereby supported as a radicalisation of liberal commitments, not as an outright rejection of them.

In order to appreciate the general line of reasoning for why holism is relevant to matters of political value, we must make a distinction between political values that are social and political values that are non-social. A political value will be social just in case it is a social property; it will be a non-social value just in case it is a non-social property. A social property, in the wide sense introduced earlier, is a property whose realisation requires that a number of people display intentional attitudes or perform intentional actions. Hence a social value will require that there are a number of people who are intentionally active in certain ways: in effect, that there are people who are intentionally involved with one another. A non-social value will not require anything of the kind. It will be the sort of value that the wholly isolated individual, even the lone occupant of a world, can enjoy.

There is a great variety of social values invoked in discussions of politics. They range from personal goods enjoyed in intimate relations to goods that are enjoyed only in the public forum. Personal goods include family and friendship, fraternity and citizenship, status and power, protection and equitable treatment, and participation. Social values also include goods that do not inhere in individuals, but in the institutions that individuals constitute; the personalist can countenance such values too, as we have seen, for their realisation will have an impact on the well-being of persons. These sorts of social values include cultural harmony, social order, political stability, and the rule of law. They also include the value that, by some interpretations, is what contractualists are really anxious to defend: the value involved in an institution's being publicly justifiable and justified (Gaus 1990).

There are also a variety of non-social values that are invoked in discussions of politics. Material welfare is an obvious example, since it is clear that the isolated individual may logically enjoy that sort of good without any involvement with other people. Another example is happiness or utility, in the sense in which this is associated, as it is in the utilitarian tradition, with the balance of pleasure over pain or the absence of frustrated preferences and desires. A third example is liberty in the sense in which this requires just the absence of intentional interference by others; the isolated individual is likely to score particularly well on this count, since there won't be any others intentionally involved with her as potential interferers. And a final example is liberty in the more positive sense in which it requires not just the absence of interference by others but also a high degree of autonomy or self-rule.

Apart from clearly social and clearly non-social values, there are also some values invoked in political discussion that can be interpreted either way. A good example here is the value of equality. This may be understood as an active sort of equality that presupposes that people are intentionally involved with one another and which requires that they recognise one another as equals in certain ways: say, as equals before the law. Alternatively it may be taken in a purely passive mode, as a comparative value that someone may enjoy relative to others with whom she has no dealings whatsover and, therefore, as a value that someone may enjoy in total isolation; she is equal, say equal in material respects, with the other human beings who happen to exist, or exist in a certain region, but that is a brute fact about her, as it were, not a fact that has social significance.

Now that we have drawn the distinction between social and non-social values, I can state my thesis about the likely impact of holism on our thinking about political values. The thesis is that if we are atomists then we must regard only non-social values as the primes in radical political evaluation whereas if we are holists, then we are as likely to assign that criterial status to social values as to non-social values. Radical political evaluation is evaluation that is meant to apply not just to a sub-set of the possible alternatives—say, feasible alternatives—but to every arrangement conceivable. The primes in such evaluation are the basic criteria by which the different arrangements are to be assessed. I assume that we are certainly personalists, and indeed universalistic personalists, in the sense explained earlier: that we think it is ultimately only individuals who count in political assessment and that we take them to count equally. My claim now is that atomism will prompt us to be non-social personalists—to think that it is only non-social values that count with individuals—whereas holism will encourage us to explore the possibility that social values may matter just as much, and just as primitively.

Now to the line of reasoning that supports this thesis. Anyone who is an atomist is bound to take the possibility of the isolated individual to be a relevant alternative in radical political evaluation: in evaluation that covers all conceivable alternatives. It may be enough in casual political discussion to argue for the superiority of an arrangement over the status quo, and over other salient alternatives, but in radical political thought the arrangement must also be shown to be superior to the lot of the isolated individual; otherwise, as the atomist sees things, the business of political evaluation will not be logically complete. Unsurprisingly, then, we find that the many thinkers in the atomist tradition that goes back to Hobbes have emphasised that the isolated individual gives us a relevant perspective on political arrangements. They have implicitly or explicitly assumed that we should judge the attraction of political arrangements, at least in part, from the point of view of that individual: from the point of view, as it is often articulated, of a state of nature in which isolation is the norm. They have assumed, to put the matter otherwise, that part of the job of supporting any political arrangement is to show what there is in it for individuals who could logically have enjoyed a solitary existence instead: what there is about that arrangement that makes it superior for such individuals to a solitary existence.

Suppose then that the atomist takes the possibility of the solitary individual as a relevant scenario and thinks that the task of political evaluation is to rank various social arrangements, not just against one another, but also against the scenario where individuals remain isolated. It follows that the terms of evaluation he uses in the course of this ranking must be terms that apply not just under different social arrangements but also in the condition of the isolated individual. The values that serve as his ultimate criteria of assessment must be capable of being realised, not just in the social alternatives assessed, but also in that more or less solipsistic condition. This is to say that the values that serve as evaluative primes for the atomist must all be non-social values, for only such values can be realised in the circumstances of the solitary individual as well as under social arrangements. And so we reach the conclusion that atomism forces the political theorist to restrict his attention to non-social values.

This line of reasoning may be evaded by the atomist who chooses to regard the scenario of the isolated individual as politically irrelevant; such an atomist will resist the aspiration to the radical form of political evaluation that would bear on that possibility as well as on more familiar arrangements. What the reasoning establishes is that if the isolated individual is allowed to figure among the alternatives to be evaluated in political theory, as it has traditionally figured, then atomism filters social values out of the set of ultimate political criteria; it drives the theorist to countenance only non-social values as the ultimate terms of political assessment.

Holism undermines the tendency to focus political theory on non-social values. The holist dismisses the possibility of the isolated individual and the only situations that he sees as options to be ranked against each other in political evaluation are intrinsically social arrangements. Thus he will be quite open to the thought that the ultimate terms of political assessment, the ultimate criteria by which to judge political alternatives, are social values, or at least include social values. If we begin by being atomists then there is a very good chance that we will feel a pressure to restrict our attention to non-social values. If we go over to the sort of holist perspective defended in this book, then we will cease altogether to register such a pressure. We will be open-minded in wondering about which values to cast as the fundamental criteria of political judgment.

I think that the general line of reasoning from atomism to non-social values may help to explain a feature of political thought in the English-speaking tradition of the last couple of hundred years, at least outside the idealist circles that were influenced by German ideas. It is striking that among the many different strands of radical political thought that emerged in that tradition, all of them tended to emphasise distinctively non-social values as the ultimate criteria of judgment. Some appealed to utility as the basic good of individuals, others to the enjoyment of non-interference, others to the realisation of a degree of personal autonomy, and others again to the attainment of a certain level of material welfare or a passive sort of material equality. This is a particularly striking feature in the tradition, given that there were many factors that might have been expected to lead the tradition towards the articulation of certain social values as the basic terms of political evaluation.

Democracy became a rallying point for many radicals in the tradition, for example, yet few of them thought of democratic participation or the democratic resolution of differences—the achievement of public deliberation—as a fundamental criterion of political assesment: democracy was valuable, if at all, for its effects in promoting utility or liberty or whatever. Again, the rule of law was hailed by all as one of the great features of English common law institutions but no one advanced the rule of law as an ultimate value by which to judge a system; on the contrary, as we shall see, the fashion among radicals like Bentham was to see law as a mixed good, as a form of interference that was justified, if at all, by the other forms of interference that it inhibited. Finally, while the chartist and trade union movements emphasised the importance of solidarity and comradeship, none of the theorists of those movements ever really argued that whether such a value would be realised was a basic test to administer in assessing a proposed political arrangement. Socialism may have pushed many thinkers in that direction but mostly the push was resisted.

Consider how different were the approaches to politics that emerged in the same period in continental circles. Think of Rousseau on the general will and on the value of democractic deliberation. Think of Herder on the cultivation of the self in relation to the *Volksgeist*. Think of Kant on the kingdom of ends or Hegel on the realisation of *Geist* in the world. Think of juridical ideals like that of the *Rechtstaat* or sociological ideals like the overcoming of *anomie*. In all of these cases we see a spontaneous tendency to assume that the basic values for the assessment of political structures are essentially social in nature. There is no evidence of the imperative that ruled most English-speaking circles: the imperative to go back to properties that could be enjoyed even by a solitary individual in the search for basic political criteria.

Why should the English-speaking tradition have proved so resistant, over such a long period, to the idea that social values might offer the basic terms of political assessment? The atomistic presuppositions that Hobbes bequeathed to the tradition may go some way towards explaining the resistance. The general line of reasoning from atomism to a concentration on non-social values may have registered, however implicitly, on the tradition that took Hobbes as one of its founders. The Hobbesian endowment matured in the nineteenth century, as so many of his ideas—and not least, as we shall see, his ideas on liberty—became part of the English-speaking intellectual tradition. It would have been difficult for anyone who was heir to that tradition not to take on board the fundamentally atomistic perspective developed by Hobbes and not to shy away from the idea that the basic terms of political evaluation might involve social values.

So much for the general line of reasoning in favour of the view that the adoption of holism is likely to have an impact on valued-centred thinking about politics: in particular, that it is likely to motivate a rethinking about the relevance of social values in political evaluation. I now want to look at the likely impact of holism on liberal commitments in particular. The dominant, value-centred political philosophy in the West today is probably liberalism. Liberals come in two main varieties, depending on whether they are deontologists or consequentialists: deontologists will usually focus on the rights of liberty that all individuals and agencies are to respect; consequentialists will present liberty as a goal that ought to be promoted, and in particular promoted by the state. Whatever the differences between them, all liberals share a commitment to the value of negative liberty, arguing that it is the unique, or at least the primary, value in politics.[8] I now propose to argue that the espousal of holism is likely to affect the conception of negative liberty itself, giving that value a republican rather than a more characteristically liberal cast, and that it thereby facilitates a transformation of liberal commitments.[9]

The republican tradition that I have in mind here goes back to Roman times when its greatest exponent was Cicero. By recent accounts it was of enormous importance, not just in classical Rome, but from the period of the Renaissance down to the beginning of the last century (Oldfield 1990, Pocock 1975, Skinner 1990b). It shaped the language of political debate in the northern Italian cities of the quattrocento; in England of the seventeenth and eighteenth centuries; and in the circles that inspired and were inspired by the American and French revolutions. It gave way eventually to liberalism, when the conception of negative liberty that it represented—so, at least, I see things—lost ground to the alternative, liberal conception.

I should stress that the strategy of beginning with the liberal commitment to negative liberty and showing how it might be transformed under the influence of holistic individualism is pragmatically dictated. There are a wide range of effects that holistic individualism is likely to have on thinking about the values hailed in political discussions. It is likely to have us rethink the requirements and attractions of democracy, as it emphasises the fundamental connection between rational deliberation and the incorporation of potentially dissident parties; here it connects with the recent interest in deliberative democracy (Barber 1984, Cohen and Rogers 1983). It is likely to make us sensitive to many of the points emphasised by communitarians, in particular points associated with the role of shared understandings in a community (Sandel 1982, Walzer 1983). And it is likely to connect with feminist critiques of political theory, as it gives the lie to the image in which self-sustaining and distinctively male individuals come together—out of solitude, as it were—to form and run society (Gatens 1991, Sunstein 1988). It would be impossible for me to survey the full range of such effects and so I concentrate on some that I think are particularly important. Most people, certainly most people in the west, have something of the liberal in them. It should be a matter of interest if, as I allege, holistic individualism has the capacity to facilitate a transformation of liberal commitments.

How to argue my claim that the adoption of holism may prompt a liberal to understand negative liberty in a republican rather than a liberal fashion? First, I will describe the received account of negative liberty, indicating an indeterminacy covered up in that account, and show that whereas the liberal tradition makes the concept determinate in one way, the republican tradition makes it determinate in another and more fetching manner; the traditions offer different conceptions of one and the same concept of negative liberty. Then, having done this groundwork, I will argue that whereas atomism would force us to understand negative liberty in the liberal mould, holism would allow us to feel the attractions of the republican conception: the conception, otherwise described, of liberty as franchise. Finally, I will indicate why I think that espousing the conception of liberty as franchise would lead to a transformation of liberal commitments: to a very different view of what the polity should be doing.

The concept of negative liberty, in particular as it applies to individual persons, involves two distinctive elements (Pettit 1989b). First, it picks out certain action-types as important to the enjoyment of liberty. These are usually identified as those independent activities—those activities not directly involving others—that the normal agent in any society is capable of pursuing in normal conditions without others having to help her to do so. They include activities like following your thoughts where they lead, speaking your mind, moving where you will, and associating with anyone who will have you: in a word, exercising the traditional liberties.

The second feature of the negative concept of liberty will be equally familiar. As it picks out certain independent activities as the relevant domain of freedom, so it identifies a certain sort of constraint on those activities as the relevant danger to freedom. The sort of constraint that it sees as the threat to freedom is the interference of others, in particular the kind of interference that is sufficiently intentional, negligent, reckless, or indifferent to count as blameworthy. To enjoy liberty under the negative concept is to be given an area of non-interference in which you can pursue relevant, independent activities.

What non-interference requires is a matter of some debate. At the least it requires that you are not prevented from pursuing those activities. Under most interpretations of non-interference, it requires in addition that you are not frustrated in your pursuit of those activities through the imposition of certain penalties; and further, that you are not coerced to avoid those activities by the credible threat of prevention or punishment (Pettit 1989b). In what follows I shall assume the richer interpretation of non-interference, though my argument would also go through under the more austere one.

But whether we go with the richer or more austere interpretation of what interference involves, there is one further question that needs to be settled. Two individuals perform the same independent actions, without interference, over their lives. But one of them, unlike the other, is in such a position that had she tried to perform certain other independent actions, then she would have been blocked. Did they enjoy the same level of non-interference? I shall assume that they did not. In other words, I shall assume that for a given level of non-interference, it is required both that others do not interfere with the actions actually performed and that they would not have done so with certain counter-factual initiatives. The assumption is reasonable, since otherwise we will have to say, paradoxically, that a person can increase the level of non-interference she enjoys, just by refraining from activities that would attract interference.

We may relate these comments on negative liberty to the triadic structure which Gerald MacCallum (1967) claims to find in every concept of liberty. According to MacCallum, liberty or freedom is always *of* something, *from* something, *to* do, not do, become, or not become something. This triadic structure may not be found in every concept of freedom (Baldwin 1984; C. Taylor 1985, chap. 8). But it is certainly present in the concept of negative liberty. Negative liberty involves the freedom *of* individual persons, *from* the interference of others, *to* perform independent activities: to exercise the traditional liberties. What makes the concept in question one of negative—or at least negative personal—liberty is the fact that the three variables allowed by the scheme are fixed in the pattern given.

But though our comments fix the concept of negative liberty fairly well, they leave one important matter indeterminate; they leave one further variable unfixed. They present negative liberty as a form of freedom or liberty that is marked by who has it, from what, to do what, but they say nothing on the freedom itself; they say nothing on what it is to be free from something, in particular free from the interference of others to perform independent activities.

There are two ways in which this indeterminacy has been traditionally resolved, as I read the relevant history. Under the one approach, to be free of interference is simply to be blessed by its absence, so that the measure of freedom is the quantity of the non-interference one enjoys. Under the other approach, to be free of interference is to be more or less secure against it, so that the measure of freedom is the quality of the protection one enjoys. The ordinary language of freedom allows both of these readings. In certain contexts the absence of an irritation will be enough to count as freedom from it, whereas in other contexts freedom from the irritation will be taken to require that one is truly rid of it: that one is no longer at its mercy.

The two ways of reading the concept of negative liberty correspond, I suggest, with what may reasonably be described as the liberal and the republican conceptions of freedom. The liberal conception puts a premium on quantity of non-interference, the republican conception on quality of protection against interference. Liberal freedom requires that one is undamaged by the interference of others; republican freedom requires that one is not susceptible, at least beyond a certain level, to such interference: the emphasis is not on escaping damage but rather on not being fragile. I will support this claim, first with reference to the republican conception of freedom, and then with reference to the liberal.

Perhaps the best way of making out the claim about the republican conception is to go back to a distinction that, we are told, was of some importance in republican Rome. Although they conceived of liberty in essentially a negative fashion, with the concept of non-interference to the fore, the Romans insisted that there was a great difference between the *liber*, the free person, and the *servus sine domino*, the slave without a master (Wirszubski 1968). The reasoning behind the distinction is obvious. The *liber* may be unfortunate enough to suffer a great deal of interference and the *servus sine domino* may be lucky enough, or cunning or fawning enough, to avoid such interference. But the *liber* is nonetheless the free one among the two, for freedom is determined not by fortune but by the standing that one has within the community, and especially before the law. To be a free person, to be a *liber*, is to be in a position to call the law out against anyone who interferes or tries to interfere; it is to have the backing of the law against such interference, whether the threat comes from other private individuals or from the officials of the state itself. The *servus sine domino* lacks that protected status, even if the *servus* manages to escape interference. And so the *servus* is not free.

Under this republican reading, freedom means having security against interference, and the measure of freedom is the quality of protection provided. In this sense of freedom, the all-powerful autocrat will enjoy the greatest freedom imaginable. But if freedom is to be something that everyone may have, at least in principle, then it comes to nothing more or less than full citizenship in the sort of society that provides its citizens with protection: it is constituted by full citizenship in a free society. Freedom becomes equivalent to the freedom given by a city, the franchise—in the old, rich sense—given by a state.

Freedom as it was understood in the Roman republican tradition was explicitly equated with citizenship. 'At Rome and with regard to Romans full *libertas* is coterminous with *civitas*' (Wirszubski 1968, p. 3). The equation was explained by at least one thirteenth-century writer on the ground that *civitas* derives from *civium libertas*, the freedom of the citizens (Skinner 1990a, p. 134). The equation continued in common mediaevel usage. '*Freedom* can mean simply "citizenship". . . . The meaning is fossilised in the surviving English use of *franchise*' (C. Lewis 1967, p. 125). The theme, or something close to the theme, is reiterated in Machiavelli, the main figure in the revival and modern development of republican thought. 'For Machiavelli . . . the law is in part justified because it ensures a degree of personal freedom which, in its absence, would altogether collapse' (Skinner 1983, p. 13).

This reading of republicanism may be found surprising, for many commentaries represent republicans as fundamentally concerned, not with protection from non-interference—not with any version of negative liberty—but with the positive liberty of participation in collective self-determination. But I follow a different reading of the tradition. According to Machiavelli, as Quentin Skinner (1983, 1984) has argued, the active participation of a large number of people was essential for the general enjoyment of liberty or *liberta*; his idea was that without general participation, an elite would be likely to commandeer the apparatus of the state for its own ends, thereby destroying the rule of law. But even in Machiavelli it is clear that not only was freedom a distinct notion from that of active participation in politics, the individual's own participation was not even necessary for her personal freedom. One commentator suggests that in this regard Machiavelli may have made a break with an older republican emphasis, though that itself is probably an exaggeration. '*Liberta* here does not mean active participation in the government of the state as it did in the republican or communal tradition, but rather, as it will come to do more and more commonly in subsequent centuries, the passive enjoyment of a condition in which the security of single individuals is guaranteed by the law' (Guarini 1990, p. 28).

The equation of freedom with citizenship has some very distinctive results in the republican tradition. Citizenship, like any social status, naturally involves awareness: it means that the citizen, and those with whom she deals, are aware of her standing, and it means that this awareness is itself a matter of common recognition. But if citizenship or freedom involves this sort of awareness, then it also means being able to live without fear or deference; freedom connotes frankness, where 'frankness' is etymologically related to 'franchise'. The theme, once again, has Roman origins. 'Security of life and property, sanctity of hearth and home, inviolability of civic rights were the chief elements of Roman *libertas.* . . . People who lived under oppression in danger of their lives came gradually to conceive *libertas* as meaning, primarily, order, security, and confidence' (Wirszubski 1968, p. 159). The theme is notoriously absent in the work of Machiavelli but it received considerable emphasis in later republican thought. 'The political liberty of the subject is a tranquillity of mind, arising from the opinion each has of his safety. In order to have this liberty it is requisite the government be so constituted as one man need not be afraid of another' (Montesquieu 1977, p. 202).

The republican notion of freedom as citizenship has a second distinctive connotation, apart from this association with confidence. Citizenship gives protection and constitutes freedom only so far as the law is appropriately framed, respected, and applied and only so far, therefore, as people can be relied upon to make themselves available for public office and to behave properly while in office. But this means that citizenship amounts to freedom only in a society of the virtuous: only in a society where cultural and institutional pressures are such that even if people are not always lovers of the good, still they can be relied upon to do their bit in the civic realm. The theme is familiar and does not need much labouring. Quentin Skinner (1983, p. 4) documents it nicely with regard to Machiavelli. 'In the classical oxymoron that Machiavelli is restating, freedom is thus a form of service, since devotion to public service is held to be a necessary condition of maintaining personal liberty. If we wish to maximise our freedom to control our private affairs without anxiety or interference, the moral is that we must first turn ourselves into wholehearted servants of the public good. Machiavelli's way of summarising these claims is to say that *liberta*, both personal and public, can only be maintained if the citizens as a whole display *virtu*' (see also Skinner 1984, 1990b).

Republicans are sometimes accused of requiring people to be spontaneously good: requiring them to be lovers of the common weal. But it is worth emphasising that all that freedom as citizenship requires is reliably beneficent behaviour, whether the reliability be a result of character or circumstance. Machiavelli was clear, for example, that spontaneous beneficence was not generally available and that republics must make do with the civic beneficence secured by the pressures of law and opinion. 'By the force of law', as Quentin Skinner (1983, p. 11) writes in commentary, 'the people were liberated from the natural consequences of their own *corruzione* and transformed in effect into *virtuosi* citizens'. Again, it is clear in someone like Montesquieu that what he cherished was reliable beneficence, however procured, not just reliable beneficence of the upright sort. Tocqueville writes of Montesquieu on virtue. 'We must not take Montesquieu's idea in a narrow sense. . . . When this triumph of man over temptation results from the weakness of the temptation or the consideration of personal interest, it does not constitute virtue in the eyes of the moralist, but it does enter into Montesquieu's conception, for he was speaking of the effect much more than the cause'.[10]

So much for the republican way of concretising the concept of negative freedom. Freedom from interference comes to mean the security against interference that is provided by citizenship in the sort of society that protects its citizens. And freedom in this sense comes to connote enjoyment of a subjective degree of confidence and membership of a society where civic virtue is the norm. The other, liberal way of determining the concept of negative freedom construes freedom from interference as the absence of interference, rather than as any sort of security against it.

It was probably Hobbes who first clearly rejected the equation of liberty with citizenship and took the measure of liberty to consist in the quantity of non-interference, however that non-interference is secured. We see this line appearing in his infamous remark that a resident of despotic Constantinople can enjoy the same degree of non-interference as a citizen of republican Lucca and, this being so, can be equally free. Never mind that the person in Constantinople may enjoy non-interference only in virtue of her own cunning or fawning or fortune, and not by the sort of right that accrues to the citizen of Lucca. 'Whether a Commonwealth be Monarchicall, or Popular, the Freedome is still the same' (Hobbes 1968, p. 266).[11] This line, unsurprisingly, attracted the wrath and derision of Hobbes's English republican contemporary, James Harrington (1977, p. 70): 'to say that a Lucchese hath no more liberty or immunity *from* the laws of Lucca than a Turk hath from those of Constantinople, and to say that a Lucchese hath no more liberty or immunity *by* the laws of Lucca than a Turk hath by those of Constantinople, are pretty different speeches'. For Harrington freedom is essentially freedom or immunity by the law—that is, a protected status conferred by law—and in that republican sense only the Lucchese is free.

Yet it was Hobbes, not Harrington, who prevailed with the later English-speaking tradition of political thought, in particular with the classical liberals of the early nineteenth century. Perhaps the best test of whether a thinker goes along with the Hobbesian line is to see what his attitude is to the law whereby people are protected from interference. If one assumes that coercion is a form of interference, the person who follows the Hobbesian line is bound to see the law itself as an invasion of people's freedom, albeit an invasion that typically does more good than harm, whereas the republican will think of that law as constitutive, at least in part, of what freedom is. Applying this test, we find that the English-speaking tradition comes out decisively in favour of the Hobbesian approach (Benn and Peters 1959, p. 213; Pettit 1989a, p. 160). Isaiah Berlin (1958, p. 8) sums up the refrain of the tradition, which he traces explicitly to Hobbes. 'Law is always a "fetter", even if it protects you from being bound in chains that are heavier than those of the law, say, arbitrary despotism or chaos'.

The rejection of the republican equation of liberty with citizenship led also, within the liberal tradition, to a rejection of the two connotations of republican freedom. The notion that freedom might be invaded just so far as someone has to live in fear or deference is foreign to the nineteenth-century liberal ways of thinking which came to dominate political theory. It is a frequent complaint among those who identify with this development that it is confused to think that freedom is lost just in virtue of certain attitudes being forced on people. 'The difference between being free and feeling free is clear enough, but not always noted' (Weinstein 1965, p. 156). There is a clear divide drawn between being free and having such a status that one is not cowed by others. Although there is a recognition that we all desire that sort of standing, the desire is separated off from the desire for liberty (Shklar 1989). As Berlin (1958, p. 43) says: 'it is not with liberty, in either the "negative" or the "positive" senses of the word, that this desire for status and recognition can easily be identified'.

The second republican connotation also disappeared in the republican tradition. The idea that virtue is needed if people are to enjoy liberty, the idea that the greatest enemy of liberty is corruption, found little resonance in the thought of early nineteenth-century liberals. Those liberals were raised on the notion introduced in the early eighteenth century by Bernard Mandeville, and made respectable in the works of Adam Smith and others, that there are invisible-hand arrangements under which people do best by one another as a whole, people do best by the public interest, if they look only to their own advantage. (See Hayek 1976, pp. 185–86, for references.) Private vice, as it was said, is public virtue. This ideology made it impossible to argue that liberty required virtue in the republican fashion. In any case, it seemed that invisible-hand arrangements represented the very consummation of liberty, for those arrangements involved no kind of interference whatsoever with the agents involved. The idea quickly grew that the real threat to liberty was not corruption, but the intervention of the state in areas where invisible hands were otherwise ready to reign.

If I am right about the different ways in which the republican and liberal traditions understand negative liberty, then a natural question to ask is why the republican conception gave way so completely to the liberal in the nineteenth century. The prevalence of atomism may have been relevant, for reasons to appear shortly. But there is also another factor worth mentioning. Where there is a shift from one conception of some subject-matter to another, that shift is often occasioned by the conception coming to be opposed to a different antonym (see Holmes 1990). I believe that the shift from a republican conception of liberty to a liberal one could have been occasioned by just such a shift of antonym.

The liberal conception is a conception of liberty under which the antonym is any form of restraint or interference. If unfreedom consists in being restrained, then freedom involves not being restrained: it involves non-interference, pure and simple. The republican conception of liberty, on the other hand, is a conception under which the antonym is slavery or subjection or, more generally, any condition in which a person is vulnerable to the will of another (Patterson 1991). If unfreedom consists in being vulnerable in this way, then freedom involves not being vulnerable. And in order to enjoy such relative invulnerability it is necessary not to be anyone's slave or subject and, more than that, it is necessary, as the Romans realised, to be the very opposite of a slave: to be a *liber* who is equally protected with the best, not just a *servus sine domino*.

It is understandable why, in the republican tradition, the antonym of liberty should have been slavery or subjection or vulnerability. The aim of republican theorists was to identify the characteristics of a society in virtue of which its citizens—its citizens as distinct from residents who do not enjoy citizenship—are marked off from those who are the victims of despotic rule, corrupt officialdom, external control, and the like. They used the concept of liberty to serve this purpose of demarcation and so it is no surprise that they should have conceived of liberty as the antithesis of slavery or subjection. The approach is clearly in place, for example, in the eighteenth century republican tract, *Cato's Letters*. 'Liberty is, to live upon one's own terms; slavery is, to live at the mere mercy of another; and a life of slavery is, to those who can bear it, a continual state of uncertainty and wretchedness, often an apprehension of violence, often the lingering dread of a violent death' (Trenchard and Gordon 1971, Vol 2, 249–50).

As it is intelligible why republican theorists should have taken vulnerability to be the antonym of liberty, so it is understandable why liberal thinkers in the late eighteenth and the nineteenth centuries should have begun to think of liberty as something primarily opposed, not to subjection, but to restraint. Liberal thinkers in that period, especially liberal thinkers in Britain, were concerned to argue that the interference of the state was undesirable: that it hampered commerce and trade and, ultimately, the well-being of all. Liberals were the prophets of the *laisser faire* economics advocated in Adam Smith's *Wealth of Nations* and they were, in effect, the advocates of the rising commercial classes. The language of liberty offered them a rhetoric with which to combat the pretensions of the state—it was probably the only rhetoric that could have served their purposes adequately—and in adopting that language, they reforged the existing conception of negative liberty. When they proclaimed the glories of liberty, they were not heaping scorn, as republicans did, on conditions of slavery or subjection or vulnerability. The other side of their devotion to liberty was an antipathy, not to such traditional ills, but to restraint, in particular to restraint imposed by the state: 'all restraint, qua restraint, is an evil', as John Stuart Mill (1972) expressed the new orthodoxy.

I have sketched a more or less standard account of the concept of negative liberty, identified an indeterminacy in that concept, and argued that whereas the republican tradition resolved the indeterminacy in one fashion, the liberal tradition came to resolve it in the other. It is time now to make a connection back to the issue between atomism and holism. I wish to argue that whereas atomism makes it natural to construe negative liberty just as non-interference, the adoption of holism facilitates a shift to the older, republican interpretation.

It will be clear why atomism makes it natural to construe negative liberty in the liberal way, taking quantity of non-interference to be the measure of liberty. Construed in this way negative liberty is a non-social property that may be enjoyed by the isolated, even the totally solitary, individual. Indeed, construed in this way, negative liberty is perhaps most perfectly enjoyed out of society, where there are no others around to get in a person's way; its finest flowering, as we might put it, is in the freedom of the heath. Construed in the republican fashion, however, negative liberty is an inherently social property. It amounts to the freedom of the city and it is something that a person can enjoy only if she has a certain status in the society of others.

The line of reasoning rehearsed earlier explains why atomism will make it natural for a political theorist to prefer the liberal to the republican rendering. The radical political theorist will want to be able to assess all the options and, if he is an atomist, these will include the condition of the isolated individual. But if that solitary condition is to be compared in political terms with the various institutional arrangements that can be provided in society, then the currency of comparison must be legitimate tender on both sides; it must be constituted by a set of values, a set of criteria, that can be realised both in the condition of the isolated individual and under various social set-ups. And the only terms that can satisfy this condition are non-social values: that is, valued properties that can logically be enjoyed by the isolated individual. Suppose then that a radical political theorist embraces the value of negative liberty as a criterion of political assessment. If he is an atomist, then that means that he is bound to go for the liberal rather than the republican conception of negative liberty. Atomism plus radicalism plus a commitment to negative liberty combine to require liberalism: they combine to require the belief that the liberal version of negative liberty offers a basic criterion for assessing political alternatives.

So much for the tie between atomism and the liberal conception of liberty. Why do I say that the adoption of a holist perspective is likely to facilitate a transfer of allegiance from the liberal to the republican conception of freedom and thereby a transformation in liberal commitments? Mainly, because the republican conception is intuitively much more attractive and more attractive, in particular, to someone of a broadly liberal disposition. Mainly, in other words, because I think that once the republican conception of liberty comes into view, the assumption of atomism is the main stumbling block that is likely to stop the liberal from approving of that conception.

The Romans were quite clear that however fortunate we might count the *servus sine domino*, especially if things went prosperously for her, we could not regard such a master-less slave as free. A parallel intuition ought to be just as compelling for us, especially those of us with a broadly liberal outlook. Although the liberal conception of freedom came to be forged, if my historical conjecture is correct, as a tool for advancing anti-government causes like that of free trade, still liberals continue to be concerned about securing the sort of invulnerability that republicans stressed. At least when they shift focus from narrowly economic matters, they continue to emphasise the importance to individuals of enjoying an assured independence from the whim of others: the importance to individuals of enjoying a status that allows them to be their own men and women.

Consider an ordinary resident of Haiti in the worst days of Papa Doc's rule and imagine that she managed to escape the attentions of the police, and even perhaps that her affairs prospered. She would have lived in dread of interference, of course, and she would have avoided interference only by a mixture of luck and cunning. But still she would have enjoyed a large quantity of non-interference over the period in question. Do we really want to say that such a person may have been free to the same extent as the citizen of a well-ordered society: say, a society like Sweden? Do we really want to say, as we may have to do under the liberal conception, that she may even have been more free, not having been subject to the same amount of taxation or whatever? Surely not. However we want to describe the ordinary residents of Haiti under Papa Doc, we can hardly say that any of them enjoyed freedom. How can someone be free, we want to protest, if she has to live in dread of an irresistible power and has to depend on her wits and her charms to avoid interference?

These intuitions suggest strongly that when we think about what makes for freedom in concrete settings, even what makes for negative freedom, we naturally look for the sort of protected status, the sort of objective and subjective assurance of non-interference, which requires a certain sort of law and a certain sort of culture. There is no freedom under a reign of terror, no matter how successful someone is at giving her persecutors the slip. There is no freedom exercised under the threat of an autocrat's wrath, no matter how well someone does at not incurring that wrath. Freedom, by all the intuitions that come to us, requires more than a patch of ground where others do not happen to trespass. Freedom also requires, and requires by the very logic of the concept, that that patch of ground is protected territory: it is territory where others may not trespass, on pain of suffering public rebuff and exclusion.

Faced with these intuitions, the defender of the liberal conception of freedom may shift ground slightly. He may say that what the Haitian example shows is that freedom should be measured, not by the actual quantity of non-interference enjoyed, but by the expectation of non-interference that accrues to the person, under reasonable criteria of expectation. The person we imagined may actually have enjoyed non-interference over the period in question, it will be said, but the reasonable expectation over that period would have been that she would suffer a lot of interference. This shift of ground on the liberal's part is not very significant, as it still directs us to quantity of non-interference—expected quantity rather than actual quantity—and the value involved remains a non-social one. So will it do to explain away the intuitions rehearsed?

No, it will not. Imagine an island that is autocratically ruled but ruled by someone of a wholly benign disposition, and not in the malevolent manner associated with Papa Doc. Imagine that we are given a choice between leaving the island in the hands of this benign dictator, at least until the time of her death, and instituting at some considerable expense the usual apparatus of republican protection. Imagine, in particular, that since the republican apparatus will require extra taxation for its funding, there is a higher expectation of non-interference associated with leaving the island under the dictator's control. The advocate of the revised liberal conception of freedom will have to say that under the benign dictator people enjoy more freedom than they would have under the apparatus of republican protection. And that is downright counterintuitive.

It is counterintuitive, because it offends against the intuition that someone is unfree if they are at the mercy of another's whim. Although there is only a very low probability that the benign dictator will interfere with anyone, still it remains that if she chose to interfere, then she could do so with impunity. Everyone else in the society depends for non-interference on the good will of that dictator; no one of them is protected against her. They may not be under much threat from the dictator but the important point is that they are defenceless against any intervention she might choose to make in their lives. It goes against the grain to think of someone as enjoying freedom if they are vulnerable in this manner. And that means that even the revised liberal conception of liberty fails to fit our intuitions.

The difference between the republican conception of liberty and this revised liberal conception is important to stress. Someone enjoys a protected status only if she is defended, not just against probable dangers, but against possible dangers: in particular, against the possible threats that other individuals may pose. The revised liberal conception associates liberty with being defended against likely threats, whereas the republican conception associates it with being defended against possible threats. Under the revised liberal conception, the residents of the benignly ruled island enjoy non-interference, and the expectation of non-interference, but only conditionally on the will of another. And that, so far as the republican conception goes, is not to enjoy liberty at all. Certainly it is unlikely that the dictator will take against any individuals and submit them to her will. But the fact remains that she could take against them and that if she did, she could do so without impediment or rebuff. People in such a situation do not enjoy freedom by our intuitions, or not at least by all our intuitions, and only the republican conception can explain why.

We can draw on a concept introduced in the preceding chapter to articulate the feature of the republican conception that enables it to support our intuitions in this matter. Under that conception, it is not enough for liberty that interference is actually absent, and it is not enough that it is probably absent.

What is required is that interference is resiliently absent. Remember the ball that rolls resiliently on a straight path, because there are posts on either side that will tend to inhibit perturbing forces and that will help to return the ball to the original path, if its course is disturbed. The ball holds resiliently to a straight path, because it is more or less secure against a range of relevant perturbances, regardless of their degree of probability. By analogy, the person who enjoys negative liberty in the republican sense, the person who enjoys the resilient absence of interference, will have to be secure against the whole range of relevant forms of interference, including some forms of interference that may be unlikely to occur (Pettit forthcoming b).

I have been suggesting that the intuitions associated with a broadly liberal disposition ought to support a preference for the republican over the liberal conception of negative liberty: that the only thing that might stand in the way of that preference is an atomistic resistance to giving criterial status to a social value like the republican notion of franchise. It is worth mentioning in this connection that even in the English-speaking liberal tradition, there has always been a tendency to slip back into the republican way of thinking, seeing the rule of law as constitutive of freedom, rather than as itself an invasion of freedom.

John Locke is usually seen as the seventeenth-century precursor of liberalism and he set the precedent for this backsliding when, no doubt under the influence of republican ideas, he made his famous remark on law and liberty: 'that ill deserves the name of confinement which hedges us in only from bogs and precipices. . . . For in all the states of created beings capable of laws, where there is no law, there is no liberty' (Locke 1960, section 57). Bentham is perhaps the most insistent, in the later liberal tradition, that all law is itself an invasion of liberty, albeit a justified invasion. But even he slips on occasion into a republican way of speaking: 'personal liberty is security against a certain species of injury which affects the person; whilst, as to political liberty, it is another branch of security—security against the injustice of the members of the Government' (Bentham 1843, p. 302). In this loose fidelity to the republican way of expressing things he is joined by other liberals. For example, Lord Acton (1985, p. 7): 'By liberty I mean the assurance that every man shall be protected in doing what he believes his duty against the influence of authority and majorities, custom and opinion'. One commentator, discussing this habit of backsliding, says that there is an ambiguity in the liberal attitude to state power and to the law: 'although law may always be a fetter, liberals readily accept that such fetters are necessary, and have made "the rule of law" one of their most conscious slogans' (Arblaster 1984, p. 74; see too Cranston 1967, p. 49 and Hayek 1960, p. 143).

Perhaps I have said enough to indicate why I think that the adoption of holism can facilitate a move from a liberal to a republican perspective. But I have described the move as a transformation of liberal commitments and it remains, finally, to comment on why the shift should be represented in such dramatic terms. It may seem that the difference between the liberal and the republican commitments is extremely subtle, turning on a nuance in the interpretation of liberty. Why should a shift in the construal of negative liberty make for anything like a transformation of liberal commitments?

Distinctive liberal commitments, in particular the commitments of someone who recognises no value other than negative liberty—the libertarian or classical liberal—are generated in three stages by the liberal conception of freedom. The first stage is the recognition that under the quantity-centred conception of freedom any legislation and, more generally, any intervention by the state always impacts negatively on people's freedom. At the least, any intervention by the state will require taxation and a degree of fiscal interference. The second stage is the observation that if the state does intervene in some area, say with a view to protecting some individuals from interference, then that is justified only so far as the amount of interference associated with the state's initiative is outbalanced by the amount of interference it inhibits. If freedom is determined by quantity of non-interference, and if every state initiative involves some interference, then the arithmetic of freedom requires that the state should intervene only when the interference it prevents is greater than the interference it displays. The third stage in the generation of liberal commitments comes with spelling out what this observation should mean in practice. The obvious way to spell it out would be to require that the state should intervene if and only if reasonable estimates of probability show that the level of interference in the society as a whole is likely to be decreased by the intervention. But since the interference associated with any state initiative is certain to occur, and the interference it may help to reduce is only a matter of probability, this requirement easily strengthens into a presumption against state initiative: a presumption that the state should not intervene in any area until the case for intervening is more or less overwhelming.

This liberal—or, better, libertarian—presumption sums up what I describe as distinctively liberal commitments. Many liberals, particularly those on the left of centre, recognise other, balancing commitments, but these will be associated with values like equality or utility, not with liberty as such: not with the core liberal value. I now want to show that if a liberal succumbs to the attractions of the republican conception of negative liberty, as the espousal of holism may lead him to do, then this will undermine the libertarian presumption and transform his distinctive commitments. It turns out that as the libertarian conception of liberty generates the libertarian presumption, so the republican conception generates quite a different set of commitments.

The process of generation can also be mapped in three stages. The first stage is the recognition that if freedom is measured by the quality of protection against interference that is provided under the law, and more generally under the culture, then a protective intervention by the state does not necessarily take from anyone's freedom; it will take from that freedom only at the point where the state's powers grow so strong that the rule of law is jeopardised and the guardians of freedom become themselves a threat. The second stage is the observation that if the state is to concern itself with the cause of freedom, as republicans think it should, then the question for political theory is how the state can increase the protection that it provides for its citizens, even in cases where threats are not particularly imminent. The third stage in the generation of republican commitments comes with the attempt to answer this question. Traditionally, republicans have concerned themselves mainly with how to stop the protective state becoming itself a threat to freedom, the focus being placed on the best checks and balances to introduce in public life. But if republicans are serious about the cause of franchise, then there is every reason why they should also look to where the state can improve the defences that it provides for its citizens against possibilities of interference. This would require an investigation of the various initiatives in criminal justice, in educational and medical provision, and in the provision of social security, whereby the value of citizenship can be increased. It represents a significant research program for political theory (see Braithwaite and Pettit 1990 and Pettit 1989a, 1991d).

The contrast between the liberal and republican perspectives can be nicely illustrated by reference to a fanciful example. Consider the situation where there is only one local employer and many employees, so that the bargaining power on the two sides is significantly different. In such a situation we may expect that the employer would certainly call out the law against any employee who interfered or tried to interfere with her but we may not expect that of any employee with whom the employer interfered or tried to interfere: we might well think that the employee would choose to ignore such acts for fear of losing her job on calling out the law against the employer. In such a situation the employer and an employee may enjoy equal non-interference, actual and expected, but they would not enjoy it with equal security. How might we increase security against interference in such a case, ensuring something like equal franchise on the two sides? One way would be by introducing a form of social insurance that would make the prospect of losing a job less than wholly intolerable. So should we think of introducing this?

The liberal is likely to argue that we should back off from any scheme of the kind proposed, if liberty is his only concern. His grounds will be that it would require taxation—itself an assault on liberal freedom—and that it is not likely that the employer will interfere with the employee: after all, it is scarcely going to be in the employer's economic interest to interfere with employees. But the republican will not be particularly moved by any of these arguments, at least if he assumes that the level of taxation required is not going to have serious flow-on effects. He is likely to think that under the situation described, the employee is not sufficiently empowered relative to the employer to be able to enjoy equal franchise and that the demands of franchise, however interpreted, suggest the need for the introduction of a social insurance system. Such a system may be necessary to allow the employee to stand eye-to-eye with her employer. Although he starts out from a negative concept of liberty, as the liberal does, the angle that the republican develops in his particular conception of negative liberty means that he is likely to have quite a different response to the situation described.

The example is meant to be illustrative. I hope it will show how, across a range of policy questions, the republican is likely to go a different way from the libertarian. Concerned with the quality of protection, the quality of liberty, which individuals enjoy, the republican is hardly going to be satisfied with the hands-off policies that libertarians prefer. He is more likely to go in the direction of those left-of-centre liberals who introduce the claims of equality or the concerns of the worst-off to supplement the demands of the quantity-centred conception of negative liberty; he will side with them in being prepared to consider a more than minimal role for the state. But notice that the republican may differ from such liberals on precisely what sorts of initiatives and institutions the state should foster. We have not said anything on exactly what is going to be required for advancing the quality-centred, republican conception of liberty. It would be inappropriate to anticipate the results of investigating that matter by associating republicanism too closely with left-of-centre liberalism.

Institution-Centred Thought

In the previous two sections we charted the effect of our holistic individualism on political theory: first, its effect on contract-centred thinking and, second, its effect on thinking of a value-centred kind. These two tasks discharged, there remains one issue that we must address in conclusion.

It is one thing to maintain that a property like liberty or franchise, or indeed any other property, is attractive and valuable and that a basic structure that promotes the realisation of that value will have certain features. It is quite another thing to say that the value is capable of being reliably institutionalised, that the structure that would promote it is feasible. In this final section the issue I want to address is whether our holistic individualism leaves us any room for arguing that one or another structure can meet this sort of feasibility condition; whether it leaves us any room for the pursuit of feasibility analysis. It may seem that our holism will make trouble for feasibility analysis. The holist claim that people conduct themselves in the light of their shared concepts, and that those concepts may vary across societies, appears to be hostile to the project of judging different arrangements by common standards of feasibility: by standards that apply independently of the concepts deployed in the culture. But I shall argue, to the contrary, that our holistic individualism does leave us room for feasibility analysis and that it pushes that analysis in a particular direction. The argument connects closely with the discussion of rational choice theory in the preceding chapter.

It will be useful if we consider the question about feasibility analysis with some concrete political proposal in view. I propose that we train our sights on the sort of republican proposals that we have just been discussing. I mentioned in the previous section that the promotion of franchise requires the promotion of virtue among the citizens. A person can enjoy non-interference resiliently at the hands of others only if those others are reliably or resiliently beneficent: only if they can be relied upon, across an indefinite range of contingencies, to refrain from interference themselves and, if it falls to their official brief, to deal appropriately with anyone who does interfere. The feasibility challenge in this case is to show that there are institutional means whereby we can induce such personal and official virtue in people, even in people who are fundamentally self-interested. We must show that there is some hope, for example, of keeping officials honest or virtuous: of keeping corruption at bay (Goodin 1992, Oldfield 1990).

The problem is even more challenging for the republican tradition of thought than this suggests. One of the recurrent and plausible themes in that tradition is that there is no hope of keeping officials honest or virtuous, no hope of avoiding corruption, if people generally are apathetic about political life: if they are willing to leave the political arena to the hacks, being too lazy to take an active part themselves or even to be conscientious in the exercise of their vote (Skinner 1983, 1984, 1990b). If this proposition is allowed, then it becomes particularly important to be able to show that we can design institutions that will generate the required virtue. Indeed most republican writing is concentrated on precisely this problem of institutional design (Oldfield 1990).

If people could be relied upon to be lovers of the good, if they could be relied upon to internalise and satisfy the demands of promoting franchise, then there would be no problem of institutional design. But of course people are not all going to be spontaneously virtuous in this way. However many are naturally virtuous, we can be sure that some will fail of such high-minded motives. Thus we have to design institutions that will elicit virtuous behaviour in those who are not spontaneously virtuous and we have to do so, presumably, in a way that does not undermine the spontaneous virtue of those who need no persuading (Ayres and Braithwaite 1992).

Any proposals for institutional design are bound to make some assumptions in social theory. In particular they are bound to make some assumptions about how people are going to behave under the introduction of the measures recommended. Such assumptions will have to be guided by a background picture in which certain controls are identified as effective for producing suitable forms of individual response. The institutional measures introduced will be triggers designed to activate those controls and produce the desired sorts of behaviour. What background picture should we adopt, then, as we think about questions of institutional design? Our ecumenism in the area of social theory means that this question can be answered on more or less pragmatic grounds. There are a variety of plausible controls, operative at different levels, to which we might look in wondering how to facilitate the appearance of civic virtue. We will naturally want to focus, however, on those controls that can be most easily triggered by institutional measures and that can be most readily made to generate suitable behaviour (Brennan and Pettit 1991).

The controls that institutional design generally assumes, the controls that look most susceptible to institutional triggering, are those the rational choice tradition countenances. As we interpreted that tradition in the preceding chapter, the basic assumption is that most people are more or less rational about their self-regarding interests in economic gain and social acceptance: specifically, that they do not generally act in a way that flouts those interests, reducing their satisfaction below a certain culturally variable level. The relevant level, I suggested, is probably fixed by the reference group to which an individual belongs. The basic assumption is that only the rare individual is likely to make a martyr of herself in regard to economic interest and social acceptance; only the rare individual is likely to be self-sacrificial to the point where she cannot maintain the life-style associated with her reference group.

The task before us then is this. We aspire to have people behave in the virtuous manner required for the establishment of a republican regime but we assume that many of them conform to the rational choice model of more or less self-regarding creatures. The question is whether we can design things institutionally, so that the assumption gives us grounds for being optimistic about the aspiration. Can we put institutions in place that will make it reasonably probable that rational, self-regarding agents behave in the virtuous fashion desired? And can we do this in a way that will not undermine the motivation and dedication of those who are spontaneously virtuous?

If we are to persuade people to be reliably beneficent then one possible way of doing so is to educate them, in particular drill and train them as children, in the overall benefits of such virtue and in the desirability of each doing her virtuous part. In confronting the task just described we assume that all measures of a kind that are likely to increase the level of spontaneous virtue have already been put in place. We assume that nonetheless there is still a proportion of the populace who cannot be relied upon always to be beneficent. The question is how we may motivate them, and motivate them without demoralising the others.

There are two possible ways we might go in designing a solution to the problem. One is to look for institutional sanctions that will make it rational in self-interested terms—which will make it prudent—for the unvirtuous to behave, and behave reliably, in a beneficent way. The other is to look for institutional filters that will ensure that only those who are spontaneously virtuous in any area of behaviour, or only those who can be moulded by certain sanctions into a virtuous profile, get into a position to display that behaviour. Institutional filters will certainly play a role in any institutional arrangement for the promotion of civic virtue but they will probably be of marginal rather than central importance. Civic virtue must involve all the citizenry, even if it is more important in some—those occupying public offices—than in others. Thus we cannot rely on filters to play a major role in developing civic virtue. Or at least we cannot do so, short of resorting to radical filters like ostracism, incarceration, or elimination of the morally recalcitrant!

The upshot is that any institutions for promoting civic virtue are going to have to depend in most part on putting suitable sanctions into operation. Sanctions come in negative and positive forms, as penalties or rewards. Such penalties and rewards may enter into the calculations of agents and give them prudential reason to commit themselves to the virtuous path. Alternatively the agents may be spontaneously committed to that path and the sanctions may have an effect—besides reassuring them about others—only in the event of the agents beginning to deviate. In the first sort of case the presence of the sanctions will explain the civic virtue of the agents in question. In the second it will not explain the virtue itself, since this is spontaneously generated, but it will explain the resilience of that virtue. In either case we may think of the sanctions as promoting civic virtue among the population.

What sort of sanctioning system might we envisage for the job of promoting civic virtue? The most salient possibility is a centralised system, under which an arm of the state, or some such corporate authority, confers rewards or imposes penalties on suitable persons. Centralised sanctions will include, in the reward camp, the remuneration or promotion of public servants and the conferring of public honours on selected citizens; in the penalty camp, they will include the rigours of the criminal law, such as the fines and terms in prison that the courts may impose. Centralised sanctions represent the rule of the iron hand.

It is undoubtedly going to be necessary in any society to have sanctions of this centralised kind, in particular sanctions of the criminal law, in place. Such sanctions help to guarantee and signal the status of each individual as someone worthy of protection. But the question for us now is whether the presence of such sanctions can serve not just to protect potential victims against interference, and not just to mark their protected status, but to promote civic virtue: to make potential offenders into reliably beneficent agents, into agents who will be beneficent across a variety of possible contingencies.

There are three problems that counsel against depending, or at least depending exclusively, on centralised, iron-hand institutions for the promotion of civic virtue. A first is a regress problem, well expressed in an old question: *Quis custodiet custodes*? Who will guard the guardians? Who will mind the minders? The problem is that if we centralise the sanctioning system in the manner envisaged, then it is not clear how that system can effectively serve to control the behaviour and dispositions of those at the top who run the system. The problem may not be overwhelming but it should at least give us pause.

Even if the regress problem is solved, however, there is another difficulty in the way of the centralised sanctioning approach. This is a version of what is more widely known among economists as the principal-agent problem. Think of the central sanctioners—those at the top—as the principal in an arrangement in which the principal has to rely on other people to be its agents. The principal's aim is to get civic virtue realised, and the agents whom it recruits in pursuit of that aim are the different individuals who are required to be virtuous: the officials lower down the hierarchy, for example. The principal-agent problem is that even if the principal is wholly dedicated to the aim, its agents often won't be, and that while the principal can put sanctions in place in order to get the agents to pursue the aim, the agents will often be able to serve their own ends, while presenting themselves to the principal as serving the common goal. In other words, the agents will often diverge in motivation from the principal and will often have the presentational resources to satisfy their divergent ends, while appearing to pursue the goal appointed for them. The problem is familiar and scarcely needs documenting.

But suppose that we could devise a centralised sanctioning system that avoided both the regress problem and the principal-agent problem. It turns out that there may still be a difficulty in relying on such a system for the promotion of virtue. The difficulty is that in order to be effective, the system envisaged would have to have a police officer on every corner, and in every committee room. It would have to put a regime of surveillance in place that would hammer out the lesson that crime doesn't pay, that honesty is mere prudence. Morever, in order to persuade potential offenders to conform to the law on occasions when the chances of getting away with deviance seem good, it would have to be able to call on an effective system of investigation and detection and it would have to be able to threaten severe punishments. But if the system does this, then it is all too likely that it will transform people who would otherwise be spontaneously virtuous into resentful and recalcitrant agents (Ayres and Braithwaite 1992). As a result, it may undermine people's self-regulation and end up making things worse rather than better.

It appears that if we are to have any hope of promoting civic virtue by institutional means, then we cannot rely exclusively on the centralised, iron hand. We must also be able to rely on an informal, decentralised system of sanctioning. We must be able to rely on a system in which everyone is a police officer for everyone else, and indeed everyone is also a potential judge and a potential source of rewards or penalties. So what are the prospects for such a decentralised system?

There is a standard line in rational choice theory, according to which there is no prospect of setting up a decentralised system of sanctions in any situation where the system is necessary: in any situation where at least some people are sufficiently self-regarding not to be spontaneously virtuous. If people are self-regarding to that extent, so the argument goes, then why should they bother with sanctioning others? If we cannot rely on them to be lovers of virtue, as we might put it, how can we rely on them to be enforcers of virtue?

The problem is sometimes described as the enforcement dilemma. The dilemma, as it arises for us, is this: either people are spontaneously virtuous, in which case there is no need for the enforcement of virtue; or they are not spontaneously virtuous, in which case there is no possibility of getting them to enforce virtue. The problem derives from the alleged fact that if compliance involves costs sufficient to inhibit suitable behaviour, there are costs associated with enforcement that ought equally to be sufficient to inhibit suitable en-forcement. James Buchanan (1975, pp. 132–33) puts the crucial premise suc-cinctly. 'Enforcement has two components. First, violations must be discovered and violators identified. Second, punishment must be imposed on violators. Both components involve costs'.

The enforcement dilemma is often presented in the terms of collective action theory (M. Taylor 1987, p. 30). Take a collective action predicament in which everyone is better off if everyone is virtuous than if no one is, but where some find it prudentially rational to free ride. With widespread civic virtue, that predicament would be resolved. But if civic virtue is supposed to be produced by decentralised enforcement, then there must be something else in play to produce such enforcement. For decentralised enforcement itself raises a collective action predicament. We are each better off if everyone is an enforcer than if no one is, but at least some of us will find it prudentially rational to free ride: we will bank on being able to enjoy the fruits of other people's virtuous enforcement without ourselves paying the costs associated with enforcement.

The enforcement dilemma constitutes a serious challenge, as we look for institutions to promote civic virtue. We have seen reason to think that a centralised sanctioning system would not be very effective in promoting such virtue and the enforcement dilemma suggests that there is no possibility of establishing a decentralised system either. So is the enforcement dilemma an insurmountable problem?

No, it is not (Pettit 1990c). The dilemma raises a problem—and, we may assume here, an insurmountable problem—for a decentralised sanctioning system, in which the sanctioning is intentionally pursued: in which the enforcer has to make intentional efforts to identify, or to punish or reward, the relevant agent. But sanctioning need not be intentional, and non-intentional sanctioning will not incur the sorts of costs that might inhibit its exercise. Thus we need not be demoralised by the enforcement dilemma. There is still some hope of finding institutions capable of promoting civic virtue. There are two different ways in which we might try to set up a decentralised system with non-intentional sanctioning. The first of these possibilities receives a lot of attention in the current literature, despite some difficulties with the proposals canvassed; the second is a more promising line but few contemporary theorists have attended to it.

The first possibility for decentralised, non-intentional sanctioning is the sort illustrated by the invisible hand. Consider the way in which producers in the ideally competitive market are allegedly drawn to sell their goods at the lowest price at which they can stay in business, thus benefitting everybody. Such producers are drawn or driven towards the competitive price, because any one of them who tries to sell above it will be punished by consumers: she will lose business to competitors who undercut her price. Thus there is a system of decentralised sanctioning in place that keeps the producers on track. But the sanctioning involved is not intentionally practised by consumers. Sure, they do something intentionally: they seek out the best price. But it is not intentional on their part that in doing this, they sanction or discipline those who would sell above the competitive price. They pursue a sanctioning line but they do so because it is independently in their interest, not because it involves sanctioning.

Can we identify invisible hand mechanisms that would promote civic virtue? Can people be put in such a situation relative to one another, can they be subjected to such an institutional arrangement, that they would each have an independent motive for following a suitable, sanctioning line?

We saw in the preceding chapter that, if David Lewis (1969) is right, then certainly people can find themselves in such a situation that they will be motivated each to act in a way that will punish anyone who defects from a convention like that of driving on the left. If driving on the left is in process of becoming a convention, then each will be motivated to act in a way that punishes anyone who defects from that convention: each will be motivated to drive on the left and thereby to threaten any defector with a serious accident.

Being virtuous in the sense that is relevant here will not usually involve just abiding by conventions of this kind. It will more typically involve abiding by a regularity of the kind that solves a free-riding problem: a regularity of such a kind that even when it is in place, each may be disposed to try to break it; a regularity of such a kind that deviance has independent attractions. Being virtuous will involve respecting the persons of others, respecting their property, faithfully carrying out any assigned role or brief, and the like. In such cases everyone is better off if everyone conforms than if no one does so but each is tempted by the prospect of getting away without conforming themselves. Conventional behaviour solves a coordination predicament and there is no temptation to defect from the behaviour. Virtuous behaviour is more typically called on to solve a free-riding predicament and so there is such a temptation. The question then is whether there is a possibility of an invisible hand generating or sustaining typically virtuous behaviour in the way in which the mechanism described by Lewis (1969) supports conventional behaviour.

A number of possibilities have been identified in the past, particularly within the republican tradition. One is the sharing of power arrangement, under which individuals or agencies that are likely to lean in different directions, if they lean away from virtue, are forced to come to common decisions. Another is the splitting of power device, under which different individuals or agencies are set up as mutual checks, with each reponsible for a different range of decisions. Under such systems, we can see how agents might be manipulated into sanctioning one another for any behaviour that is not, as it happens, in the common interest. These devices are worthy of serious investigation but they are not likely to be feasible outside a narrow range of cases. Thus we cannot look to them with any general optimism.

Many contemporary authors have explored another invisible-hand arrangement for generating a certain sort of cooperation or virtue (Axelrod 1984, R. Hardin 1982, Sugden 1986, M. Taylor 1976, 1987). Take the indefinitely iterated prisoner's dilemma involving two subjects: that is, the sequence of dilemmas such that at no stage does either party know when the last round is. If both subjects defect in each round, then they will both do worse than if they each choose a conditional strategy like tit-for-tat: under this strategy, each cooperates in round one and then does whatever the other party did in the previous round. That is to say that tit-for-tat represents a Pareto-superior outcome to permanent defection. But tit-for-tat also represents an equilibrium outcome, because neither party will be in a position to benefit by departing unilaterally from the outcome under which both tit-for-tat. The defector may win an extra reward on the occasion of defection but she will have to pay for that reward, so far as the other punishes her by defection in the next round. Furthermore, in order to get the other to cooperate again she will have to be willing to cooperate in some round in which the other defects.

In view of these features, it has been suggested that two parties in the situation imagined, and indeed many parties in corresponding many-party situations, will be driven by an invisible hand towards something like virtuous behaviour: towards tit-for-tat or towards some such conditionally cooperative strategy. Each will be in a position where any defection will rationally attract the punitive defection of the others and so each will be policed by the others, albeit unintentionally, into behaving appropriately. All that is necessary for a tit-for-tat outcome to emerge and stabilise is that each can persuade the others that she is a tit-for-tatter, so that it is to the advantage of the others to tit-for-tat with her.

There is at least one major problem with this sort of arrangement for institutionally promoting civic virtue. It does not look likely to be feasible in what is perhaps the most important sort of case: the common sort of free-rider predicament, where there are many parties involved and where defection by one has only a very marginal impact on the welfare of others. This predicament is illustrated by the littered beach, the polluted city, the overfished lake, and the like. Why would others be willing to put at risk the degree of cooperation they have achieved in order to punish the single, barely irritating defector? Why would others be ready to fulfill the threat of retaliation that is implicit in a tit-for-tat strategy, against the lone defector? There may be no problem with the foul-dealer predicament, as I have described it elsewhere (1986b), in which the lone defector will put at least one other party beneath the baseline of universal defection. But in the more common free-rider sort of predicament, it is not clear that anyone would be disposed to implement the threat of retaliation against the lone defector. And, if that is so, then equally it is not clear that anyone would be able to persuade others that she is a tit-for-tatter: in particular, that she is prepared to defect from a general cooperative line in order to punish the lone defector. Thus it is not clear that the tit-for-tat solution can ever get off the ground (Pettit 1986b; for discussions see Axelrod 1988, Pettit 1989d, Schmidtz 1988, and Tuomela 1988).

So much for the first possibility of decentralised, non-intentional sanctioning. The second has not been explored much in the literature but it is very promising (Pettit 1990c). Here the sanctioning is non-intentional in a different way from that which obtains in the invisible hand case. In that case people perform certain actions intentionally—the buyers take their custom elsewhere, in the market example—but it happens not to be intentional on their part that the actions serve to sanction someone. In the case envisaged now people do not do anything intentionally in the course of exercising their sanctioning role and so, necessarily, their sanctioning is non-intentional. Sanctioning in the invisible-hand case is contingently non-intentional: it happens not to be intentional on the agents' part that their intentional actions have a sanctioning effect. Sanctioning in this other case is essentially non-intentional: the responses that have a sanctioning effect are not intentional actions at all.

This second sort of decentralised, non-intentional sanctioning deploys attitudes rather than actions as sanctioning responses and, sticking with the established metaphor, we may describe it as employing an intangible hand (Brennan and Pettit forthcoming). In our discussion of rational choice theory we distinguished between two sorts of goods that agents are assumed by the theory to want for themselves: action-dependent goods, summed up in the notion of economic gain; and attitude-dependent goods, summed up in the notion of social acceptance. Either sort of good can be effective as a reward and either of the corresponding sorts of bad can be effective as a penalty. Where the invisible hand deploys actions as sanctioning responses and imposes economic costs or benefits, the intangible hand deploys attitudes and imposes social costs or benefits.

The social costs or benefits involved in the intangible hand may affect the appreciation, affection, gratitude, trust, or esteem of others. Equally, the costs or benefits may be imposed individual by individual, or in a manner marked by collective recognition. I abstract from these matters here. The crucial point is that the good or bad involved, being attitude-dependent, is not conferred via an intentional action on the part of the sanctioners, though intentional action may be involved in the stage-setting. And so such a mode of sanctioning will not be intentional and need not require sanctioners to incur any costs.

Buchanan mentions two apparently inevitable costs of sanctioning: that involved in identifying those who behave appropriately or inappropriately, and that associated with conferring a suitable reward or punishment. But people may not have to do anything to identify the relevant agent: they may just be around in sufficient numbers to make it highly likely that some of them will notice the doer and the deed. And equally people may not have to do anything to make the agent suffer or rejoice in such a case. The fact that they notice the agent, in particular notice what she has done under circumstances where this is bound to attract their favourable or unfavourable regard, may be penalty or reward in itself. You are punished for looking through the keyhole, as you smart under the gaze of someone who comes upon you doing so. That person doesn't need to say anything or do anything. She punishes you without having to undergo any costs in the exercise of that role.

It will be readily conceded that given sufficient numbers, enforcement need not involve intentionally seeking out the violators of a norm or whatever. What will come as a shock to many, however, is the claim that a violator can be punished—or of course a conformer rewarded—by the attitudes of others, even when those attitudes are not intentionally expressed, say, in censure or praise. Yet the point is obvious, in the light of our discussion of social acceptance in the preceding chapter. We care not just about the rebukes and commendations we receive from others but also about whether they take a negative or positive view of our actions and persons: look at the eagerness with which we search for cues as to the view they actually take. We care about their dispositions to rebuke or commend us, even if the costs—say, the costs of social embarrassment—mean that those dispositions are not much exercised. Adam Smith (1982, p. 116) puts the point nicely. 'We are pleased to think that we have rendered ourselves the natural objects of approbation, though no approbation should ever actually be bestowed upon us: and we are mortified to reflect that we have justly merited the blame of those we live with, though that sentiment should never actually be exerted against us'.[12]

In the discussion of social acceptance in the preceding chapter I argued that there was reason to think that we desire this good for its own sake, and not just so far as it promotes our expectation of economic gain. Accordingly, I shall assume here that the good has intrinsic rather than instrumental status. But it is worth noting that were we to concede that the good was desirable only as a means to economic gain, still this would not require major qualifications to our line of argument. It would introduce some limits on the social acceptance that is likely to count but not limits of a drastic kind. Suppose that the only things we are ultimately interested in getting from others are action-dependent goods. And suppose that we seek the acceptance of others, because we get more action-dependent goods as more and more other people grant us acceptance. These assumptions still compel us to recognise that we are punished or rewarded just by the attitudes that others come to have towards us; we do not remain indifferent up to the point when the others act. That someone comes to think ill of me, even if various costs stop her saying or doing anything then and there, means that there is a decrease in the probability of her favouring me in any future situation where she is in a position of power or patronage. If I care about whether she will favour me in such a situation, then I must care about a decrease in the probability of receiving such favour. Imagine that as others looked at us, a dial on their foreheads indicated our acceptance rating. Even if we only cared about acceptance on an instrumental basis, we would still be concerned at any drop in that rating; it would bode ill for our economic fortunes.

These considerations establish that if we are looking for sanctioning systems to encourage civic virtue, then we should explore the prospects of a system under which the pressures of shame and fame, alienation and acceptance, keep people reliably on the virtuous path. We can look with some reasons for optimism to what I call the intangible hand. It would have been no surprise to traditional republicans that this possibility is live, since there is a recurrent emphasis in the tradition on the role of such pressures in sustaining the sort of political and social life required for liberty. Cicero, who is one of the founding fathers of republican theory, puts the point as follows. 'The sages, taking nature as their guide, make virtue their aim; on the other hand, men who are not perfect and yet are endowed with superior minds are often incited by glory, which has the appearance and likeness of *honestas*' (Lovejoy 1961).

But it is hardly enough from our point of view to be able to argue that such a non-intentional sanctioning system—such an intangible hand—is a live possibility. We should do something to show that it is positively likely, under appropriate circumstances, that people will use the system to pressure and police one another into a life of civic virtue, a life of reliable, civic beneficence. It is positively likely that they will allow themselves to be guided by an intangible hand.

There are three conditions under which we may expect people to be persuaded by an intangible hand system to display a certain sort of behaviour. One is that it is common knowledge in the community that the type of behaviour involved is encouraged by all, its opposite discouraged. Each knows that it is encouraged, each knows that each knows this, and so on perhaps to other levels: there may be ignorance at a higher level about what holds lower down, but certainly there is no error (D. Lewis 1969, chap. 2). A second condition is that it is also common knowledge that anyone displaying or failing to display that behaviour is likely to be noticed and, if noticed, will suffer a loss of acceptance. A third is that the type of behaviour does not carry costs in terms of economic gain that, by whatever standards are taken as relevant—I can only hope that we will be able to identify such standards—would make it prudentially irrational for most people to pursue it.

If such conditions are fulfilled with any type of behaviour then we may expect that behaviour to be displayed resiliently in the community. The behaviour may be produced by many or most people for deliberative reasons that have nothing to do with the acceptance it earns; it may be produced, for example, for reasons of upright virtue. But the fact that departing from the behaviour carries net costs for an agent should mean that if people consider departing from it, then they will be faced in deliberation with those costs and will be induced to resume the behaviour. Each will be confronted with the fact that if she fails to perform appropriately then her failure is likely to be discovered by at least some people and she will in that case incur a serious net cost: the loss of a certain degree of acceptance. It is in the self-interest of each, whether self-interest has an actual or virtual presence, to behave in the manner that virtue requires.

If the sanction of acceptance is the sole source of someone's beneficence, then she will be beneficent, it appears, only in cases where it pays. Does this mean that the intangible hand cannot produce virtue proper: that is, reliable or at least relatively reliable beneficence? No. There is a reason why even an agent moved solely by the desire for acceptance may be more or less reliably beneficent. In every situation there will be some chance of her being discovered failing to do an encouraged sort of action; indeed things can usually be organised so as to keep that chance fairly high. And the cost of having her failure in such a situation discovered is relatively large. To be caught out behaving unsuitably in the sort of case where one might have expected to get away with it is to reveal oneself as the sort of person who performs encouraged acts only for the sake of acceptance. And to reveal oneself as such a person is to lose a good deal of the acceptance otherwise accruing: it is to establish that one is not worthy of the appreciation, affection, gratitude, trust, or honour that might otherwise have been forthcoming. 'The general axiom in this domain', as Jon Elster (1983b, p. 66) puts it, 'is that nothing is so unimpressive as behaviour designed to impress'. Thus there is some ground for thinking that even if people are moved by acceptance alone, still they will display virtuous behaviour fairly resiliently. They will be loath to run risks, even in cases where the risk is low, for those are precisely the cases where their acceptability to others is put under the most crucial test; those are the cases where they have a chance to prove themselves truly likeable, admirable, trustworthy, honourable, or whatever.

This point can be formulated also as follows. The intangible hand is not designed to change people's character, at least in the first instance; unlike moral education, for example, it does not aim at transforming them into spontaneously virtuous agents. But neither is the intangible hand designed merely to change people's behaviour, as the invisible hand does. The device, if it operates properly, serves to do something in between these two extremes: it serves to change the persona that agents put forward, the image of themselves they project, as they enjoy the approval and favour of others. We may well expect someone who is otherwise disciplined by an invisible hand to take advantage of any opportunity to break from the pattern of behaviour that the discipline imposes: we may expect the seller in the market to take advantage of the foreigner and charge above the going rate. But we should not necessarily expect this with someone who is subject to the discipline of the intangible hand. The seller who exploits the foreigner will not suffer any invisible-hand penalty if her act is discovered later by others. But the person who is later discovered to have behaved badly in a situation where she thought no one would notice will suffer, and suffer seriously, under the intangible hand.

The upshot of this discussion is that there is a live possibility of promoting civic virtue by means of an acceptance-centred system of decentralised sanctioning, by means of intangible-hand institutions. There is a possibility, however abstract, of institutionally mobilising the forces of social acceptance so as to encourage people, both as ordinary citizens and as public officials, in the ways of virtue that the enjoyment of franchise requires. There is no prospect of proving in the present context that this possibility is really fruitful and, for example, that it is sufficiently fruitful for the republican proposals canvassed in the preceding section to look like feasible recommendations. But we can at least illustrate the possibility and show what its realisation would come to in a concrete, institutional setting.

Any likely republican dispensation will involve a heavy reliance on committees. In particular it will rely on the sorts of committees that have to make a judgment in accordance with some publicly assigned brief. One such committee is the jury. It is important for the franchise of citizens that jurors be conscientious, and be known to be conscientious, in assessing the impact of evidence on guilt: in discharging their publicly assigned duty. But conscientiousness may not be motivating for all jurors; they may lack the spontaneous virtue to pursue it. Conscientiousness takes time and trouble and while everyone is better off if every juror is conscientious than if none is, everyone is better off in self-interested terms if she can be the one juror who gets away with being less than conscientious. Hence it is important for a republican dispensation that we be able to put sanctions in place that will promote conscientiousness among jurors.

The same lesson goes, by parallel reasoning, for a variety of public committees: the public hearing or inquiry, the appointments or promotions committee, the economic advisory board, and so on. Indeed an analogue of the lesson may also go for the forum of public debate in which the representatives of the people allegedly try to determine what is for the public good. If people are to enjoy franchise, if they are each to be resiliently secure against interference, then such committees must be constituted by conscientious members; they must not be stacked by those who will pursue sectarian interests at whatever costs to others.

If the committee arrangement generally works in our societies, then it probably does so, in some part, because of the operation of an intangible hand. The arrangement offers a good illustration of such a device at work. Consider the arrangement in the case of the jury. First, it is common knowledge among the jurors that conscientiousness is the only approved and acceptable type of conduct, for it should be manifest to all that each of them is vulnerable to manipulation by others if jurors are not generally conscientious; besides, the requirement of conscientiousness is highlighted as they are sworn in and as they are given their instructions by the judge. Second, it is common knowlege among the jurors that if anyone is not conscientious then this is likely to be noticed by others and to attract a degree of rejection; it is common knowlege, after all, that jurors can ask each other questions and that this interrogation can reveal whether someone is being casual or biassed in their judgment. Finally, it is also the case under the jury arrangement that conscientiousness ought not to be economically irrational: members of the jury are vetted so that they do not have a special interest in the outcome and their deliberations are covered by a veil of secrecy, so that they ought not to be exposed to intimidation (but see Thompson 1986).

These three conditions being fulfilled, we may reasonably expect that jurors will police one another into the display of conscientiousness, even if they are not spontaneously virtuous in this respect. The vetting and confidentiality measures ideally ensure that they are not strongly motivated away from conscientiousness. And the other aspects of the arrangement mean that this fertile motivational ground is seeded with incentives to adopt a conscientious profile, doing one's best to determine whether the evidence puts guilt beyond reasonable doubt. The juror who fails to be conscientious is liable to look silly or unsavoury, as her cavalier or prejudiced attitudes are revealed. With nothing else to care about, this may be sufficient to elicit the desired sort of conscientiousness in even the most hardened soul.

This may be sufficient, I say. But of course there is no guarantee that it will be; the intangible hand is not a foolproof remedy for the lack of spontaneous civic virtue. The committee arrangement may fail to work, whether in the jury case or in any of the other examples mentioned, in a variety of ways. The committee may be stacked, despite the best vetting procedures, with those who have a special interest in the outcome; or the effect of the desire for acceptance, in particular the acceptance of the other jurors, may be neutralised, despite efforts at confidentiality, by outside intimidation. Again, the committee may be vitiated by an unwillingness to pursue mutual interrogation as to the reasons for any judgment made and by the absence of a belief that those who are not being conscientious will be exposed. Or, more basically still, the arrangement may be undermined by the appearance of divisions in society that make some members of the committee willing to present themselves as 'shameless' to others. A society can become so divided that committee members care only for the acceptance of those in their own particular subgroup, be the grouping one of colour, creed, gender, or whatever.

I mention such problems to highlight the qualified nature of the position I wish to defend. The position is that while there is some hope for the institutional promotion of civic virtue, in particular some hope for its promotion by intangible-hand devices, that hope will only be realised under careful and informed institutional design. One of the most important tasks for anyone interested in political ideals, especially rich and demanding ideals like the republican one, is to develop an account of the detailed arrangements that may be expected to promote the sort of compliance the ideals require: in the republican case, to promote what we have been describing as civic virtue. Institutional design is a research challenge for social theory and I mention the problems confronting the committee arrangement in order to underline the fact that the program requires more than the cursory sort of observation deployed here.

Whatever its problems, however, the arrangement has great appeal within the perspective developed in this book. Not only does it represent a possibility of decentralised, intangible-hand sanctioning and hold out some hope for those political theorists, like republicans, who favour a relatively rich state. The arrangement lays great emphasis on the role of discourse and deliberation as a discipline on individuals; in this, indeed, it is a distinctively republican device (Sunstein 1988). And in laying this emphasis on discourse and deliberation, it makes a connection with the deliberative view of intentional subjects that we have been defending as part of our holistic individualism. There are a number of reasons why we should be interested in exploring this particular possibility of establishing civic virtue. Not only does it hold out the promise of being effective. It also answers in a particularly satisfying manner to our image of the human subject.

But even if we think we can design or redesign the committee arrangement in such a way as to promote the associated sort of civic virtue—conscientiousness—there are other major problems of institutional design we will also have to confront if we are to be serious about republican ideals. We must be able to design a voting system that can motivate individuals to be conscientious at the polls and can discipline politicians appropriately, imposing costs on them for the pursuit of anything other than what is at least arguably in the public interest (see Brennan and Lomasky 1992, Brennan and Pettit 1990). We must be able to design a bureaucracy that can remain alive to the demands of republican ideals, and not get bogged down in routines of small-mindedness, inflexibility, or empire-building. We must be able to design a police force that can pursue the difficult tasks assigned to it without becoming a self-serving, mutually protective elite; the police force represents the sort of problem for contemporary republicans that the standing army was taken to be in earlier days (Braithwaite and Pettit 1990). In general, we must be able to design legislative, executive, and judicial systems that pursue their respective goals, and carry on their allotted business, in such a way that people's franchise is facilitated and furthered. I see here a program for extended research, in the grand tradition of Machiavelli's *Discourses*, Montesquieu's *Spirit of Laws*, and Mill's *Representative Government*.

There is one final question that I would like to raise about the acceptance-centred, intangible-hand system of sanctioning. I said earlier that any system of sanctioning that is designed to promote civic virtue should not have the effect of diminishing or undermining the motives of those who are spontaneously virtuous and who need no institutional persuasion. The question then is how far the intangible hand would support the stance of those who display such spontaneous civic virtue. I said before that a centralised system of sanctioning might well introduce such sticks and carrots as would drive the spontaneously virtuous person to assume a resentful and more self-regarding mentality. The question now is whether the decentralised, acceptance-centred system that I favour might have the same effect. Would a resort to the pressures of shame and fame, ignominy and honour, alienation and incorporation, drive out spontaneous virtue?

No, it would not. As mentioned previously, an acceptance-based system of sanctioning rewards individuals for the persona they display, in particular for a persona of concern for the relevant public good. But this means that the person who is most reliably rewarded under an intangible hand is the spontaneously virtuous agent. If anyone else is to be rewarded, indeed, then they have to pass as virtuous. As we noticed, no one gains acceptance for being the sort of person who seeks acceptance. A person gains acceptance only for being someone who appears to be independently disposed to do acceptable things: only for being someone who acts virtuously by dint of independently sourced inclinations or efforts.

There is another way of supporting the claim that intangible-hand devices ought not to undermine spontaneous virtue. Under an acceptance-centred system we should expect suitable norms of behaviour to get established: norms of behaviour of the kind that the spontaneously virtuous person will honour and hail. That the system is in place will mean that suitable cultural norms are established. And that the system is in place will support moralisation about those norms. Or so at least I argue (Pettit 1990c).

Here is a plausible definition of a cultural norm.

> A regularity, R, in the behaviour of members of a population, P, is a cultural norm if and only if, in any instance of S among members of P:
> 1. Nearly everyone conforms to R.
> 2. Nearly everyone approves of nearly anyone else's conforming and disapproves of their deviating.
> 3. The fact that nearly everyone approves and disapproves on this pattern helps to ensure that nearly everyone conforms.

This definition may require some further clauses, say one requiring common knowledge of what obtains, but it will do for our purposes here (for a defence, see Pettit 1990c). Let approval and disapproval be understood in the broadest possible sense, so that to approve of an action is to find it a ground for the greater social acceptance of an agent, to disapprove is to find it a ground for her lesser social acceptance. Now suppose that a certain pattern of behaviour is sustained in a community through an acceptance-centred system of sanctioning; many people may spontaneously and virtuously display the behaviour but the acceptance it attracts helps to support the behaviour. It turns out that in any such case the regularity of behaviour involved will constitute a cultural norm. Nearly everyone will conform, by hypothesis. Nearly everyone will approve of nearly anyone else's conforming and disapprove of their deviating, since the behaviour is of a kind to win acceptance. And this pattern of approval and disapproval, this connection with social acceptance, will help to ensure the general conformity, since it will help to support the behaviour; it will keep the non-virtuous on track and it will guard against the contingency of a spontaneously virtuous agent breaking her habits.

That a regularity is a cultural norm in this sense does not mean that it is necessarily something about which people moralise. People who moralise in the relevant sense will censure deviance and praise conformity, at least in some contexts, on grounds of the kind that weigh with the spontaneously virtuous agent: that deviance is unfair, takes from the common good, or whatever. There is no reason so far given to think that cultural norms will attract that sort of defence. We may find a norm of conscientiousness on committees, for example, without anyone waxing moralistic on the topic. But it turns out that under an acceptance-centred system of sanctioning, under the rule of the intangible hand, we can understand why such moralisation might emerge and stabilise; we can provide something like a rational choice derivation of the activity.

Suppose that there is a certain cultural norm in place in a society, such as a norm of conscientiousness on committees. In such a situation it will be in the self-interest of individuals to moralise about the norm; specifically, it will increase the social acceptance they get. If people moralise in the society, most will moralise in a spontaneous way. But the fact that moralising has this social pay-off gives us reason to think that moralising may become universal and will be resilient.

Why will moralising about a norm win acceptance in a society? Moralising about the norm is likely, by common assumption, to increase the degree of conformity to the norm. Since conformity wins acceptance, the moralising that is taken to promote it will presumably win acceptance too. There will sometimes be costs associated with this moralising that make it an unattractive option overall. But the pay-off in acceptance ought to mean that there are some situations where it will appeal to the most self-interested person. Thus we can see why any type of behaviour that is promoted by an acceptance-centred system of sanctioning, any type of behaviour that constitutes a cultural norm, ought also to assume the status of a moralised norm.

This rational choice account of norms is sketchy and may not be found persuasive. But even if the account is rejected, I hope that the point it is designed to support will not be contested. This is that if we have an acceptance-centred system of sanctioning in place to support any type of behaviour, this sort of sanctioning is unlikely to diminish or undermine spontaneous virtue. Even if it does not lead to the appearance of a moralised norm of the kind that the spontaneously virtuous agent will endorse, there is no reason to think that it will undermine the motivation of that sort of agent. On the contrary, the motivation of the spontaneously virtuous agent should be reinforced by the presence of intangible-hand sanctioning; such a form of sanctioning will provide extra incentives for that agent to remain the sort of person she is.

I have argued in this final section that our holistic individualism gives us an interesting line on feasibility analysis: on the analysis of political values for their institutional feasibility. We discussed this line with particular reference to republican values. The realisation of the republican ideal, the advancement of franchise, requires the presence of civic virtue among the citizens and officials of the society. We cannot assume that such virtue will appear naturally in all parts, even if many people do spontaneously display it. And so we must look to the design of social and political institutions to see if we can artificially elicit civic virtue. Are there any likely candidates on offer? I argued that the prospects are not entirely dim, in particular that there is some hope of exploiting people's interest in social acceptance to get them to be reliably beneficent in appropriate ways: to get them to do the decent, civic thing. There is some hope of exploiting intangible-hand institutions, as there is some hope of exploiting certain invisible-hand arrangements, to generate the virtue that republicanism requires.

This completes our investigation of the impact of holistic individualism on the project of political theory. The doctrine raises serious questions about contract-centred thinking, it introduces powerful influences on the nature of our thinking about values, and now, so we have just seen, it leaves room for institution-centred thinking too. In fact, given the way in which it leads us to interpret rational choice theory, holistic individualism directs us towards a particular mode of institution-centred thinking: it suggests a more or less novel way of approaching the task of feasibility analysis. The doctrine serves us well in the project of political evaluation, as it serves us well in thinking about social explanation.

Notes

1. The dictum is ascribed by Mill in his *Utilitarianism* (Mill 1969, p. 257) but Bentham does not use those words in the reference to his work provided by Mill's editors. I am grateful to Knud Haakonssen for his help in this connection.

2. A certain sort of subjectivism would support this commitment. I have in mind the assumption that nothing is good for a person unless it would have been equally good for her in any situation where everything was the same from her conscious point of view, including situations where there are vast, unregistered differences in the world around her: including situations, indeed, where she is a brain in a vat and is deluded about the world she inhabits. See Jackson 1990. I do not provide an argument against such solipsism here but it is worth mentioning that it does not fit well with the externalism about intentionality and thought that was defended in Chapters 1 and 4. Such solipsism may have been influential but it is highly counterintuitive: we may value the situation where someone enjoys friendship with her colleagues, while recoiling from the situation where she enjoys the delusion, even the permanent delusion, of such friendships.

3. The characterisation of Ackerman's position is difficult, because he has the neutrality condition—and other associated constraints—do double duty. Not only is the best state the state that would be supported by neutral dialogue. The best state is

the state that promotes neutral dialogue, in the sense of allowing the holders of power and other goods to justify themselves to others without offending against neutrality.

4. I have benefitted from exchanges with Fred d'Agostino on this point.

5. Scanlon (1982, p. 110) may seem to come close to endorsing the accommodation conception when he speaks of the principles to be sought in contractual thinking as principles 'which no one could reasonably reject as a basis for informed, unforced general agreement'? But the word 'reasonably' argues against that interpretation. The most straightforward reading of his proposal is that we are to look at what principles contractors would find intellectually compelling as they look for principles to organise social affairs: to organise the affairs of people who wish to live in peace and agreement.

6. This latter version is more plausible, because it satisfies a plausible condition on any right-making property: the condition that the property not refer us to particulars. See Hare 1981. The property will make one basic structure right for one community, another for a different community, but it will make each right on a common basis. See Scanlon 1982, p. 112.

7. What if it is said that the deontologist has no truck with abstract, universal values: that this is already a consequentialist way of framing things (see Foot 1985)? What if it is said that all the deontologist does is to prescribe certain actions or institutions on the grounds that they are, quite simply, right? I reply that if a deontologist holds it is right of X in circumstances C to take option F, then he must be prepared to universalise and hold that it is right of any agent or agency in C-like circumstances to take the F-like option. See Hare 1981. But that universal state of affairs—the choice of F-like options by any agents or agencies in C-like circumstances—counts then as something that the deontologist values and, more specifically, it counts as a value that he thinks that X should honour rather than promote: he does not think, presumably, that if X could better promote the realisation of that state of affairs by proselytising about its attractions, rather than by taking option F, then X should proselytise.

8. I here ignore many of the so-called modern liberals of the late nineteenth century—liberals in the T. H. Green mould—who defended a different concept of liberty. Significantly, they were greatly influenced by the continental tradition of thought, in particular the Hegelian. See Gaus 1983.

9. I am grateful to David Neal, who first drew my attention to the similarities between the notion of freedom that I had been trying to extract from holistic individualism and the republican notion of freedom, at least as that had begun to be explicated by Quentin Skinner. I am also grateful to Quentin Skinner for his encouragement and advice. If there is a difference between my reading of the republican notion of freedom and that articulated by Skinner, it is that the tie between liberty on the one side, and law and institutions on the other, is constitutive under my account, whereas Skinner sometimes suggests that it is causal: the enjoyment of liberty is seen as causally rather than constitutively requiring a decent legal and institutional life, and the civic virtue that that sort of life presupposes.

10. Quoted from the preparatory notes to volume 2 of *Democracy in America* in Aron 1968, p. 201.

11. See Tuck 1989, pp. 74–75, for a discussion. Hobbes on liberty is extremely elusive, however, as appears in the detailed discussion in Skinner 1990c.

12. Here and elsewhere Smith wants to be able to say that we desire not only to be such that others are disposed to praise us but also to be such that others are rightly disposed to praise us. I suspect that he illicitly uses the first claim to make the second, intuitively stronger claim seem plausible.

References

Abel, Theodore (1948). 'The Operation Called *Verstehen*'. *American Journal of Sociology*, Vol. 54, pp. 211–18.

Ackerman, Bruce A. (1980). *Social Justice in the Liberal State*. New Haven: Yale University Press.

Acton, Lord (1985). *Essays in the History of Liberty*, ed. J. Rufus Fears. Indianapolis: Liberty Classics.

Alchian, A. A. (1950). 'Uncertainty, Evolution and Economic Theory'. *Journal of Political Economy*, Vol. 58, pp. 211–21.

Althusser, Louis (1970). *Reading Capital*. London: New Left Books.

Anscombe, G. E. M. (1957). *Intention*. Oxford: Basil Blackwell.

Apel, Karl-Otto (1976). *Sprachpragmatik und Philosophie*. Frankfurt: Suhrkamp Verlag.

Arblaster, Anthony (1984). *The Rise and Decline of Western Liberalism*. Oxford: Basil Blackwell.

Armstrong, David (1968). A *Materialist Theory of the Mind*. London: Routledge.

Aron, Raymond (1968). *Main Currents in Sociological Thought*. Vol. 1. Harmondsworth: Penguin.

Axelrod, Robert (1984). *The Evolution of Cooperation*. New York: Basic Books.

—— (1988). 'The Further Evolution of Cooperation'. *Science*, Vol. 242, pp. 1385–90.

Ayres, Ian, and John Braithwaite (1992). *Responsive Regulation*. New York: Oxford University Press.

Baldwin, Tom (1984). 'MacCallum and the Two Concepts of Freedom'. *Ratio*, Vol. 26, pp. 125–42.

Barber, Benjamin (1984). *Strong Democracy*. Berkeley: University of California Press.

Barker, Ernest, ed. (1947). *Social Contract*. Oxford: Oxford University Press.

Barry, Brian (1965). *Political Argument*. London: Routledge.

—— (1989). A *Treatise of Social Justice*, Vol 1: *Theories of Justice*. Berkeley: University of California Press.

—— and Russell Hardin, eds. (1982). *Rational Man and Irrational Society?* Beverly Hills: Sage Publications.

Barwise, Jon, and John Perry (1983). *Situations and Attitudes*. Cambridge, Mass.: MIT Press.

Becker, Gary (1976). *The Economic Approach to Human Behavior*. Chicago: University of Chicago Press.

Benn, S. I., and R. S. Peters (1959). *Social Principles and the Democratic State*. London: Allen and Unwin.

Benn, Stanley I. (1988). A *Theory of Freedom*. Cambridge: Cambridge University Press.

Bennett, Jonathan (1976). *Linguistic Behaviour*. Cambridge: Cambridge University Press.
—— (1990). 'Why Is Belief Involuntary?' *Analysis*, Vol. 50, pp. 87–107.
Bentham, Jeremy (1843). *The Works of Jeremy Bentham*, Vol. 1. Edinburgh: William Tait.
Berlin, Isaiah (1958). *Two Concepts of Liberty*. Oxford: Oxford University Press.
—— (1976). *Vico and Herder*. London: Hogarth Press.
Bishop, John (1989). *Natural Agency: An Essay on the Causal Theory of Action*. Cambridge: Cambridge University Press.
Blackburn, Simon (1984a). *Spreading the Word*. Oxford: Oxford University Press.
—— (1984b). 'The Individual Strikes Back'. *Synthese*, Vol. 58, pp. 281–301.
—— (1991). 'Losing your Mind'. In John Greenwood, ed., *The Future of Folk Psychology*. New York: Oxford University Press pp. 196–225.
Block, Ned (1980). 'What is Functionalism?' In Ned Block, ed., *Readings in Philosophy of Psychology*, Vol. 1. London: Methuen, pp. 171–84.
—— (1981). 'Psychologism and Behaviorism'. *Philosophical Review*, Vol. 90, pp. 5–43.
—— (1990a). 'Can the Mind Change the World?' In George Boolos, ed., *Meaning and Method: Essays in Honor of Hilary Putnam*. Cambridge: Cambridge University Press, pp. 137–70.
—— (1990b). 'The Computer Model of the Mind'. In D. Osherson and E. E. Smith, eds., *An Invitation to Cognitive Science*, Vol. 3. Cambridge, Mass.: MIT Press, pp. 247–89.
Boghossian, Paul (1989). 'The Rule-Following Considerations'. *Mind*, Vol. 98, pp. 504–50.
Bradley, F. H. (1876). *Ethical Studies*. 2nd ed. (1927). London: Oxford University Press.
Braithwaite, John, and Philip Pettit (1990). *Not Just Deserts: A Republican Theory of Criminal Justice*. Oxford: Oxford University Press.
Bratman, Michael (1987). *Intention, Plans, and Practical Reason*. Cambridge, Mass.: Harvard University Press.
Brennan, Geoffrey (1987). 'The Buchanan Contribution'. *Finanzarchiv*, Vol. 45, pp. 1–24.
——, and Loren Lomasky (1984). 'Inefficient Unanimity'. *Journal of Applied Philosophy*, Vol. 1, pp. 157–63.
——, and Loren Lomasky (1992). *Democracy and Decision*. Cambridge: Cambridge University Press.
——, and Philip Pettit (1990). 'Unveiling the Vote'. *British Journal of Political Science*, Vol. 20, pp. 311–33.
——, and Philip Pettit (1991). 'Modelling and Motivating Academic Performance'. *The Australian Universities' Review*, Vol. 34, pp. 4–9.
——, and Philip Pettit (forthcoming). 'Hands Invisible and Intangible'. *Synthese*.
Broome, John (1990). 'Irreducibly Social Goods'. In G. Brennan and C. Walsh, eds., *Rationality, Individualism and Public Policy*. Canberra: ANUTECH, pp. 80–85.
—— (1991). *Weighing Goods*. Oxford: Basil Blackwell.
Brown, Robert (1984). *The Nature of Social Laws*. Cambridge: Cambridge University Press.
Bubner, Rüdiger (1981). *Modern German Philosophy*. Cambridge: Cambridge University Press.
Buchanan, James (1975). *The Limits of Liberty*. Chicago: University of Chicago Press.

—, and Gordon Tullock (1962). *The Calculus of Consent*. Ann Arbor: University of Michigan Press.

Burge, Tyler (1979). 'Individualism and the Mental'. *Midwest Studies in Philosophy*, Vol. 4, pp. 73–121.

Campbell, John (forthcoming). 'A Simple View of Colour'. In John Haldane and Crispin Wright, eds., *Reality, Representation and Projection*. Oxford: Oxford University Press.

Candlish, Stewart (1978). 'Bradley on My Station and Its Duties'. *Australasian Journal of Philosophy*, Vol. 56, pp. 155–70.

Cherniak, Christopher (1986). *Minimal Rationality*. Cambridge, Mass.: MIT Press.

Churchland, Patricia (1986). *Neurophilosophy*. Cambridge, Mass.: MIT Press.

Churchland, Paul (1979). *Scientific Realism and The Plasticity of Mind*. Cambridge: Cambridge University Press.

Cohen, G. A. (1979). *Karl Marx's Theory of History*. Oxford: Oxford University Press.

Cohen, Joshua, and Joel Rogers (1983). *On Democracy*. Harmondsworth: Penguin.

Coleman, James (1990). *Foundations of Social Theory*. Cambridge, Mass.: Harvard University Press.

Coleman, James S. (1974). *Power and the Structure of Society*. New York: W. W. Norton.

Collins, John (1988). 'Belief, Desire and Revision'. *Mind*, Vol. 97, pp. 333–42.

Craig, Edward (1982). 'Meaning, Use and Privacy'. *Mind*, Vol. 91, pp. 341–64.

Crane, Tim, and D. H. Mellor (1990). 'There is No Question of Physicalism'. *Mind*, Vol. 99, pp. 185–206.

Cranston, Maurice (1967). *Freedom: A New Analysis*. 3rd ed. London: Longmans.

Cummins, Robert (1983). *The Nature of Psychological Explanation*. Cambridge, Mass.: MIT Press.

—— (1989). *Meaning and Mental Representation*. Cambridge, Mass.: MIT Press.

Currie, Gregory (1984). 'Individualism and Global Supervenience'. *British Journal for the Philosophy of Science*, Vol. 35, pp. 345–58.

Dallmayr, Fred R., and T. A. McCarthy, eds. (1977). *Understanding and Social Inquiry*. Notre Dame: University of Notre Dame Press.

Dancy, Jonathan (1985). *An Introduction to Contemporary Epistemology*. Oxford: Basil Blackwell.

Danto, Arthur (1973). *Analytical Philosophy of Action*. Cambridge: Cambridge University Press.

Davidson, Donald (1980). *Essays on Actions and Events*. Oxford: Oxford University Press.

—— (1984). *Inquiries into Truth and Interpretation*. Oxford: Oxford University Press.

—— (1985). 'Rational Animals'. In Ernest Le Pore and Brian McLaughlin, eds., *Action and Events*. Oxford: Basil Blackwell, pp. 473–80.

Davies, Martin (1990). 'Thinking Persons and Cognitive Science'. *AI and Society*, Vol. 4, pp. 39–50.

—, and Lloyd Humbertstone (1981). 'Two Notions of Necessity'. *Philosophical Studies*, Vol. 48, pp. 1–30.

Dawkins, Richard (1976). *The Selfish Gene*. London: Granada Publishing.

—— (1986). *The Blind Watchmaker*. Harlow: Longman Scientific & Technical.

Dennett, Daniel (1979). *Brainstorms*. Brighton: Harvester Press.

—— (1987). *The Intentional Stance*. Cambridge, Mass.: MIT Press.

Devitt, Michael (1984). *Realism and Truth*. Oxford: Basil Blackwell.

Dretske, Fred (1988). *Explaining Behavior.* Cambridge, Mass.: MIT Press.
Dumont, Louis (1986). *Essays on Individualism.* Chicago: University of Chicago Press.
Durkheim, Émile (1933). *The Division of Labour in Society.* New York: Macmillan.
—— (1938). *The Rules of Sociological Method.* New York: Free Press.
—— (1951). *Suicide.* New York: Free Press of Glencoe.
Dworkin, Ronald (1978). *Taking Rights Seriously.* London: Duckworth.
Eells, Ellery (1982). *Rational Decision and Causality.* Cambridge: Cambridge University Press.
Elster, Jon (1976). 'A Note on Hysteresis in the Social Sciences'. *Synthese,* Vol. 33, pp. 371–91.
—— (1979). *Ulysses and the Sirens.* Cambridge: Cambridge University Press.
—— (1983a). *Explaining Technical Change.* Cambridge: Cambridge University Press.
—— (1983b). *Sour Grapes.* Cambridge: Cambridge University Press.
—— (1985). *Making Sense of Marx.* Cambridge: Cambridge University Press.
—— (1986a). 'The Market and the Forum: Three Varieties of Political Theory'. In J. Elster and A. Hilland, eds., *Foundations of Social Choice Theory.* Cambridge: Cambridge University Press, pp. 103–28.
—— ed. (1986b). *Rational Choice.* Oxford: Basil Blackwell.
—— (1989). *Nuts and Bolts for the Social Sciences.* Cambridge: Cambridge University Press.
Evans, Gareth (1982). *The Varieties of Reference.* Oxford: Oxford University Press.
Finnis, John (1980). *Natural Law and Natural Rights.* Oxford: Oxford University Press.
Fodor, Jerry (1975). *The Language of Thought.* Cambridge, Mass.: Harvard University Press.
—— (1985). 'Fodor's Guide to Mental Representation'. *Mind,* Vol. 94, pp. 76–100.
—— (1987). *Psychosemantics.* Cambridge, Mass.: MIT Press.
—— (1990). *A Theory of Content.* Cambridge, Mass.: MIT Press.
Fogel, Robert William, and Stanley L. Engerman (1974). *Time on the Cross: The Economics of American Negro Slavery.* Boston: Little, Brown.
Fogelin, Robert (1987). *Wittgenstein.* London: Routledge.
Foot, Philippa (1985). 'Utilitarianism and the Virtues'. *Mind,* pp. 196–209.
Frank, Rober (1985). *Choosing the Right Pond.* New York: Oxford University Press.
French, Peter A. (1984). *Collective and Corporate Responsibility.* New York: Columbia University Press.
Friedman, Debra, and Michael Hechter (1988). 'The Contribution of Rational Choice Theory to Macrosociological Research'. *Sociological Theory,* Vol. 6, pp. 201–18.
Garfield, Jay L. (1988). *Belief in Psychology.* Cambridge, Mass.: MIT Press.
Gatens, Moira (1991). *Feminism and Philosophy.* Cambridge: Polity Press.
Gaus, G. F. (1983). *The Modern Liberal Theory of Man.* London: Croom Helm.
—— (1990). *Value and Justification.* Cambridge: Cambridge University Press.
Gauthier, David (1986). *Morals by Agreement.* Oxford: Oxford University Press.
Geach, Peter (1957). *Mental Acts.* London: Routledge.
Giddens, Anthony (1976). *New Rules of Sociological Method.* London: Hutchinson.
—— (1984). *The Constitution of Society.* Cambridge: Polity Press.
Godfrey-Smith, Peter (1991). 'Signal, Decision, Action'. *Journal of Philosophy,* Vol. 88, pp. 709–22.
Goffman, Erving (1975). *Frame Analysis: An Essay on the Organization of Experience.* Harmondsworth: Penguin.

Goldfarb, Warren (1985). 'Kripke on Wittgenstein on Rules'. *Journal of Philosophy*, Vol. 82, pp. 471–88.

Goodin, Robert (1992). *Motivating Political Morality*. Oxford: Basil Blackwell.

Goodman, Nelson (1968). *Languages of Art*. London: Oxford University Press.

—— (1978). *Ways of Worldmaking*. Sussex: Harvester Press.

Gouldner, A. W. (1969). *The Coming Crisis in Western Sociology*. London: Heineman.

Guarini, E. F. (1990). 'Machiavelli and the Crisis of the Italian Republics'. In G. Bock, Q. Skinner, and M. Viroli, eds., *Machiavelli and Republicanism*. Cambridge: Cambridge University Press, pp. 17–40.

Habermas, Jürgen (1973). 'Wahrheitstheorien'. In H. Fahrenbach, ed., *Wirchlichkeit und Reflexion: Walter Schulz zum 60 Geburtstag*. Pfullingen: Neske, pp. 211–65.

—— (1984, 89). *A Theory of Communicative Action*. Vols. 1 and 2. Cambridge: Polity Press.

Hacking, Ian (1980). 'Grounding Probabilities from Below'. *Proceedings of the Philosophy of Science Association*, Vol. 1, pp. 110–16.

—— (1981). 'The Taming of Chance by an Army of Statistics'. *The Times Higher Education Supplement*, 30.10.81, pp. 12–13.

—— (1990). *The Taming of Chance*. Cambridge: Cambridge University Press.

Hamlin, Alan (1989). 'Liberty, Contract and the State'. In A. Hamlin and P. Pettit, eds., *The Good Polity: Normative Analysis of the State*. Oxford: Basil Blackwell, pp. 87–101.

——, and Philip Pettit (1989). 'The Normative Analysis of the State: Some Preliminaries'. In A. Hamlin and P. Pettit, eds., *The Good Polity: Normative Analysis of the State*. Oxford: Basil Blackwell, pp. 1–13.

Hampton, Jean (1986). *Hobbes and the Social Contract Tradition*. Cambridge: Cambridge University Press.

Hardin, C. L. (1988). *Color for Philosophers*. Indianapolis: Hackett.

Hardin, Russell (1982). *Collective Action*. Baltimore: Johns Hopkins University Press.

Hare, R. M. (1981). *Moral Thinking: Its Levels, Method and Point*. Oxford: Oxford University Press.

Harman, Gilbert (1986). *Change in View*. Cambridge, Mass.: M.I.T. Press.

—— (1991). 'Justification, Truth, Goals, and Pragmatism', *Philosophy and Phenomenological Research*, Vol. 51, pp. 195–200.

Harrington, James (1977). *The Political Works of James Harrington*, ed. J. G. A. Pocock. Cambridge: Cambridge University Press.

Harsanyi, John (1969). 'Rational Choice Models of Behavior versus Functionalist and Conformist Theories'. *World Politics*, Vol. 22, pp. 513–38.

Hayek, F. A. (1960). *The Constitution of Liberty*. London: Routledge.

—— (1976). *Law, Legislation and Liberty*. Vol. 2. London: Routledge.

—— (1978). *New Studies in Philosophy, Politics, Economics and the History of Ideas*. London: Routledge.

Hilbert, David R. (1987). *Color and Color Perception*. Stanford, Calif.: Center for the Study of Language and Information.

Hill, Christopher (1972). *The World Turned Upside Down*. London: Temple Smith.

Hindess, Barry (1987). *Politics and Class Analysis*. Oxford: Basil Blackwell.

—— (1988). *Choice, Rationality, and Social Theory*. London: Unwin Hyman.

Hobbes, Thomas (1968). *Leviathan*, ed. C. B. MacPherson. Harmondsworth: Penguin.

Hobsbawm, E. J. (1990). *Nations and Nationalism since 1780: Programme, Myth, Reality*. Cambridge: Cambridge University Press.

Hollis, Martin (1987). *The Cunning of Reason*. Cambridge: Cambridge University Press.

Holmes, Stephen (1989). 'The Permanent Structure of Antiliberal Thought'. In N. L. Rosenblum, ed., *Liberalism and the Moral Life*. Cambridge, Mass.: Harvard University Press, pp. 227–53.

—— (1990). 'The Secret History of Self-Interest'. In J. J. Mansbridge, ed., *Beyond Self-Interest*. Chicago: University of Chicago Press, pp. 267–86.

Holton, Richard (1992). 'Response-dependence and Infallibility'. *Analysis*, Vol. 52.

Hookway, Christopher (1978). 'Indeterminacy and Interpretation'. In C. Hookway and P. Pettit, eds., *Action and Interpretation: Studies in the Philosophy of the Social Sciences*. Cambridge: Cambridge University Press, pp. 17–42.

Horgan, Terence, ed. (1984). *Spindel Conference 1983: Supervenience*. Supplement to Southern Journal of Philosophy, Vol. 22.

Hornsby, Jennifer (1980). *Actions*. London: Routledge.

Humphrey, Nicholas (1986). *The Inner Eye*. London: Faber.

Hurley, Susan (1989). *Natural Reasons*. New York: Oxford University Press.

Jackson, Frank (1977). 'A Causal Theory of Counterfactuals'. *Australasian Journal of Philosophy*, Vol. 55, pp. 3–21.

—— (1984). 'Internal Conflicts in Desires and Morals'. *American Philosophical Quarterly*, Vol. 22, pp. 105–14.

—— (1990). 'Irreducibly Social Goods: Comment'. In Geoffrey Brennan and Cliff Walsh, eds., *Rationality, Individualism and Public Policy*. Canberra: Federalism Research Centre, Australian National University, pp. 86–90.

—— (1992). 'Block's Challenge'. In K. Campbell, J. Bacon, and L. Rhinehart, eds., *Ontology, Causality, and Mind: Essays on the Philosophy of David Armstrong*. Cambridge: Cambridge University Press.

——, and Philip Pettit (1988). 'Functionalism and Broad Content'. *Mind*, Vol. 97, pp. 381–400.

——, and Philip Pettit (1990a). 'In Defence of Folk Psychology'. *Philosophical Studies*, Vol. 57, pp. 7–30.

——, and Philip Pettit (1990b). 'Program Explanation: A General Perspective'. *Analysis*, Vol. 50, pp. 107–17.

——, and Philip Pettit (1990c). 'Causation in the Philosophy of Mind'. *Philosophy and Phenomenological Research*, Vol. 50, pp. 195–214.

——, and Philip Pettit (1992 a). 'Some Content is Narrow'. In J. Heil and A. Meile, eds., *Mental Causation*. Oxford: Oxford University Press.

——, and Philip Pettit (1992 b). 'In Defence of Explanatory Ecumenism'. *Economics and Philosophy*, Vol. 8, pp. 1–21.

——, and Philip Pettit (1992c). 'Structural Explanation and Social Theory'. In David Charles and Kathleen Lennon, eds., *Reductionism and Anti-reductionism*. Oxford: Oxford University Press.

James, Susan (1984). *The Content of Social Explanation*. Cambridge: Cambridge University Press.

Jeffrey, Richard C. (1983). *The Logic of Decision*. 2nd ed. Chicago: University of Chicago Press.

Johnston, Mark (1989). 'Dispositional Theories of Value'. *Proceedings of the Aristotelian Society*, Supp. Vol. 63, pp. 139–74.

—— (forthcoming). 'Objectivity Refigured: Pragmatism with Verificationism'. In John Haldane and Crispin Wright, eds., *Reality, Representation and Projection*. Oxford: Oxford University Press.

Kim, Jaegwon (1974). Non-causal Connections. *Nous*, Vol. 8, pp. 41–52.

—— (1984). 'Epiphenomenal and Supervenient Causation'. *Midwest Studies in Philosophy*, Vol. 9, pp. 257–70.

—— (1990). 'Supervenience as a Philosophical Concept'. *Metaphilosophy*, Vol. 21, pp. 1–27.

Kiser, E., and M. Hechter (forthcoming). 'The Role of General Theory in Comparative-Historical Sociology'. *American Journal of Sociology*.

Kripke, Saul A. (1980). *Naming and Necessity*. Rev. ed. Oxford: Basil Blackwell.

—— (1982). *Wittgenstein on Rules and Private Language*. Oxford: Basil Blackwell.

Kuhn, Thomas (1970). *The Structure of Scientific Revolutions*. 2nd ed. enlarged. Chicago: University of Chicago Press.

Kukathas, Chandran, and Philip Pettit (1990). *Rawls: A Theory of Justice and Its Critics*. Cambridge: Polity Press, and Stanford, Calif.: Stanford University Press.

Kymlicka, Will (1988). 'Rawls on Teleology and Deontology'. *Philosophy and Public Affairs*, Vol. 17, pp. 173–90.

—— (1990). *Contemporary Political Philosophy*. Cambridge: Cambridge University Press.

—— (1991). 'The Social Contract Tradition'. In Peter Singer, ed., *A Companion to Ethics*. Oxford: Basil Blackwell, pp. 186–96.

Larmore, Charles (1987). *Patterns of Moral Complexity*. New York: Cambridge University Press.

Lessnoff, Michael (1986). *Social Contract*. London: Macmillan.

Lévi-Strauss, Claude (1971). *L'Homme Nu*. Paris: Plon.

Lewis, C. S. (1967). *Studies in Words*. Cambridge: Cambridge University Press.

—— (1983). *That Hideous Strength*. London: Pan Books.

Lewis, David (1969). *Convention*. Cambridge, Mass.: Harvard University Press.

—— (1973). *Counterfactuals*. Oxford: Basil Blackwell.

—— (1983a). 'New Work for a Theory of Universals'. *Australasian Journal of Philosophy*, Vol. 61, pp. 343–77.

—— (1983b). *Philosophical Papers*. Vol. 1. Oxford: Oxford University Press.

—— (1986). *Philosophical Papers*. Vol. 2. Oxford: Oxford University Press.

—— (1988). 'Desire as Belief'. *Mind*, Vol. 97, pp. 323–32.

Locke, John (1960). *Two Treatises of Government*, ed. P. Laslett. Cambridge: Cambridge University Press.

Lovejoy, Arthur O. (1961). *Reflections on Human Nature*. Baltimore: Johns Hopkins University Press.

Luce, R. D., and Howard Raiffa (1957). *Games and Decisions*. New York: John Wiley.

Lukes, Steven (1973). *Émile Durkheim: His Life and Work: A Historical and Critical Study*. Harmondsworth: Penguin.

Lyons, David (1982). 'Utility and Rights'. *Nomos*, Vol. 24, pp. 107–38.

MacCallum, Gerald C., Jr. (1967). 'Negative and Positive Freedom'. *Philosophical Review*, Vol. 76, pp. 312–14.

Macdonald, Cynthia, and Graham Macdonald (1986). 'Mental Causes and Explanation of Action'. *Philosophical Quarterly*, Vol. 36, pp. 145–58.

Macdonald, Graham, and Philip Pettit (1981). *Semantics and Social Science*. London: Routledge & Kegan Paul.

MacIntyre, Alasdair (1988). *Whose Justice? Which Rationality?* London: Duckworth.

Mackie, J. L. (1977). *Ethics*. Harmondsworth: Penguin.

Maitland, F. W. (1900). Introduction to *Political Theories of the Middle Age* by Otto Gierke. Cambridge: Cambridge Univesity Press.

Mansbridge, Jane (1990). *Beyond Self-Interest*. Chicago: University of Chicago Press.

McCullagh, C. Behan (1991). 'How Objective Interests Explain Action'. *Social Science Information*, Vol. 30, pp. 29–54.

McDowell, John (1980). 'Meaning, Communication, and Knowledge'. In Z. van Straaten, ed., *Philosophical Subjects*. Oxford: Oxford University Press, pp. 117–39.

—— (1981). 'Anti-realism and the Epistemology of Understanding'. In H. Parret and J. Bouveresse, eds., *Meaning and Understanding*. Berlin: de Gruyter, pp. 225–48.

—— (1983). 'Aesthetic Value, Objectivity and the Fabric of the World'. In E. Schaper, ed., *Pleasure, Preference and Value*. Cambridge: Cambridge Univesity Press, pp. 1–16.

—— (1987). 'In Defence of Modesty'. In B.Taylor, ed., *Michael Dummett: Contributions to Philosophy*. Dordrecht: Reidel, pp. 59–80.

——, and Philip Pettit (1986). 'Introduction'. In Pettit and McDowell, eds., *Subject, Thought and Context*. Oxford: Oxford University Press, pp. 1–15.

McGinn, Colin (1982). *The Character of Mind*. Oxford: Oxford University Press.

—— (1984). *Wittgenstein on Meaning*. Oxford: Basil Blackwell.

—— (1989). *Mental Content*. Oxford: Basil Blackwell.

Mellor, D. H. (1977). 'Conscious Belief'. *Proceedings of the Aristotelian Society*, Vol. 78, pp. 87–101.

—— (1982). 'The Reduction of Society'. *Philosophy*, Vol. 57, pp. 51–75.

Menzies, Peter (1987). 'Against Causal Reductionism'. *Mind*, Vol. 97, pp. 551–74.

Milgram, Stanley (1974). *Obedience to Authority*. New York: Harper & Row.

Mill, John Stuart (1969). *Essays on Ethics, Religion and Society* (Collected Works, Vol. 10). London: Routledge.

—— (1972). *On Liberty*. London: Dent.

Miller, David (1990). 'Equality'. In K. Hunt, ed., *Philosophy and Politics*. Cambridge: Cambridge University Press, pp. 98–108.

Miller, Richard (1978). 'Methodological Individualism and Social Explanation'. *Philosophy of Science*, Vol. 45, pp. 387–414.

Millikan, Ruth (1984). *Language, Thought and Other Biological Categories*. Cambridge, Mass.: MIT Press.

Montesquieu, Baron de (1965). *Considerations on the Causes of the Greatness of the Romans and Their Decline*. G. D. Lowenthal, ed. Ithaca: Cornell University Press.

—— (1977). *The Spirit of Laws*, ed. D. W.Carrithers. Berkeley: University of California Press.

Morton, Adam (1990). *Disasters and Dilemmas*. Oxford: Basil Blackwell.

Nagel, Ernest (1961). *The Structure of Science: Problems in the Logic of Scientific Explanation*. London: Routledge & Kegan Paul.

Nagel, Thomas (1986). *The View from Nowhere*. Oxford: Oxford University Press.

Nelson, R., and S. Winter (1982). *An Evolutionary Theory of Economic Change*. Cambridge, Mass.: Harvard University Press.

Nietzsche, Friedrich (1956). *The Birth of Tragedy* and *The Genealogy of Morals*. Trans. F. Golffing, New York: Doubleday.

North, Douglass (1981). *Structure and Change in Economic History*. New York: Norton.

Nozick, Robert (1974). *Anarchy, State, and Utopia*. Oxford: Basil Blackwell.

—— (1981). *Philosophical Explanations*. Oxford: Oxford University Press.

O'Neill, John, ed. (1973). *Modes of Individualism and Collectivism*. London: Heinemann.

Oldfield, Adrian (1990). *Citizenship and Community: Civic Republicanism and the Modern World*. London: Routledge.

Olson, Mancur (1965). *The Logic of Collective Action*. Cambridge, Mass.: Harvard University Press.

Ostrom, Elinor (1991). 'Rational Choice Theory and Institutional Analysis: Toward Complementarity'. *American Political Science Review*. Vol. 85, pp. 237–43.

Papineau, David (1987). *Reality and Representation*. Oxford: Basil Blackwell.

Parfit, Derek (1984). *Reasons and Persons*. Oxford: Oxford University Press.

Patterson, Orlando (1991). *Freedom in the Making of Western Culture*. New York: Basic Books.

Peacocke, Christopher (1983). *Sense and Content*. Oxford: Oxford University Press.

—— (1988). 'The Limits of Intelligibility: A Post-Verificationist Proposal'. *Philosophical Review*, Vol. 97, pp. 463–94.

—— (1989). 'What are Concepts?'. *Midwest Studies in Philosophy*, Vol. 14, pp. 1–28.

—— (1990). 'Contents and Norms in a Natural World'. In E. Villaneuva, ed., *Information, Semantics and Epistemology*. Oxford: Basil Blackwell.

—— (1992). A *Study of Concepts*. Cambridge, Mass.: MIT Press.

Pettit, Philip (1980). *Judging Justice*. London: Routledge.

—— (1982). 'Habermas on Truth and Justice'. In G. H. R. Parkinson, ed., *Marx and Marxisms*. Cambridge: Cambridge University Press, pp. 207–28.

—— (1983). 'Wittgenstein, Individualism and the Mental'. In *Epistemology and the Philosophy of Science: Proceedings of the Seventh International Symposium*, pp. 446–55. Vienna: Holder-Pichler-Tempsky.

—— (1984). 'In Defence of "A New Methodological Individualism"'. *Ratio*, Vol. 26, pp. 81–87.

—— (1985). 'The Prisoner's Dilemma and Social Theory'. *Politics*, Vol. 20, pp. 1–11.

—— (1986a). 'Broad-minded Explanation and Psychology'. In P. Pettit and J. McDowell, eds., *Subject, Thought and Context*. Oxford: Oxford University Press, pp. 17–58.

—— (1986b). 'Free Riding and Foul Dealing'. *Journal of Philosophy*, Vol. 83, pp. 261–79.

—— (1986c). 'A Priori Principles and Action Explanation'. *Analysis*, Vol. 46, pp. 39–45.

—— (1987a). 'Humeans, anti-Humeans and motivation'. *Mind*, Vol. 96, pp. 530–33.

—— (1987b). 'Inference and Information'. *Brain & Behavioural Science*, Vol. 10, pp. 727–29.

—— (1987c). 'Universalizability without Utilitarianism'. *Mind*, Vol. 96, pp. 74–82.

—— (1987d). 'Rights, Constraints and Trumps'. *Analysis*, Vol. 47, pp. 8–14.

—— (1988a). 'The Paradox of Loyalty'. *American Philosophical Quarterly*, Vol. 25, pp. 163–71.

—— (1988b). 'The Consequentialist Can Recognise Rights'. *Philosophical Quarterly*, Vol. 35, pp. 537–51.

—— (1989a). 'The Freedom of the City: A Republican Ideal'. In A. Hamlin and P. Pettit, eds., *The Good Polity*. Oxford: Basil Blackwell, pp. 141–68.

—— (1989b). 'A Definition of Negative Liberty'. *Ratio*, Vol. n.s.2, pp. 153–68.

—— (1989c). 'Consequentialism and Respect for Persons'. *Ethics*, Vol. 99, pp. 116–26.

—— (1989d). 'Foul Dealing and an Assurance Problem'. *Australasian Journal of Philosophy*, Vol. 67, pp. 341–44.

—— (1990a). 'The Reality of Rule-Following'. *Mind*, Vol. 99, pp. 1–21.

—— (1990b). 'Affirming the Reality of Rule-Following'. *Mind*, Vol. 99, pp. 433–39.

—— (1990c). '*Virtus Normativa*: Rational Choice Perspectives'. *Ethics*, Vol. 100, pp. 725–55.

—— ed. (1990d). *Contemporary Political Theory*. New York: Macmillan.

—— (1991a). 'Decision Theory and Folk Psychology'. In Michael Bacharach and Susan Hurley, eds., *Essays in the Foundations of Decision Theory*. Oxford: Basil Blackwell, pp. 147–75.

—— (1991b). 'Realism and Response-dependence'. *Mind*, Vol. 100, pp. 587–626.

—— (1991c). 'Consequentialism'. In Peter Singer, ed., *A Companion to Ethics*. Oxford: Basil Blackwell, pp. 230–40.

—— (1991d). 'Liberty in the Republic'. In G. Oddie and R. Perrett, eds., *Justice, Ethics and New Zealand Society*. Auckland: Oxford University Press, pp. 171–91.

—— (1992). 'Naturalism'. *Proceedings of the Aristotelian Society*, Supp. Vol. 66, pp. 245–66.

—— (1993). 'Liberal-Communitarian: MacIntyre's Mesmeric Dichotomy. In J. Horton and S. Mardys, eds., *After MacIntyre*. Cambridge: Polity Press.

—— (forthcoming a). 'Social Holism without Collectivism'. In Edna Ullmann-Margalit, ed., *The Israel Colloquium: Studies in the History, Philosophy and Sociology of Science*. Vol. 5. Dordrecht: Reidel.

—— (forthcoming b). 'Negative Liberty, Liberal and Republican'. Mimeo, Australian National University.

—— (forthcoming c). 'Towards Interpretation.' *Philosophia*.

——, and Huw Price (1989). 'Bare Functional Desire'. *Analysis*, Vol. 49, pp. 162–69.

——, and Michael Smith (1990). 'Backgrounding Desire'. *Philosophical Review*, Vol. 99, pp. 565–92.

——, and Michael Smith (1992). 'Parfit's P'. In J. Dancy, ed., *Reading Parfit*. Oxford: Basil Blackwell.

——, and Michael Smith (forthcoming). 'Practical Unreason'. *Mind*.

——, and Robert Sugden (1989). 'The Backward Induction Paradox'. *Journal of Philosophy*, Vol. 86, pp. 169–82.

Plamenatz, John (1960). 'The Use of Political Theory'. *Political Studies*, Vol. 8, pp. 37–47.

Platts, Mark (1979). *Ways of Meaning*. London: Routledge.

Plekhanov, Georgi (1976). 'On the Question of the Individual's Role in History'. In *Selected Philosophical Works*, Vol. 2, pp. 283–315. Moscow: Progress Publishers.

Pocock, John (1975). *The Machiavellian Moment: Florentine Political Theory and the Atlantic Republican Tradition*. Princeton: Princeton University Press.

Popper, K. R. (1960). *The Poverty of Historicism*. 2nd ed. London: Routledge & Kegan Paul.

Price, Huw (1983). "Does 'Probably' Modify Sense?". *Australasian Journal of Philosophy*, Vol. 61, pp. 396–408.

—— (1988). *Facts and the Function of Truth*. Oxford: Basil Blackwell.

—— (1989). 'Defending Desire-as-belief'. *Mind*, Vol. 98, pp. 119–27.

Prigogine, Ilya, and Isabelle Stengers (1984). *Order out of Chaos*. Toronto: Bantam Books.

Putnam, Hilary (1975). *Mind, Language and Reality, Philosophical Papers*. Vol. 2. Cambridge: Cambridge University Press.

—— (1978). *Meaning and the Moral Sciences*. London: Routledge.

—— (1982). *Reason, Truth and History*. Cambridge: Cambridge University Press.

Quine, W. V. O. 1974. *The Roots of Reference*. La Salle: Open Court Publishers.

—— (1970). *Word and Object*. Cambridge, Mass.: MIT Press.

Rabinowicz, Wolodzimierz (1979). *Universalizability: A Study in Morals and Metaphysics*. Dordrecht: Reidel.

Ramsey, W., S. Stich, and J. Garon (1990). 'Connectionism, Eliminativism and Folk Psychology'. *Philosophical Perspectives*, Vol. 4, pp. 499–533.

Rawls, John (1971). *A Theory of Justice*. Oxford: Oxford University Press.

—— (1978). 'The Basic Structure as Subject'. In A. Goldman and J. Kim, eds., *Values and Morals*, pp. 47–71. Boston: Reidel.

Raz, Joseph (1986). *The Morality of Freedom*. Oxford: Oxford University Press.

Resnik, Michael D. (1987). *Choices: An Introduction to Decision Theory*. Minneapolis: University of Minnesota Press.

Ricoeur, Paul (1966). *Freedom and Nature: The Voluntary and the Involuntary*. Evanston, Ill.: Northwestern University Press.

Rorty, Richard (1980). *Philosophy and the Mirror of Nature*. Oxford: Basil Blackwell.

Rose, Margaret (1991). *The Post-Modern and the Post-Industrial*. Cambridge: Cambridge University Press.

Rosenberg, Alexander (1988). *The Philosophy of Social Science*. Oxford: Oxford University Press.

Rousseau, Jean-Jacques (1973). *The Social Contract and Discourses*. London: J. M. Dent.

Ruben, David-Hillel (1985). *The Metaphysics of the Social World*. London: Routledge & Kegan Paul.

Rudder Baker, Lynne (1987). *Saving Belief*. Princeton: Princeton University Press.

Runciman, W. G. (1972). *Relative Deprivation and Social Justice*. Harmondsworth: Penguin.

Ryan, Alan (1970). *The Philosophy of the Social Sciences*. London: Macmillan.

Salmon, Nathan U. (1982). *Reference and Essence*. Oxford: Basil Blackwell.

——, and S. Soames, eds. *Propositional Attitudes*. Oxford: Oxford University Press.

Sandel, Michael J. (1982). *Liberalism and the Limits of Justice*. Cambridge: Cambridge University Press.

—— (1988). 'The Political Theory of the Procedural Republic'. *Revue de Metaphysique et de Marale*, Vol. 93, pp. 57–68.

Sartre, Jean-Paul (1963). *The Problem of Method*. London: Methuen.

Scanlon, T. M. (1982). 'Contractualism and Utilitarianism'. In A. Sen and B. Williams, eds., *Utilitariansism and Beyond*. Cambridge: Cambridge University Press, pp. 103–28.

Schelling, T. C. (1969). 'Models of Segregation'. *American Economic Review*, Vol. 59, pp. 488–93.

Schick, Frederic (1984). *Having Reasons*. Princeton: Princeton University Press.

—— (1991). *Understanding Action*. Cambridge: Cambridge University Press.

Schmidtz, David (1988). 'Pettit's "Free Riding and Foul Dealing"'. *Australasian Journal of Philosophy*, Vol. 66, pp. 230–33.

Searle, John R. (1983). *Intentionality*. Cambridge: Cambridge University Press.

Sen, Amartya (1970). *Collective Choice and Social Welfare*. Edinburgh: Oliver and Boyd.

Shklar, Judith (1989). 'The Liberalism of Fear'. In N. L. Rosenblum, ed., *Liberalism and the Moral Life*. Cambridge, Mass.: Harvard University Press, pp. 21–38.

Skinner, Quentin (1983). 'Machiavelli on the Maintenance of Liberty'. *Politics*, Vol. 18, pp. 3–15.

——— (1984). 'The Idea of Negative Liberty'. In R. Rorty, J. B. Schneewind, and Q. Skinner, eds., *Philosophy in History*. Cambridge: Cambridge University Press, pp. 193–221.

——— (1989). 'The State'. In T. Ball, J. Farr, and R. Hansen, eds., *Political Innovation and Conceptual Change*. Cambridge: Cambridge University Press, pp. 90–131.

——— (1990a). 'Pre-humanist Origins of Republican Ideas'. In G. Bock, Q. Skinner and M.Viroli, eds., *Machiavelli and Republicanism*. Cambridge: Cambridge University Press, pp. 121–42.

——— (1990b). 'The Republican Ideal of Political Liberty'. In G. Bock, Q. Skinner, and M. Viroli, eds., *Machiavelli and Republicanism*. Cambridge: Cambridge University Press, pp. 293–309.

——— (1990c). 'Thomas Hobbes on the Proper Signification of Liberty'. *Transactions of the Royal Historical Society*, Vol. 40, pp. 121–51.

Skyrms, Brian (1980). *Causal Necessity*. New Haven: Yale University Press.

Slote, Michael (1989). *Beyond Optimizing: A Study of Rational Choice*. Cambridge, Mass.: Harvard University Press.

Smart, J. J. C. (1959). 'Sensations and Brain Processes'. *Philosophical Review*, Vol. 68, pp. 141–56.

——— (1963). *Philosophy and Scientific Realism*. London: Routledge.

——— (1987). *Essays Metaphysical and Moral*. Oxford: Basil Blackwell.

Smith, Adam (1982). *The Theory of the Moral Sentiments*, ed. D. D. Raphael and A. L. Macfie. Indianapolis: Liberty Classics.

Smith, John Maynard (1986). *The Problems of Biology*. Oxford: Oxford University Press.

Smith, Michael (1987). 'The Humean Theory of Motivation'. *Mind*, Vol. 96, pp. 36–61.

——— (1988). 'On Humeans, Anti-Humeans and Motivation: A Reply to Pettit'. *Mind*, Vol. 97, pp. 589–95.

——— (1992). 'Valuing: Desiring or Believing?' In David Charles and Kathleen Lennon, eds., *Reductionism and Anti-reductionism*. Oxford: Oxford University Press.

Sober, Elliott (1984). *The Nature of Selection*. Cambridge, Mass.: MIT Press.

——— (1985). 'Methodological Behaviorism, Evolution, and Game Theory'. In James H. Fetzer, ed., *Sociobiology and Epistemology*. Dordrecht: Reidel, pp. 181–200.

Sperber, Dan, and Deirdre Wilson (1986). *Relevance: Communication and Cognition*. Oxford: Basil Blackwell.

Sosa, Ernest (1980). 'Varieties of Causation'. *Grazor Philosophische Studies*, Vol. 11, pp. 93–103.

Stalnaker, Robert C. (1984). *Inquiry*. Cambridge, Mass.: MIT Press.

Sterelny, Kim (1990). *The Representational Theory of Mind*. Oxford: Basil Blackwell.

———, and Philip Kitcher (1988). 'The Return of the Gene'. *Journal of Philosophy*, Vol. 85, pp. 339–61.

Stich, Stephen (1983). *From Folk Psychology to Cognitive Science*. Cambridge, Mass.: MIT Press.

——— (1990). *The Fragmentation of Reason*. Cambridge, Mass.: MIT Press.

Stoljar, S. J. (1973). *Groups and Entities: An Inquiry into Corporate Theory*. Canberra: Australian National University Press.

Sugden, Robert (1986). *The Economics of Rights, Cooperation and Welfare*. Oxford: Basil Blackwell.

——— (1990). 'Contractarianism and Norms'. *Ethics*, Vol. 100, pp. 768–86.

Summerfield, D. M. (1990). 'On taking the Rabbit of Rule-Following Out of the Hat of

Representation: A Response to Pettit's "The Reality of Rule-Following"'. *Mind*, Vol. 99.

Sunstein, Cass R. (1988). 'Beyond the Republican Revival'. *Yale Law Journal*, Vol. 97, pp. 1539-90.

—— ed. (1989). *Symposium on Feminism and Political Theory*. *Ethics*. Special Issue. Chicago: University of Chicago Press.

Taylor, Charles (1975). *Hegel*. Cambridge: Cambridge University Press.

—— (1985). *Philosophy and the Human Sciences: Philosophical Papers 2*. Cambridge: Cambridge University Press.

—— (1989). *Sources of the Self: The Making of the Modern Identity*. Cambridge: Cambridge University Press.

Taylor, Michael (1976). *Anarchy and Cooperation*. London: Wiley.

—— (1987). *The Possibility of Cooperation*. Cambridge: Cambridge University Press.

—— (1988). 'Rationality and Revolutionary Collective Action'. In Michael Taylor, ed., *Rationality and Revolution*. Cambridge: Cambridge University Press, pp. 63-97.

Tennant, Neil (1987). *Anti-Realism and Logic: Truth as Eternal*. Oxford: Clarendon Press.

Thompson, E. P. (1986). 'Subduing the Jury'. *London Review of Books*, 4 December, pp. 7-9, and 18 December, pp. 12-13.

Tocqueville, Alexis (1980). 'The Revolution of 1848'. In Tocqueville, *On Democracy, Revolution and Society*, ed. J. Stone and S. Mennell. Chicago: University of Chicago Press, pp. 250-79.

Tolstoy, Leo (1972). *War and Peace*. Trans. Constance Garnett. London: Heinemann.

Trenchard, John, and Thomas Gordon (1971). *Cato's letters*. 2 Vols. New York: Da Capo.

Tuck, Richard (1989). *Hobbes*. Oxford: Oxford University Press.

Tully, James (1988). *Meaning and Context: Quentin Skinner and His Critics*. Cambridge: Polity Press.

Tuomela, Raimo (1988). 'Free-Riding and the Prisoner's Dilemma'. *Journal of Philosophy*, Vol. 85, pp. 421-27.

Turner, Jonathan H., and Alexandra, Maryanski (1979). *Functionalism*. Menlo Park, Calif.: Benjamin/Cummings Publishing Co.

Ullmann-Margalit, Edna (1977). *The Emergence of Norms*. Oxford: Oxford University Press.

—— (1978). 'Invisible-hand Explanations'. *Synthese*, Vol. 39, pp. 263-91.

Van Parijs, Philippe (1981). *Evolutionary Explanation in the Social Sciences*. London: Tavistock.

Veyne, Paul. (1984). *Writing History*. Middletown, Conn.: Wesleyan University Press.

Walzer, Michael (1983). *Spheres of Justice*. Oxford: Martin Robertson.

Webb, Leicester C. (1958). *Legal Personality and Political Pluralism*. Melbourne: Melbourne University Press.

Weber, Max (1949). *The Methodology of the Social Sciences*. New York: Free Press.

Weinstein W. L. (1965). 'The Concept of Liberty in Nineteenth Century English Political Thought'. *Political Studies*, Vol. 13, pp. 156-57.

Weiskrantz, L., et al. (1974). 'Visual Capacity in the Hemianopic Field Following a Restricted Occipetal Ablation'. *Brain*, Vol. 97, p. 709.

Wells, G. A. (1987). *The Origin of Language*. La Salle, Ill.: Open Court.

West, David (1990). *Authenticity and Empowerment: A Theory of Liberation*. Brighton: Wheatsheaf.

Wiggins, David (1980). 'What Would Be a Substantial Theory of Truth?' In Z. van Straaten, ed., *Philosophical Subjects*. Oxford: Oxford University Press, pp. 189–221.

—— (1987). *Needs, Values, Truth*. Oxford: Basil Blackwell.

Williams, Bernard (1973). *Problems of the Self*. Cambridge: Cambridge University Press.

—— (1978). *Descartes*. Harmondsworth: Penguin.

—— (1984–85). 'Formal and Substantial Individualism'. *Proceedings of the Aristotelian Society*, Vol. 85, pp. 119–132.

Wirszubski, Ch. (1968). *Libertas as a Political Ideal at Rome*. Oxford: Oxford University Press.

Wittgenstein, Ludwig (1958). *Philosophical Investigations*. 2nd ed. Oxford: Basil Blackwell.

—— (1978). *Remarks on the Foundations of Mathematics*. 3rd ed. Oxford: Basil Blackwell.

Wokler, Robert (1987). *Rousseau on Society, Politics, Music and Language*. New York: Garland.

Wright, Crispin (1980). *Wittgenstein on the Foundations of Mathematics*. London: Duckworth.

—— (1986). 'A Cogent Argument against Private Language?' In P. Pettit and J. McDowell, eds., *Subject, Thought and Context*. Oxford: Oxford University Press, pp. 209–66.

—— (1988). 'Moral Values, Projection and Secondary Qualities'. In *Proceedings of the Aristotelian Society*, Supp. Vol. 63, pp. 1–26.

Wynne-Edwards, V. C. (1986). *Evolution through Group Selection*. Oxford: Blackwell.

Index

beliefs (*continued*)
  consciousness of, 64
  contents of, 61, 63, 64, 74, 106n1
  evaluative, 240, 241
  formation of, 41, 56, 59, 60, 70, 73, 93,
      98, 138, 195, 234, 283n2
  functionalist account of, 62
  requirements on, 18, 55
  sensitivity of, 69
Benn, S. I., 120, 314
Bennett, J., 60, 64
Bentham, J., 287, 307, 319
Berkeley, G., 197
Berlin, I., 166, 194, 314
Bishop, J., 107n2
Blackburn, S., 24, 32, 99, 179, 247
Block, N., 32, 45, 53n6, 53n9, 164n7,
      238n8
Boghossian, P., 83
boundary conditions, 124, 125, 135, 149,
      151, 154, 155
Bradley, F. H., 167, 170, 174
brain-states, 106n1
Braithwaite, J., 321, 323, 326, 335
Bratman, M., 283n2
Brennan, G., 291, 323, 329, 335
Broome, J., 52n1, 286
Brown, R., 126
Bubner, R., 189
Buchanan, J., 291, 326, 330
Buckle, T. H., 128
Burge, T., 175

Campbell, J., 208
Candlish, S., 170
capacities, human, 112, 114, 138, 162,
      169, 223
capacity to think, ix, 169–173, 178, 179,
      192, 213, 247; *see also* thought
Cartesianism, 26
causal factors, 40, 41
causal history, 256–258, 277, 283n9
causal power, 230
causal relevance, 32–42, 54, 164n7, 244,
      250, 254, 255
  program model of, ix, 4, 5, 47, 51, 219
  of properties, 147, 148, 150, 217, 230–
      234, 252, 277
cause
  general, 263

higher-order, 33
lower-order, 33
particular, 263
standby, 277
Cherniak, C., 11, 41
Churchland, Patricia, 45
Churchland, Paul, 45
Cicero, 308, 331
citizenship, 311–314
cognitive states, 23
cognitivism, 23, 24, 52n2, 240, 301,
      302
Cohen, G. A., 261
Cohen, J., 308
Coleman, J. S., 124, 126, 138
collaboration, causal, 150
collective action theory, 326, 327
collectivism, 111–114, 131, 134, 157, 162,
      163, 165, 166, 168, 253, 263
  and individualism, ix, 118, 127, 137–
      143; *see also* individualism and
      collectivism
  atomistic, 172, 173
  confused with holism, 175
  extreme, 143–146, 155
  holistic, 172
  limited, 146–152, 155
  source of, 126
Collins, J., 23, 283n3
committees, public, 333–335
commonability
  of rules, 114, 118, 181, 182–184, 186–
      189
  of thought, 181, 182–184, 191, 209–
      211, 213, 237
  thesis, 181–183, 186, 189, 190
communication, 193, 211; *see also*
      language
communitarianism, 308
community, 9, 114, 115, 181, 189, 190,
      210, 211, 237, 238
community of knowledge, 210
compatibilism, 45, 46
competence, 198–200
competition
  economic, 159; *see also* market,
      competitive
  group, 158
  levels of, 157, 158
Comte, A., 128